THE NEW PHYSICS
OF
«DYSMETRIC»
SPACES

J. M. ARNAIZ
cfejma@gmail.com

In this work, the false hypothesis of the International System of Units is saved with the First Algebra of Magnitudes in history or *dyadic algebra* and the innovative evidence is discovered that the natural thing is the «dysmetry», which allows to represent the infinite areas of empty space, characterizing its physical properties tensorically.

To curious spirits who yearn to understand the principles of science, to facilitate their work; and also in honor of the absent friends who inspired everything with altruistic spirit. And especially to Maria, my wife, without whose encouragement «dysmetry» would not exist.

PART I
FIRST ALGEBRA OF MAGNITUDES (DYADIC ALGEBRA)

EXORDIUM

EXORDIUM

*«Arithmetizing» Physics suffocates our minds, incapable of
to understand what magnitudes and «dysmetry» are,
a fundamental physical-mathematical truth*

The 19th century witnessed a radical reform of mathematics, which transformed it from an «arithmetized» discipline serving practical purposes to one that became autonomous and abstract. This period was characterized by unprecedented rigor applied to the invention of abstract algebraic structures, based on deep introspection about their foundations. Évariste Galois introduced *general group theory*. William Rowan Hamilton invented *quaternions*, a new number system where multiplication is not commutative, proving that arithmetic rules were not unique but could be extended. Mathematicians such as Carl Friedrich Gauss, János Bolyai, and Nikolai Lobachevsky saw that non-Euclidean geometries could be defined, demonstrating that geometry was not an absolute truth about space, but a logical structure susceptible to modification, which represented an unprecedented mathematical revolution. Georg Cantor developed *set theory*, which became a universal language and a basic tool for underpinning all of mathematics. George Boole's contribution to this modernization was crucial, connecting logic with algebra and creating the basis for symbolic logic and computation.

The fruits of this modernization were not long in coming. Boolean algebra has been the basis of all digital computer logic since its inception. Abstract number theory is fundamental to current cryptography, e-commerce security, and many other applications. Algorithms and programming solve highly complex problems that were previously intractable. In short, our knowledge society has only been possible thanks to the modernization of mathematics. The ability to abstract and

11

generalize mathematical concepts is the tool with which scientists and technologists of the 20th and 21st centuries have provided us with the prodigious advances we enjoy. But, paradoxically, such evident progress today had to overcome, at its origins, a strong resistance to the adoption of abstraction, which is also slowing the transition of Physics to abstract modernity.

It is therefore a historical fact that mathematics eventually overcame this clumsy resistance to abstraction, but Physics has not yet managed to overcome it even slightly, remaining anchored in the archaic «artimetization» of the 19th century and deprived of its own algebraic structure, which would allow it to legitimately operate with quantities of magnitudes and allow abstraction to bear its natural fruits, following the successful and proven example of mathematics. Since there is no excuse not to do so, this abstract reform of the physical foundations is what is promulgated with this work, in the certainty that it will produce countless advances that are now unthinkable, as undoubtedly guaranteed by the triumphant transition from mathematics to modern abstract algebra.

Our initial motivation for modernizing Physics was to answer such basic questions as: How do you multiply a kilogram by a meter? What is the multiplier, the kilogram or the meter? How do you square a second? How do you divide the product kg x m by a second squared? What is the full meaning of compound units such as the newton, a unit of force, or the joule, a unit of energy? What are the inverses of quantities? Do inverses of units such as s^{-1}, m^{-1} or kg^{-1} exist? Does empty space have physical properties? Are standard units constant, or can their quantity vary in space and time? Do universal physical constants really exist? And others like them.

Since we began our elementary study of Physics, we have been accustomed to using operations with entities that indicate concrete quantities of magnitudes, and, due to the subliminal influence of arithmetic operations with abstract numbers such as the real numbers, we naturally believe that concrete operations should follow the same calculation rules. Thus, for example, if a moving object travels a distance of 100 meters, taking 25 seconds to do so, we say that the speed of the movement is $100/25=4$ meters per second, and we will abbreviate this speed with the graph 4 m/s. Likewise, if in 3 seconds the speed changes from an initial value of 10 m/s to 70 m/s, we firmly affirm that the moving object would have experienced an acceleration of $(70-10)/3=20$ meters per second and per second, writing it as 20 m/s^2. With this, we will have unconsciously assumed that unit symbols, such as the meter or the second example, must be operated with the same laws of composition that we have established for abstract real numbers, that is, as if the units were mere formal algebraic variables. A careless assumption, given the absence of any motivation to justify it. Without fully clarifying this point, one should not skip to the second page of any Physics textbook; but it turns out that the texts themselves induce this mess, because they all forget to plumb a fundamental pillar of science: the abstract algebra of magnitudes.

So what we unscrupulously get used to and are unconsciously taught to do from childhood, seeming so natural to us, is in reality not only not at all obvious, but completely incorrect, since it omits something crucial: the epistemic definitions of the laws of composition between entities that represent measurements or quantities of magnitudes and their units. So it is not surprising that the effects of this omission have been a source of concern among experts. Perhaps the first to recognize this was Fourier in his Analytical Theory of Heat (1822), where he introduced the important concept of dimension inherent to all physical magnitudes. Another distinguished thinker who warned about this issue was Clerk Maxwell, author of the unification of electromagnetism, who alluded to the typical case of teachers at all educational levels, who explain the unit of work simply by referring to the formula: one joule equals one kilogram multiplied by one square meter and divided by one second squared, according to the well-known expression:

$$1\,joule = \frac{1\,kg \times 1\,m^2}{1\,s^2}$$

And he continued saying that teachers do not realize that they would be in a difficult position if an inquisitive student asked them what it means to multiply a kilogram by a square meter and how to divide the result by a second squared, or even what the meaning of a second squared is.

And it is that, indeed, we happily admit that $1\,s \times 1\,s = 1\,s^2$, because it seems elementary algebra; however, we must take a closer look at expressions like this. It is clear that any multiplication of abstract numbers such as 3×4 means, by definition of multiplication, adding the multiplicand 3 as many times as indicated by the multiplier 4, that is, we have the abbreviated sum $3 \times 4 = 3 + 3 + 3 + 3 = 12$. So, if we applied this same algorithm to the product $3\,s \times 4\,s$, establishing as multiplying the first factor and the second as multiplier, or vice versa, we would have $3\,s \times 4\,s = 3\,s + 3\,s + 3\,s + 3\,s = 12\,s$. Why, then, do we so easily admit that $3\,s \times 4\,s = 12\,s^2$? Just because it has the written form of an algebraic expression, do we ignore the arithmetic definition of multiplication? Because obviously $12\,s$ cannot be the same as $12\,s^2$, since we can easily understand what a second is, since we have clocks that measure that quantity of time; Now, what kind of entity will a second squared be? It does not seem that such a thing can be observed in nature, which would seem to delegitimize composite units like this, which appear everywhere in the study of physical phenomena.

There is therefore a gap pending to be resolved in this of operations with physical magnitudes, which causes the proliferation of diverse and contradictory opinions regarding their nature and formulation, discussions that would be ended simply by defining the appropriate composition laws. A group of authors such as RC Tolman attribute to the symbols of dimensional expressions a certain impenetrable or mystical character and consider that «the true essence of magnitudes, from the physical point of view, is represented by their dimensional formula»

(Physics Review, p 25, 1917). This hypothesis does not seem to be true, because it would suppose that such disparate magnitudes as the momentum of a force and its work, which can both be expressed in «newton×meter», were essentially manifestations of the same magnitude, energy, which seems clearly a madness, as we justify in section XXVI of the Dyadic Algebra of Magnitudes. Great authors such as Planck indicate that «it is as meaningless to speak of the "real" dimension of a magnitude as it is of the "real" name of an object», which would mean that physical magnitudes should be hidden from the understanding. Planck seems to indicate to us that we must not forget that physical quantities are mental entities and that, like any other name that indicates an extramental object, they are the result of the arbitrariness of thought. The positivist faction of the Vienna Circle, led by Bridgman, states that «dimensions do not have absolute value at all, but must be defined precisely from the process used to measure the respective magnitude», which makes us to suggest that, indeed, in the field of magnitudes there must be a good deal of arbitrariness; which bothered Planck so much that he criticized positivism thus: «The views of the positivists cannot be fought from a purely logical point of view. And yet a careful examination of them reveals that they are inadequate and sterile, because they dispense with a circumstance that is of decisive importance for scientific progress. As much as positivism boasts of being free from prejudice, it has to start from a fundamental premise if it is not to degenerate into an unintelligible solipsism. This premise is that every physical measurement can be reproduced in such a way that the result is independent of the observer's personality, the place and time in which the measurement is made, and any other circumstance. All this simply reveals that the decisive factor for the measurement result lies outside the observer and that consequently the measurements pose problems involving causal connections in an objective reality independent of the observer».

And in the same way as in these notable examples, different beliefs about the nature of physical quantities proliferate, forming a kind of intellectual pandemonium in this matter, so that everyone conforms to the usual way of operating with quantities of magnitudes, although each one has their own subjective notion of them and without being aware of the problem that the lack of epistemic definition of their composition laws raises, so that for the majority the vice does not exist and it is common not even to ask why it must operate with such physical entities as it is done and not otherwise. Moreover, all authors versed in dimensional analysis take it for granted that unit abbreviations operate with the same algebra of abstract numbers, and on this tacit assumption and without justifying it in any way they elaborate their respective theories, which completely omit all specific algebra. for the magnitudes. And the same happens in the educational field, where the philosophical problems related to magnitudes and their laws of composition are ignored, as if they did not exist, teaching concrete operations in an intuitive, subjective and arbitrary way, leaving students with, Even without knowing it, a residue of uncertainty that vitiates all the knowledge acquired with this gap pending to be clarified, degrading the teaching quality, because the key to perfect understanding is not to advance at all without first having precisely defined all of the foregoing, and more, if possible, in the case of something so fundamental to understand and develop natural laws, such as magnitudes, their measurements and their operations. In our view, the reason for this inertia common to science and education could be due to the intuition fed by the algebra of geometric segments, which more or less we all know to some extent since adolescence. We know that the multiplication of segments does not follow the model of the arithmetic product, but is conceived in geometry as a new magnitud, called surface, with two factors, or called volume, if there are three, so that the multiplication of lengths gives rise to two new derived magnitudes, the surface and the volume, in accordance with certain suitably defined laws of composition, and whose commutative, associative and distributive properties retain the form of the arithmetic. We will analyze all this in more detail throughout this work.

Thus, in the same way that there are algebras for numbers and abstract vectors, universally accepted, an algebra of concrete entities, representatives of quantities of magnitudes, should be established, because only in this way would the prevailing confusion be ended and they would be better clarified the meanings of the various composite magnitudes. And this is precisely what is humbly carried out in the monographic memory that we have entitled Dyadic Algebra of Magnitudes: starting as the foundation of the algebra of geometric segments, one immediately has that of lengths and, by mere generalization, it is easy to arrive at the convenient definition of the laws of composition for any magnitudes. This reveals the hidden frameworks of the derived units and the meanings that can be attributed to them can be judged more accurately. Thus we will see that such mythical magnitudes as, for example, the so-called energy, surprisingly, are rather a kind of ether fruit of the arbitrary nature of scientific thought, rather than real entities or qualities. This does not mean that the magnitudes do not correspond to some aspect of the physical realm, but rather that we must be prudent when we derive conclusions about the world from dimensional formulations, pertaining to the mental realm, in order not to end up entangled in childish disquisitions wrong. And for this it is very convenient to understand where the compositions of the magnitudes that intervene in any phenomenon have arisen, since, otherwise, the entire analysis carried out on the case under examination will be mutilated or corrupted without possible remission and subject to capricious speculation.

So the notion of dimension of all magnitude has to be considered after and not before having conceived an algebra of magnitudes, whose mathematical expression is concrete entities. Hence, the method followed in the work of the Dyadic Algebra of Magnitudes, faithful to the step-by-step sequence that characterizes us, has to be presented according to the following sequence: first, the basic concepts of physical magnitudes are established in general; then, they are assigned a mathematical entity and the concrete entities or physical dyads are created; An algebra is defined below for such special entities, which are adopted as precise representatives of the quantities of natural magnitudes; then the meaning of the definition equations, of the universal laws and of other physical entities is investigated; and, finally, the basis for dimensional analysis is finished. Without trying at all to describe this matter with perfect exhaustiveness, but with enough detail so that its unfailing character can be appreciated, leaving for successive editions the development of more abstract structures tailored to the various scientific theories.

The capital elements of Physics are the measurements or quantities of magnitudes, associated through certain invariant relations that operate with them; so, considering that simple algebraic entities are not composed, because they always carry a unit, it is not admissible that the laws of composition that define exactly how these physical features of nature should be composed are lacking. It would be something like if Mathematics established the algebraic formulas in the abstract without having specified the tables for the addition or multiplication of natural numbers. We would certainly be shocked by it. And yet, with Physics, this is what has been tacitly done for centuries with all naturalness and indifference on the part of the majority, except for a few wise men who have been upset by the tradition of simply admitting that magnitudes operate. as elements of ordinary algebra, without their concern having resulted in a definitive solution, adding to the confusion that many refer to as the «mysterious problem of dimensions» or the «puzzling character of compound units».

With the Dyadic Algebra of Magnitudes, the foundations are laid so that the forgotten fundamental pillar of science is properly installed and, when we are before a physical law such as Newton's second law $\overline{F} = m \times \overline{a}$, we do not believe that we are facing an expression of vector algebra rather, we must understand that $\overline{F}$, m y $\overline{a}$ indicate quantities of physical magnitudes or measurements, represented by concrete mathematical entities, and that to operate with them it is necessary to have defined first of all how to do it in a convenient way with epistemic composition

18

laws without the least ambiguity, following the pattern of abstract algebra.

With this, any subjective controversy about the interpretation or natural meaning of the composite magnitudes should decline, because it will be the definitions themselves that allow it to be elucidated objectively. And this will undoubtedly result in a substantial improvement in the quality of teaching and the intellectual performance of students.

A good revelation of the Dyadic Algebra of Magnitudes is the **doubling theorem**, which determines how every physical law is decomposed into its two basic elements, the pure algebraic equation and the dimensional relationship between the units of its two members, a process in which they appear the constants of homogeneity, which show the difference between the equations of the mathematical metric, in which said constants are always the unit, hence the doubling does not produce any change in their formulations, and the physical measurements, in which in general you do not have to comply with this restriction.

And with this result, in our opinion very remarkable and transcendent, culminates what with some license of the language and without much property we could consider as a certain dyadic algebra sui generis applied to physical magnitudes, since ultimately the measurements are nothing but dyads formed by pairs of closely related elements: an algebraic primary and a dimensional secondary non numerical.

For the time being, the most spectacular and striking achievement of magnitude algebra is the discovery of the «dysmetric» dimension of every magnitude and, specifically, the «dysmetric» dimension of space, which arises naturally from the rigorous logic of the dyadic algebra of magnitudes. We firmly believe that «dysmetric» spaces are destined to transform Physics. Its first basic development is set forth in the second part of this work. The «dysmetric» observation is unquestionable and inalienable, and represents a dazzling Copernican turn in modern Physics, in contrast to the elementary and invisible isometry that has prevailed since the origins of science. Dyadic Algebra of Magnitudes initially follows the dogmatic isometric tradition, limiting itself practically to discovering, describing, and resolving the «arithmetization» paradox of Physics and the impossible hypothesis of the International System of Units. However, it soon unexpectedly encounters «dysmetry», which marks a new course promising the emergence of an inexhaustible source of physical innovations, which many enterprising researchers will undoubtedly appreciate. This is an inevitable discovery for logic and an essential one for Physics.

The paradox of the «arithmetization» of Physics is perniciously concealed by the false hypothesis of the International System of Units, accepted by everyone by tacit and convenient conventionalism, usually unconscious, which gives the appearance of rigor where there is only arbitrariness: it is universally accepted by hypothesis that physical magnitudes form an abelian multiplicative group, which cannot be sustained by any rigorous algebra, which requires defining the generating external multiplicative laws, for which the inverses cannot be conceived in the same sense as for the internal laws.

Here we do not limit ourselves to simply denouncing this false hypothesis, but the problem is identified and solved with the first coherent algebra in the history of Physics. Nobody well informed and impartial will maintain after studying the matter, that this absurd hypothesis must be maintained because Physics has given great results, because they should be reproached for how much more fruit it would bear without that falsehood in its principles.

In addition, it turns out that far from seeking a rectification, the International System of Units has consolidated this false principle in its most recent regulations, granting a letter of nature to the omission of a non-arithmetic algebra for physical quantities. Indeed, in its section 2.1 the International System defines quantities of magnitudes in this way: «*The value of a quantity is generally expressed as the product of a number and a unit. The unit is simply a particular example of the quantity concerned which is used as a reference, and the number is the ratio of the value of the quantity to the unit».*

We already observe a first substantial defect: The International System speaks of the product of a number by a unit and of the relationship between the value of the magnitude and the unit without defining these operations at all, thereby tacitly admitting that they correspond to arithmetic operations , although obviously the units and quantities of magnitudes are not abstract numbers, but quantities of physical phenomena.

This error is confirmed later in section 5.2, where it says: «*Unit symbols are mathematical entities and not abbreviations*». It is surprising and certainly inadmissible that this reckless claim is made to refer to particular parts of physical phenomena, which manifestly cannot be identified with numerical entities. On the one hand, the International System speaks of symbols of physical units and at the same time affirms that these symbols are mathematical entities, not abbreviations, without specifying to which mathematical entities it refers. In the same section it continues to say: «*In forming products and quotients of unit symbols the normal rules of algebraic multiplication or division apply*». This confirms the crazy equating of physical units and ordinary numbers. In another case, these mysterious symbols would present to the International System the character of unidentified mathematical entities.

In Section 5.4.1 the International System certainly confesses its false principle and states the following: «*Symbols for units are treated as mathematical entities. In expressing the value of a quantity as the product of a numerical value and a unit, both the numerical value and the unit may be treated by the ordinary rules of algebra. This procedure is described as the use of quantity calculus, or the algebra of quantities*».

If it weren't because we are dealing with something very important, this paragraph would make you laugh. It is not serious to affirm that an arbitrary rule that allows to operate with the physical units by means of the laws of ordinary algebra constitutes nothing less than the algebra of magnitudes which is the foundation of all Physics.

Therefore, the International System of Units normalizes the «arithmeticization» of Physics and falls squarely into the trap of constructing an abstract symbology without the slightest physical meaning, so that absolutely no one, not even the most eminent physicists, is capable of explaining what a multiplication as simple in appearance as a meter by a kilogram means. All of us, as faithful vassals, follow with blind obedience that mandate of the International System, without even asking ourselves why this fictitious algebra for physical quantities, without realizing that, thinking arithmetically, it is not possible to answer the elementary question : What is the multiplier, the meter or the kilogram? Simply because that multiplier cannot be quantified with a simple number, since neither the meter nor the kilogram are numbers, but rather indeterminate quantities of physical phenomena.

Another relevant consequence of true algebra is that the inverses of the physical units do not exist, despite the fact that the SI repeatedly grants them their own entity. Thus, in section 5.4.6 he pointed out that the quotients between units can be written a/b or $a{\times}b^{-1}$, identifying b^{-1} with the non-existent quotient $1/b$. For example, let's take the meter m, and find the quotient $1/m$. If someone knows an entity that multiplied by the amount of length implicit in a *meter* gives the abstract number 1, it would deserve a very special prize. We have already shown by non-arithmetic algebra why this quotient cannot be found, therefore it is necessary to redefine the inverses of this type with proposals such as the one in our annex.

In contrast to such intellectual servitude, when observing how Newton operates in his Principia it is observed that he did not fall into that absurd trap of «arithmetizing» Physics. True to his meticulous style, he begins with some clarifying definitions. We highlight those that come on purpose: «*I. The quantity of matter is the <u>measure</u> of the same, arising from its density and bulk* *conjunctly»; «II. The quantity of motion is the <u>measure</u> of the same, arising from the velocity and quantity of matter conjunctly»; «VI. The absolute quantity of a centripetal force is the <u>measure</u> of the same proportional to the efficacy of the cause that propagates it from the centre, through the spaces round about»; «VII. The accelerative quantity of a centripetal force is the <u>measure</u> of the same, proportional to the velocity which it generates in a given time»; «VIII. The motive quantity of centripetal force, is the <u>measure</u> of the same, proportional to the motion which it generates in a given time».*

In these definitions, we can already observe a striking fact that clashes head-on with the formula of the International System. Newton speaks of measurements without units. His concept of «measure» is «the number of times, whole or fractional, that a quantity of magnitude contains its unit». Thus, every measure is the number that multiplies the unit established by measuring a certain quantity of magnitude, so measurements can be operated using the laws of arithmetic. In this way, Newton overcomes the lack of an algebra of magnitudes and uses the arithmetic of his time, but he does so with logical coherence.

Newton then gives form to his axioms or laws of motion, well known, and states them in this way: *«Law I. Every body perseveres in its state of rest, or of uniform motion in a right line, unless it is compelled to change that state by forces impressed thereon»; «Law II. The alteration of motion is ever <u>proportional</u> to the motive force impressed; and is made in the direction of the right line in which that force is impressed»; «Law III. To every action there is always opposed an equal reaction: or the mutual actions of two bodies upon each other are always equal, and directed to contrary parts».*

In Law II Newton speaks of «proportionality», but what kind of proportion is he referring to? The answer is in the Principia and it is not precisely arithmetic proportionality. Now, if he is not referring to the proportionality of arithmetic ratios, what does Newton mean by the term «proportional»?

Newton assimilates the quantities of physical magnitudes to geometric figures, as the ancient Greeks did. Therefore, when he speaks of addition, subtraction, multiplication, or ratios between quantities of magnitudes, he is referring to these operations with segments, areas, or volumes, not with numbers, which he reserves exclusively for physical measurements. Therefore, he uses arithmetic to operate with measurements and geometry to operate with quantities of magnitudes. And this is how it should be done, because measure and quantity are not the same thing. Measure is the number that represents a quantity of magnitude in relation to a certain unit, it is the multiplier of the unit; while quantity is the union of measure and unit to express specific portions of magnitudes.

It is not possible to operate seriously with magnitudes only through arithmetic, because operations with abstract numbers are internal laws of composition and, as shown throughout this text, especially starting in section XII, operations with magnitudes require new **generating external composition laws**, which produce the composite magnitudes born from the fundamental ones, the only way to provide physical meaning to the laws and equations of Physics. And these generative operations are not at all like arithmetic ones, showing very notable differences such as, for example, the nonexistence of inverse elements for the quantities of magnitudes, which is why unitary dyads or inverse and, therefore, the inverse of any unit cannot exist, as explained in section XIV. This means that, just as the inverse of the number 2 is the rational $1/2$, the inverses of a meter, a kilogram or a second cannot be found. We insist, in the annex we propose an abstract solution to configure the dyadic inverses.

Our path of abstract modernization begins with the observation that operations with magnitudes they are not «arithmetizable», but they can be «geometrized» by affinity with geometric operations with segments, areas, and volumes. Of course, these operations are not arithmetic, because they do not relate numbers but rather elements of geometry. Next, we conceive dyadic forms, mathematical elements that represent the non numerical quantities of physical magnitudes. On the sets of dyads, we define the internal and external operations to provide them with an algebraic structure. And immediately, «dysmetry» and the transcendental consequences of this **physical-mathematical truth** emerge, two of them very important: first, the properties of empty space, which is not presented as inert but as an active entity that produces physical effects by itself, characterized by the **«dysmetric» tensors** of section XXXVI; and second, the immortal *law of dyadic variation* described in section XXXVII, dedicated to differential «dysmetry», which proves the ignored fact that «dysmetry» is natural, a truth currently invisible to arithmetic Physics, and whose foundations are set out in the second part of this work. Such original results as the nonexistence of universal constants in an absolute sense are proved by analyzing the «dysmetric» variations of the number pi and the speed of light, and the work culminates with notions about «dysmetric» tensors, which characterize the properties of empty space in the absence, therefore, of all material disturbance.

And so the two main «dysmetric» tensors are identified, called the **deformation tensor** and the **density tensor of empty space**, proving in turn with tensor calculus that in an empty «dysmetric» space neither the speed of light is constant nor its trajectories are rectilinear, even if there is no matter or energy present, establishing a **new concept of empty space**, which ceases to be a passive and inert entity, becoming an active element capable of producing physical effects on its own and, likewise, with its own nature and faculties with the power to condition physical phenomena.

The complexity this entails is frightening at first glance, but it turns out that «dysmetry» can be mathematized through the fundamental property of magnitudes that we call **«dysmetric» density**, which turns out to be dimensionless and, therefore, purely numerical, which undoubtedly magically but with perfect mathematical rigor connects magnitudes with abstract numbers.

Our passion for conveying the importance of the topic may have led us to resort to redundancies or outbursts. We apologize for this and assure you that there is no malicious intent. The reason is none other than our firm belief that what matters is Physics and its integrity, taking precedence over any subjective sensitivity or interest. At the same time, we have tried to use mathematical language that is not too abstruse, because our intention is for this abstract dyadic algebra and its universal «dysmetry» to be accessible to the majority of intellects, including those at the pre-university level. To this end, we have strived to respect the episteme, applying the logical-algebraic method, which reasons step by step based solely on previously established definitions and properties, thus eliminating any blinding prejudices. We hope to have explained ourselves well, without offending anyone, and we hope to have clarified the philosophical and practical significance for the future of Physics of the dyadic algebra of magnitudes and its initial revelation, «dysmetry».

J. M. Arnaiz

Section I

DEFINITIONS OF MAGNITUDE AND QUANTITY [1]

The first obstacle to overcome in the analysis of physical phenomena is to specify what is meant by magnitude. The contemplation of nature inspires the mental concepts of length, area, volume, time, speed, acceleration, mass, force, energy, and many others. Each of these entities does not always present itself with the same intensity, resulting in relationships of «equality» or «greater than» or «less than» between their different portions. Thus, by factual comparison, we can establish whether a certain portion of length is equal to, greater than, or less than another, just as two portions of areas, volumes, or time intervals, etc., can be determined. This fact allows us to conceive the definition of **physical magnitude** as any property that allows relationships of equality and inequality to be established between its different portions. On the other hand, the **quantity implicit** in any portion of a magnitude is the extent, degree, or intensity with which it manifests itself[2]. In the case of the magnitude of length, any straight line represents it; a portion would be any segment of the straight line; the amount of length implicit in a segment would be non-numerical and refers to the true extension of the segment.

---

[1] In English, it is common to consider the terms quantity and magnitude as synonyms. This poses a problem for English speakers in understanding the substantial difference between the meanings of these two concepts in Physics. Magnitude refers to a measurable natural property. In contrast, the quantity implicit in a portion of magnitude is precisely what symbolizes that measurement. Therefore, readers of that language are advised to rigorously understand and differentiate these two fundamental concepts for dyadic algebra and «dysmetry».

[2] A more elementary version of the algebra of magnitudes, although equally significant, can be found in another title by the same author, «Lesson 3» of Matematize 3.

Since every quantity of magnitude is non-numérical, that is, it cannot be reduced to an abstract number, to express it analytically and be able to operate with various quantities, it must be symbolized with a mathematical object that indirectly represents it. This mathematical element arises from the measurement process and is formed by a pair consisting of an abstract number or other multiplying entity and a portion of any magnitude that is taken as the unit. The physical unit is non-numerical, so it is replaced by an abstract symbol. It is vital to differentiate the physical or real quantity implicit in a portion of magnitude from the mathematical pair that represents it.

Let's take an example. When we say that the distance between two points is four meters, we write it as $4\ m$. In this nomenclature we have the following meanings: the magnitude we are not referring to is the length; the abbreviation m means that **we have taken as a reference the quantity of non-numerical length implicit in the portion we call the standard unit or meter** to measure the distance, therefore m means an amount of length that is not explicit because it cannot be expressed numerically, which is why it is symbolized with a letter. In turn, the abstract number 4 acts as a multiplier of m and means that the established distance is such that it corresponds to four times the amount of length implicit in the standard unit m. It is common to confuse 4 with a quantity of length, which is a fundamental conceptual error, because 4 is only an abstract number that by itself does not indicate magnitude; only by associating it with a unit like m does the pair thus formed acquire in this case the meaning of the quantity of the magnitude length. Thus, the difference between magnitude and quantity is clear: $4\ m$ means a quantity that refers to the magnitude of length. Since the unit m is not countable, the quantity $4\ m$ is not either, but this non-numerical quantity is represented by the mathematical pair $4\ m$. If the same number is associated with another quantity of magnitude, for example, the kilogram, the pair formed by the abstract number 4 and the quantity of non-numerical mass implicit in the standard unit, symbolized kg, acquires the meaning of a quantity four times the quantity implicit in the kilogram, expressed as 4 *kg*, of the magnitude we call mass. Here, the quantity is the meaning of the pair 4 *kg*, and the magnitude referred to is mass. The quantity of mass is not the pair 4 *kg*; this is merely the mathematical symbol for the real quantity implicit in a certain portion of mass.

Why do we say that every quantity of any magnitude is non-numerical? Let's take length. The distance between two points in space is a quantity of real or true length. How can we express it? No matter how much we search, we cannot associate it with any number. Therefore, what we do is invent measurement. We take a portion of magnitude or segment that we assume includes a certain amount of length and call it a standard unit. We take it for granted that the segment implicitly contains a certain quantity of length. Since we cannot express it numerically, we assign it a symbol to represent it, for example, *m*. Thus, to measure the distance between two points, we simply need to determine how many times the standard segment fits juxtaposed within the distance to be established. We tacitly admit that the quantity of length implicit in the standard segment does not vary along the distance to be measured, and we can now say that the quantity of length existing between the two points in question is that many times the quantity of length present in the standard segment, which seems constant to us. We have thus established a measurement consisting of a pair of elements: a number and an non-numerical standard unit. We say that this pair symbolically represents the real value of the measured quantity of length, 4 *m* in the example. In this way, we overcome the fact that we cannot directly number the magnitude of length and assume that measurement replaces it. This impossibility arises for any other magnitude, which is why there is always the need for measurement and for establishing standard units associated with the magnitude whose quantities we wish to establish. Thus, we symbolically express the quantities of magnitude that seem true to us through pairs of measurements or dyads, defined in section III, which consist of a multiplier number and an non-numerical portion or physical unit that appears to produce constant measurements. It is important to recognize that $4\,m$ is the symbol or dyad associated with the quantity of physical, real, or true length that we wish to represent. **Understanding the difference between symbol and meaning, that is, between dyad and quantity of real magnitude, is crucial to appreciating the «dysmetric» phenomenon, where the same standard can indicate different quantities.**

When studying operations with physical quantities, it will be understood that the coherent definition of magnitude refers to **any physical property affine to length**, in order to provide a logical connection to the algebra of magnitudes. Throughout this work, this affinity will gradually become more specific, allowing us to define the various laws of composition that provide an algebraic structure to the quantities of physical quantities.

Anyone who carefully follows the reasoning presented in the text will gradually discover the **richness hidden in magnitudes**. For example, in the dyadic algebra of magnitudes, it is provisionally assumed, as has always been done automatically and without realizing it, that every portion of magnitude identical to the same standard unit seems to always be associated with the same amount of real magnitude, absolutely constantly and in every situation or circumstance. However, nothing prevents, and it is moreover necessary for logic and science, the formulation of the generic prediction, that is, that portions equal to the same standard unit can be identified with different quantities of real magnitude depending on various causes, for example, position, time, or the material environment. This broader option, which we will call the «dysmetric» variant, goes beyond the limits of the current, tacitly accepted isometric hypothesis and is the subject of specific study in the second part of this work, whose careful examination is highly recommended, because it introduces the reader to the original «dysmetric» spaces, a novelty born from the dyadic algebraic order and a fruitful mathematical tool that will produce inexhaustible scientific innovations.

Section II

DEFINITION OF MEASUREMENT OF
THE QUANTITY OF A MAGNITUDE

The reasoning in section I allows us to define the measurement of a quantity of magnitude in this way: **we will call measurement the application of any procedure that allows the quantity implicit in a portion of magnitude to be represented by the pair formed by two heterogeneous elements, one mathematical and the other physical, the first serving as a multiplier of the second, which is any symbol that represents a quantity of non-numerical magnitude.** The order in which these two entities are written is irrelevant, but we agree to write the mathematical element first, followed by the physical element.

By definition of magnitude, their quantities can be treated as quantities of length, given the affinity we have postulated. Therefore, the various quantities will be comparable in terms of «equality», «greater than» or «less than» as if they were geometric segments. What quantity of length is implicit in a given segment? Obviously, this quantity is not visible to direct observation, so we must quantify it indirectly; we must determine it in some relative way. Any segment can be fragmented into equal, that is, geometrically congruent, segments. If we assume that the quantity of length implicit in each of these smaller segments is the same, we have the traditional **isometric hypothesis**, as opposed to the more general option we called **«dysmetry»** in the second part of the text, which allows for different implicit lengths in congruent segments. It is always possible to arbitrarily choose any segment as a unit of length to construct multiples and submultiples of it that allow us to compare that unit with other segments and observe how many units or fractions of the unit a given segment contains. We have called this action measurement, and given the affinity of

31

magnitudes with length, we accept that it is valid for any magnitude. The result of the measurement will be a real number or other mathematical entity, indicating the number of units or fractions of a unit contained in the quantity of magnitude being measured. A heterogeneous pair is thus obtained: the physical element ϕ, which represents the reference quantity of magnitude, whose quantity is innumerable, so it is replaced by this abstract symbol; and the mathematical multiplying element μ of the multiplicand ϕ. We will say that the heterogeneous pair thus formed represents the real quantity of the magnitude considered, and we will express it using any of the forms $\mu\,\phi$, $(\mu\,\phi)$, (μ,ϕ), or $\mu\times\phi$, which we will call the significant **measurement** of that real quantity. We define **measure** as the mathematical multiplying element resulting from a measurement, coinciding with Newton. And this is not capricious, because, if we have $\mu\times\phi=\varphi$, where φ is the quantity of magnitude to be measured with the unit ϕ, the division $\varphi/\phi=\mu$ is clear. So, if the measure is defined as the whole or fractional times that the quantity φ contains its unit ϕ, it turns out that the measure is the ratio φ/ϕ, which is the multiplier μ. This division operation is detailed in section XI.

Measurement symbolizes the quantity of real magnitude implicit in the measured phenomenon, showing its relative value to a certain quantity of magnitude taken as a reference standard, whose true value is non-numerical, so it is replaced by an abstract symbol. The comparison is made by affine geometric congruence. Assuming that congruent segments implicitly include the same quantities of length is the common isometric hypothesis. Recognizing the more general observation that congruent segments can implicitly contain different quantities of length is essential to appreciating the «dysmetric» phenomenon, the subject of the second part of this work, starting with section XXX. This means that **mathematical congruence is not synonymous with physical equality**, with important consequences, which are outlined in said second part.

32

Section III

DEFINITION OF CONCRETE ENTITY OR PHYSICAL DYAD
DYADIC SETS

Let's consider the most common measurements expressed with real numbers R or vectors of R^3. We have called measurement the quantity of a magnitude expressed with the form $q\,U$, as a symbol of the times q that a unitary quantity U is present in a phenomenon, calling q multiplier or measure with the unit U of the observed magnitude. And analogously if the measurement were a vector. The term «measure» or multiplier should not be confused with the «measured quantity» corresponding to a measurement. Here the word «measure» is the participle of the verb «to measure». Measurement is the product of the measure by the unit, which can be expressed with the common multiplication symbol $q \times U$, and the quantity it represents refers to the value that the measurement yields based on the supposedly true value of the quantity of magnitude implicit in the unit U, **an non-numerical quantity but which is taken into account at all times with the abstraction of symbolizing it with a sign that replaces it.** The expression $q \times U$ as an abstract formula for any quantity of magnitude associated with the unit U reveals the presence of the pair (q, U) formed by the elements q and U, and for convenience nothing prevents it from being written more briefly as $q\,U$, representing a mathematical entity formed by a multiplier, real number, vector or other mathematical object, and a multiplicand or dimensional quantity associated with a certain magnitude. This newborn entity that alludes to physical **measurement** can also be given a mathematical name, for example, **concrete entity** or **physical dyad**, and its elements will be called primary, measure, mathematical element or multiplier q or , and unit U, secondary, physical element, dimensional part or multiplicand. The primary is the mathematical part of the dyad. The secondary is the

33

physical or dimensional part. Perhaps the name unit number or physical number, or some other suggestive title, would be appropriate for the secondary; But, since the name does not make things, we will not waste time on this trifle, but rather we will attend to what is important, which is the nature of the concrete entity, truly born from the action of counting and joining a certain number of whole or fractional times the pattern U, an operation that, we repeat, can be indicated as the product of a number by a quantity of determined magnitude noted $q \times U$ or $q \times (1\ U)$; or, if the measurement is vector, $\overline{q} \times U$ or $\overline{q} \times (1\ U)$, where $(1\ U)$, although not specified, corresponds to a supposed true quantity of the magnitude in question, established by its empirical definition and taken as an elementary unitary pattern of said magnitude. And there is no problem in admitting, by definition, that the indicated quantity does not depend on the order of writing, so the same quantity will be $q \times (1\ U)$ as $(1\ U) \times q$, which is equivalent to axiomatizing the commutative property of this symbology. In short, it is necessary to establish a principle that allows for the construction of further reasoning, and we will call it the **fundamental postulate to be kept in mind in operations with dyads**, whose statement is that the measurement symbol $q\ U$ means that the real quantity is q whole or fractional times the quantity implicit in the unit U, which is indicated by the three forms of the following definition:

$$q\ U = q \times (1\ U) = (1\ U) \times q \qquad [3.1]$$

As we will define in the following section, here we advance that the equal sign means that all members symbolize the same quantity of magnitude, so they are substitutable for each other.

In turn, in the case of a vector primary $\overline{q}$, the concrete $\overline{q}\ U$ must symbolize the quantity of a vector magnitude that has the same direction and sense as $\overline{q}$ and whose module is the number of whole or fractional times equal to the module of $|\overline{q}|$ the quantity of the magnitude contained in the unit U. As in the scalar case, the following three notations will be accepted as indicative of this meaning:

34

$$\overline{q}\ U = \overline{q} \times (1\ U) = (1\ U) \times \overline{q} \qquad \qquad [3.2]$$

Therefore, in view of [3.1] and [3.2] it is not possible to distinguish between scalar and vector units, because, for both, every unit or quantity of magnitude U must be admitted by algebraic axiom to be identified with the scalar dyad $(1\ U)$ and that the numerical element that acts as multiplier of the dyad is q or $|\overline{q}|$, depending on whether it is associated with a magnitude of scalar nature in R or vector in R^3, or with another mathematical entity that serves as multiplier.

Magnitudes whose multipliers are such that $q \in R$ and that can take any value are called continuous, whereas those in which the multipliers can only be integers, with $q \in Z$, are called discrete. It is observed that operations with discrete magnitudes are included in the continuous ones, since their primaries will be represented by integers, which is a subset of the real numbers, so that the continuous magnitudes present greater generality than the discrete magnitudes; and the continuous magnitudes will be explained in the abstract in any case by means of the affinity with length, which fictitiously represents them all, because any of them can be assimilated to the real line, resulting in any case in the same reasoning scheme.

The choice of units for any magnitude is arbitrary. Therefore, **the broad definition of a physical dyad is any pair formed by a primary mathematical multiplier, number, vector, or other, and a secondary consisting of any symbol or symbols that designate a certain quantity of unspecified and non-numerical magnitude.**

In section I, we defined magnitude as **any physical property affine to length.** We said that this means that quantities of magnitudes can be treated as if they were geometric segments. And this helps us **formulate the concept of dyad in the abstract.** Thus, any quantity f of a magnitude can be associated with a segment of arbitrary length, which can be added to itself by juxtaposition as many times as a multiplying element indicates, or decomposed into equal segments of lesser length as many times as a divisor element indicates, in accordance with the elementary

35

geometric operations, which we assume the reader is familiar with. In this way, we can form, with the quantity of magnitude ϕ, other quantities defined by a multiplying element μ, integer or fractional, which we symbolize with the multiplicative form $\mu \times \phi$. This operation is developed in section IX, assigning it the symbol of the law of composition «°». The factor μ is generally an element of the set of real numbers R or a vector of R^3. Once we have obtained the homogeneous quantity $\mu \times \phi$ of ϕ, it is obvious that $\mu \times \phi$ can be observed as a pair of heterogeneous elements μ and ϕ. Let us remember that μ in isolation is not a quantity of length, it is merely an abstract mathematical entity. Now, nothing prevents us from naming this pair with the term abstract dyad, choosing symbols such as $\mu \ \phi$, $(\mu \ \phi)$, (μ, ϕ) or $\mu \times \phi$ to denote these pairs. As we have already seen, we will say that μ is the primary, mathematical element or multiplier of the dyad, and we will call the quantity of magnitude ϕ secondary, physical unit, dimensional part, or multiplicand. The dyad is, therefore, the mathematical reflection of the measurement process, by which a unit, divided into other smaller and equal units, allows us to formulate a measurement in the form of pairs of the type (μ, ϕ). Every measurement is a dyad, but an abstract dyad does not have to be a measurement. For operations on dyads, which represent quantities of magnitudes, to verify the associative-commutative and distributive properties, we will see in the following sections that the set of multipliers $\{\mu\}$ must have a field structure, like R. If we designate μ_1 as the multiplicative unitary element of said set $\{\mu\}$, we postulate that $\mu_1 \times \phi = \phi$ for any ϕ, and we can extend the ways of expressing the same quantity of a magnitude indicated by a dyad (μ, ϕ) with $\mu \times (\mu_1, \phi)$ and $(\mu_1, \phi) \times \mu$. The set of all dyads formed with the set of multipliers $\{\mu\}$ and associated with the quantity of magnitude f could be represented abstractly with the graph $\{\{\mu\}, \phi\}$, which we will call a dyadic set. Thus, by definition, the dyadic set f over the field of multipliers $\{\mu\}$ is, by definition, $\{\{\mu\}, \phi\} = \{(\mu, \phi) \mid \mu \in \{\mu\}\}$.

Once the definition of physical dyads and dyadic sets has been established, they would be of no use if the composition laws that allow us to operate with them were not formulated, providing them with an algebraic structure. And this is the crux of the key task for resolving the gap described at the beginning of this work, as is properly justifying operations with quantities of different physical magnitudes, which is the objective of the first part of this work, in which the terms **concrete** and **physical dyad** will be used interchangeably to name the basic elements that symbolically represent every quantity of magnitude.

The name concrete is retained due to the historical weight of this concept, which has long served to differentiate abstract numbers, those that indicate a quantity without specifying any unit and formed by a single mathematical element, from classical concrete numbers, which indicate a quantity associated with the unit to which they refer. However, the name **dyad** is our preference for heterogeneous pairs associated with any physical quantity, which is why the name dyadic algebra is reserved for the various structures that arise from the laws of composition that will be defined for dyads, symbolized indistinctly with the forms $q\ U$, $(q\ U)$ o (q,U). And analogously for vector numbers, replacing q with $\overline{q}$.

To formally develop dyadic algebra, it is necessary to epistemically define the addition of dyads, the multiplication of a dyad by a scalar, for example, a real number, the subtraction of dyads, the multiplication and division of scalar dyads, and the scalar and vector products of vector dyads. Once these operations are established, other derived operations such as exponentiation, radicalization, or logarithm of scalar dyads can be deduced by logical means, for rather theoretical purposes, in the same way that subtraction and division are derived from addition and multiplication.

Particularizing for the field R, let (μ,ϕ) be a dyad, which represents a quantity of a certain magnitude, where μ and ϕ can be any; if the magnitude is a scalar, $\mu=q$ will be a real number with $q\in$R. Any quantity of the given magnitude can be taken as the unit $\phi=U$, and we will denote the universal set of all of them

37

with $\{U\}$; therefore, we will have that every unit U will be in the total set of quantities $\{U\}$ and we will write $U \in \{U\}$. It is concluded with this notation that every dyad (q,U) will be an element of the set of all possible dyads, which we will denote $\{R,U\}$ and we will call it the dyadic set of the magnitude considered relative to the quantity U; this set in turn is evidently constructed with the Cartesian product of R and $\{U\}$, that is, $\{R,U\} = R \times \{U\}$. We reiterate that U represents any quantity of the magnitude in question, which in any case could be taken as a pattern. We observe that $\{R,U\} = \{U\}$. Obviously, for any quantity $U_0 \in \{U\}$ the dyadic set constructed with it $\{R,U_0\}$ is complete, it includes all quantities of magnitude, so it coincides with $\{R,U\}$, so $\{R,U_0\} = \{R,U\}$. Dyads can be composed among themselves by means of internal composition laws, defined by establishing applications of the Cartesian product $\{R,U\} \times \{R,U\}$ in the same $\{R,U\}$; and they can also be composed with the elements of other sets, such as R, by means of external composition laws such as applications of $R \times \{R,U\}$ in $\{R,U\}$, so the task of establishing an adequate algebra for them must be addressed. We will verify that the field conditions of R are necessary for the dyadic structure to maintain the associative, commutative and distributive properties in its operations. Instead, we will see that dyadic entities lack unitary and inverse elements in the sense that refers to the internal composition laws inherent in the group structure. In summary, for scalar quantities, we have the following fundamental analytic definitions:

$\{U\}$ = {set of all quantities U of a magnitude}
Dyadic set of U over R: $\{R,U\} = \{(q,U) \,|\, q \in R\} = R \times \{U\}$
Every dyad is $(q,U) \in \{R,U\}$ with $q \in R$ y $U \in \{U\}$

In turn, the vector dyads form a set that can be symbolized $\{R^3,U\}$ or if preferred $\{V^3,U\}$ or $\{E^3,U\}$, which are equivalent notations used interchangeably. They are susceptible to being composed among themselves by means of internal composition laws, with applications of the Cartesian product $\{R^3,U\} \times \{R^3,U\}$

in $\{R^3, U\}$; and they can also be composed with the elements of other sets, such as R, by means of external composition laws, with applications of $R \times \{R^3, U\}$ in $\{R^3, U\}$, so the task of establishing an adequate algebra for them will also be addressed, which we will have to ensure is as isomorphic as possible with the structure of the vector space R^3 over R.

We will see that the above composition laws do not present too much difficulty for their analytical formulation, because they are all **additive operations** built on the same magnitude, as defined in sections V to XI, despite which they reveal important properties, resolving, for example, the historical mysteries of the dimensionless nature of angular magnitudes, and in turn promoting the development of new and important concepts such as «dysmetric» density.

We will find more resistance in the foundation of **multiplicative operations** based on two or more equal or different magnitudes with any units U_1 and U_2 for the case of two factors, whose definitions and properties are established in sections XII to XVII. However, it will be possible to establish coherent applications between sets such as $\{R, U_1\} \times \{R, U_2\}$ and $\{R, U_C\}$, where U_C indicates a new composite unit produced when operating on U_1 and U_2 for scalar magnitudes and analogously for vector magnitudes. And this notable capacity to generate new magnitudes is precisely the characteristic note of such multiplicative operations. The sets $\{R, U_1\}$ and $\{R, U_2\}$ may be equal, when they refer to the same magnitude, but **the dyadic set $\{R, U_C\}$ will always be different from the previous two, because its elements are quantities of a different magnitude, born by multiplying magnitudes in accordance with what is explained in section XII.** Thus it turns out that, even when $\{R, U_1\}$ and $\{R, U_2\}$ are the same set, for example, if both referred to the magnitude length, it will turn out that $\{R, U_C\}$ represents all the quantities that an area can take, because **the product of two lengths gives rise to a new magnitude composed with them that we call surface or area.** In this way, what we could call **generating external composition laws** appear, which have the special quality of

39

producing new magnitudes from the multiplication of any others. Once this is done, it will be understood that this greater difficulty of such multiplicative operations is what has caused everyone, including the International System of Units, to ignore them and replace them with an easy, arbitrary and illusory undesirable hypothesis of «arithmetization» of magnitudes, an error that we have assumed by allowing ourselves to be deceived by arithmetic symbology and creating erroneous concepts. For example, physical units and any quantity of magnitude cannot have multiplicative inverses as if they were internal operations, because the multiplication of magnitudes is externally generative. Thus, notations such as U^{-1} must be reformulated, as explained in section XIV and in the appendix. Symbologies such as m^{-1}, s^{-1} or kg^{-1} must be expressly defined for physical quantities. Note also that not every external law is a generating law, although every generating law must be external. For example, applications of $R \times \{R, U\}$ to $\{R, U\}$, being external, do not generate a new quantity, because the set $\{R, U\}$, the image of that external law, is one of the initial sets. Generating laws are essential laws for the algebra of quantities, since every physical formulation is built with them, giving rise to new composite quantities. For example, Newton's second law involves composing mass and acceleration, relating them to another different magnitude: force. In turn, by composing length and time, a new magnitude arises, which is speed, or by operating with mass and volume, we generate another magnitude, which is density.

In any case, it is clear that **the algebra of magnitudes must obey operational criteria with dyads,** so we must establish specific composition laws that allow us to build sui generis structures and take into account the dyadic nature of physical phenomena, avoiding illusory simplifications.

Section IV

DEFINITIONS OF HOMOGENEITY,
UNIFORMITY AND EQUALITY

Two scalar dyads $a_1 \ U_1$ and $a_2 \ U_2$, each formed by a real number and a unit, we will say that they are **homogeneous** if and only if they refer to quantities of the same magnitude, that is, if their units U_1 and U_2 symbolize quantities uncountable empirical values of the same magnitude.

In turn, the concretes or dyads whose unit is the same we will say that they are **uniform**. So all elements of a set like $\{R, U\}$ are concrete uniform scalars.

On the other hand, two concrete or homogeneous scalar dyads $a_1 \ U_1$ and $a_2 \ U_2$ we will say that they are equal if and only if, by definition, they describe the same quantity of the associated magnitude, and we will symbolize the **equality** with the usual equal sign with an expression such as the next:

$$a_1 \ U_1 = a_2 \ U_2 \qquad [4.1]$$

Together $\{R, U_1\}$ determines all quantities of the associated magnitude with reference to the unit U_1. In turn, the set $\{R, U_2\}$ also determines all the quantities of the same magnitude linked to the unit U_2. The equal elements of both sets are related by equation [4.1]. Under these conditions, the criterion of equality of two concretes or uniform scalar dyads cannot be established in a more convenient way than this: we will say that two uniform concretes $a_1 \ U$ and $a_2 \ U$ are equal if and only if they have the same numerical or measured part, that is, if the real numbers $a_1 = a_2$ are equal.

What has been said for the dyad scalars with respect to homogeneity and uniformity must be analogous for the vector dyads $\overline{a}_1 \ U_1$ and $\overline{a}_2 \ U_2$, and regarding equality it must mean

41

that both dyads refer to the same quantity of the vector magnitude in question. The analytic expression for equality must logically respond to the dyadic equation:

$$\overline{a}_1 \, U_1 = \overline{a}_2 \, U_2 \qquad [4.2]$$

If the dyads were uniform, it will result $U_1 = U_2 = U$, and it will be said that two uniform vector dyads $\overline{a}_1 \, U$ and $\overline{a}_2 \, U$ are equal if and only if the vector equality $\overline{a}_1 = \overline{a}_2$ is verified.

Given two homogeneous units U_1 and U_2, that is, associated to the same magnitude, it is necessary to axiomatize, because physical observations so advise, that there is a real number k such that:

$$U_2 = (1, U_2) = k \, U_1 \qquad [4.3]$$

We will call this statement the **axiom of continuity** and it will promote the transformation of quantities of homogeneous magnitudes, linking them to the same unit, which will allow us to add homogeneous concrete entities, as we will see later. In particular, if the units are uniform, we will have $k = 1 \in R$. We must note that the algebraic meaning of [4.3] will be completed with the definition [9.1] of multiplication of a scalar by a dyadic entity.

Note an essential condition of the definition of equality, which is that only homogeneous quantities can be compared, that is, of the same magnitude, so that **every physical equation of equality establishes a law that relates the quantities of magnitudes specified by predefined dyadic algebra operations. and both members must be homogeneous.**

At the end of sections IX and XI we complete the concept of dyadic equality once the operations that allow it to be established with algebraic rigor have been defined: the multiplication of a scalar by a dyad and the homogeneous dyadic division.

Section V

DEFINITION OF DYADIC ADDITION

A first observation to bear in mind when defining operations with dyadic entities is that certain rules of an axiomatic nature must be respected, based on rational observations of the facts. Thus, to add concretes it is required that the addends be homogeneous, that is, that they refer to the same magnitude, although it would be admissible that the units expressed in the addends were not the same. It would not make sense to add meters with kilograms, because the result could not be indicated in any of the units of the addends; but it does have coherence to add seconds and hours, because the magnitude associated with both units is time, so they can be added and express the sum in seconds or hours, although for this one of the addends should be converted to the unit of the other, otherwise the sum of different units would also be meaningless and should be rejected, because the dyadic addition consists in fact in counting the elements of the addends and those that can be said to be equal.

Therefore, it is necessary to admit as a prior and necessary axiom for the dyadic addition to be valid that in every sum of concrete entities the addends must refer to the same magnitude, that is, the addends must be homogeneous, and before adding them they must be represented in the same unit, a statement that we could call the **axiom of uniformity** of addition. Such a transformation will always be possible by virtue of the axiom of continuity.

Let's start with adding scalar dyads. The sum cannot be conceived in any other way than by establishing a law of internal composition called addition between the uniform scalar concretes, through an application of the Cartesian product $\{R,U\} \times \{R,U\}$ in $\{R,U\}$. In this way, when the addends are already expressed in the same unit U of a certain magnitude, that is, when they are uniform, the sum of two scalar dyads $a\,U$ and $b\,U$ can be written $a\,U + b\,U$, with the meaning of counting the number of units U that accept the two addends at the same time; and here there is no other option but to admit as a result of the sum the statement that it is equal to $(a+b)$ units U, which would be written with the dyadic form $(a+b)\,U$, because what is added are quantities of elements all equal to the quantity symbolized by the letter U, which is elementary arithmetic, with which it is fully based on an application of the Cartesian product set $\{R,U\} \times \{R,U\}$ in $\{R,U\}$, characterized by the abstract formula that describes the **addition definition equation for scalar diads**:

$$a\,U + b\,U = (a+b)\,U \qquad [5.1]$$

Observing the previous definition, it should be emphasized, even at the risk of seeming repetitive, that it represents the addition of elements equal to the quantity of the unit considered, so it should be read with a meaning like the following: the addition of a quantities equal to the quantity of the unit U added to b quantities equal to the quantity of the unit U equals the sum of real numbers $(a+b)$ quantities equal to the quantity of the unit U; which reduces the addition of dyads to the addition of real numbers with perfect precision.

In order not to reiterate heavily the expression «quantities equal to the quantity of the unit U», we simply substitute the letter U, and thus we will speak simply of «a units U», of «b units U» or of «$(a+b)$ units U».

We must note that the **principle of symbolic economy** leads us to identify different laws of composition with the same symbol: in effect, the definition equation of the concrete addition [5.1] includes in the first member the plus sign in $a\,U + b\,U$, with the meaning of addition of scalar dyads, while the same sign of the second member in $(a+b)\,U$ refers to the addition of the field of real numbers. In a more abstract and symbolic way but equivalent in result, it could be observed that the addition of scalar dyads behaves analytically as if U were a number, since it could be

44

considered that it reflects the distributive form, which allows to consider that to operate with the addition of scalar concretes it is enough to do it symbolically as if the symbol of the unit of the addends were one more algebraic element and then read the result with the meaning that the addition of concretes is a dyadic entity with a primary equal to the addition in R of the real parts of the addends, associated to the same unit as these.

Let's see an example of addition: if you wanted to add the amount of time of 2 minutes, abbreviated 2 min, and the amount of time of 15 seconds, expressed 15 s, sum that symbolically is 2 min +15 s, both addends are homogeneous, because they refer to the same magnitude, time, then they can be added; but first they must be expressed in the same unit of time, given the axiom of uniformity of addition; be this the second, by definition of minute, we will have 2 min =120 s, so the sum to be calculated is 120 s +15 s, and now, since the addends specify quantities of the same unit, so they are uniform, it is enough add the numerical parts according to the definition of addition to state that the sum is 135 s.

It could have been reasoned by operating symbolically with the distributive property for the letter s in the following way, first applying the axiom of uniformity of addition to put the minutes in seconds:

$$2\ min+15\ s=120\ s+15\ s=(120+15)\ s=135\ s$$

And the result would be read with the meaning that the sum of 2 minutes and 15 seconds equals 135 seconds. Therefore, the addition of uniform scalar dyads allows operating in abstract analytical terms with the unit symbols as if they were algebraic elements, although without losing sight of the proper meaning of the indicated notation. If you think about it, this circumstance is not strange, because when 15 s is indicated the meaning is $15\times1\ s$, that is, that 15 s actually represents 15 times the quantity of the time magnitude contained in one second, **a quantity that cannot be expressed numerically it is symbolized by the letter s**, that is, the product of the real number 15 by s; so the distributive behavior

45

of the symbol s in the above reasoning scheme is not illogical, but undeniable. Simply, the number 15 acts with respect to the quantiity of time s as a **multiplier**.

In the case of the addition of vector dyads, the reasoning scheme to define this operation is totally similar to that of scalars, with the exception that the supporting structure is that of the vector space R^3 or V^3 and, therefore, The addition to which it is reduced is not that of R but the vector, according to the following equation for the **definition of addition of vector dyads**:

$$\overline{a}\ U + \overline{b}\ U = (\overline{a} + \overline{b})\ U \qquad [5.2]$$

Note that the addition of the second member of [5.2] is not that of R, as in the case of [5.1], but the vector addition of R^3 or V^3; while the addition sign of the first member indicates the sum of vector dyads defined here. The same addition sign for two different composition laws.

If you want to be more precise in the differentiation of the composition laws involved, even if only for didactic purposes, to better explain the precise meanings of the definitions [5.1] and [5.2] they must be identified with those of the exact equations, which they differentiate the operations and they could be written with «⊕» for scalar or vector dyadic additions, as well as «+» for sums of scalars or vectors. This is how the explicit analytic expressions of scalar and vector dyadic additions result:

$$a\ U \oplus b\ U = (a + b)\ U \qquad [5.3]$$

$$\overline{a}\ U \oplus \overline{b}\ U = (\overline{a} + \overline{b})\ U \qquad [5.4]$$

And even so, as with the «+» sign between scalars or vectors, which refers to two different operations, we would not be distinguishing in the spellings all those involved, although they are easy to interpret due to the nature of the elements that connect, thus, if the operation sign «⊕» were placed between scalar dyads, the addition would be the scalar dyad; and, if you linked vector dyads, the sum would be the vector dyadic. And so analogously with all the operations involved.

In what precedes, in order to materialize the dyadic addition we have required that the units of the addends be the same, which we have called the axiom of uniformity; however, there is an exception that does allow the addition of non-uniform amounts without reducing them to a common unit: the case where the primaries are equal. In this case, nothing prevents the addition from being analytically expressed when the homogeneous units of the addends are different. Thus, given two units of the same magnitude U_1 and U_2, by the affinity postulate and the isomorphism with the geometric addition of segments described in section XXVIII, article 13, it turns out that adding quantities of magnitudes corresponds biunivocally by affinity with the addition of segments, with which the dyad in explicit notation $(q, U_1 \oplus U_2)$, is equivalent to the sum $(q, U_1) \oplus (q, U_2)$, so it can be indicated:

$$(q, U_1 \oplus U_2) = (q, U_1) \oplus (q, U_2)$$

Or with the classical notation that we have been using, we can also express the above with the form:

$$q \; U_1 \oplus U_2 = q \; U_1 \oplus q \; U_2$$

And the same exception can be established for vector magnitudes, resulting indistinct for both notations:

$$(\overline{q}, U_1 \oplus U_2) = (\overline{q}, U_1) \oplus (\overline{q}, U_2)$$
$$\overline{q} \; U_1 \oplus U_2 = \overline{q} \; U_1 \oplus \overline{q} \; U_2$$

Section VI

COMMUTATIVE PROPERTIES AND
ASSOCIATIVE OF THE DYADIC ADDITION

First, let's look at adding scalar dyads. Once this internal law has been defined on the set of the concrete scalars $\{R, U\}$, it is worth asking whether $a\ U$ and $b\ U$ will turn out to be commutative for any two of its elements. The definition of concrete addition [5.3] allows us to write the equality:

$$a\ U \oplus b\ U = (a+b)\ U$$

The commutative property of the addition of the real numbers in R determines that $a+b=b+a$ then, in effect, the dyad $(a+b)\ U$ is the same as $(b+a)\ U$, which in turn is $b\ U \oplus a\ U$, and with it the scalar dyadic addition verifies the commutative property:

$$a\ U \oplus b\ U = b\ U \oplus a\ U$$

Furthermore, this dyadic addition is associative because, starting from the triple addition $(a\ U \oplus b\ U) \oplus c\ U$, the definition [5.3] of dyadic addition allows us to write the equality without problems:

$$(a\ U \oplus b\ U) \oplus c\ U = [(a+b)\ U \oplus c\ U] = [(a+b)+c]\ U \qquad [6.1]$$

Since the addition of real numbers is associative, we will have the equality in the additive group of R:

$$(a+b)+c = a+(b+c)$$

Therefore, the relationship between the dyads scalars indicated below is justified:

$$[(a+b)+c]\ U = [a+(b+c)]\ U$$

49

The definition of addition [5.3] allows the second member to be decomposed into the dyadic sum of concretes:

$$[a+(b+c)]\ U = a\ U \oplus (b+c)\ U = a\ U \oplus (b\ U \oplus c\ U) \qquad [6.2]$$

So the initial dyad $(a\ U + b\ U) + c\ U$ of [6.1] is the same as the concrete of the second member in [6.2], which is $a\ U \oplus (b\ U \oplus c\ U)$, a result we can call associative property of dyadic addition, written analytically with equality:

$$(a\ U \oplus b\ U) \oplus c\ U = a\ U \oplus (b\ U \oplus c\ U)$$

Regarding the addition of vector dyads, with an identical reasoning to that of scalars, but with the only exception that, instead of the commutative and associative properties of the real numbers of R, based on the commutative and associative properties of the addition of vectors in R^3 or V^3, the commutative and associative properties of the addition [5.4] of vector dyads are concluded, whose analytical forms are:

$$\overline{a}\ U \oplus \overline{b}\ U = \overline{b}\ U \oplus \overline{a}\ U$$

$$(\overline{a}\ U \oplus \overline{b}\ U) \oplus \overline{c}\ U = \overline{a}\ U \oplus (\overline{b}\ U \oplus \overline{c}\ U)$$

Section VII

EXISTENCE OF NEUTRAL AND SYMMETRICAL
ELEMENTS FOR DYADIC ADDITION

Let's see that the addition of scalar dyads defined on $\{R,U\}$ is a law of internal composition such that there is a neutral element. Indeed, we can easily observe that the dyad $0\ U$, where 0 is the real zero, is such that added to any dyad $a\ U$ it verifies the following reasoning:

$$a\ U \oplus 0\ U = 0\ U \oplus a\ U \qquad [7.1]$$

The above equality is a consequence of the commutative property. In turn, the definition [5.3] of addition of scalar dyads allows us to write:

$$a\ U \oplus 0\ U = (a+0)\ U \text{ y } 0\ U \oplus a\ U = (0+a)\ U$$

The zero or neutral element of the real numbers is such that $a+0$ is the same as $0+a$ and equal in both cases to a, with what we have:

$$a+0 = 0+a = a$$

Therefore, the two members of the first relation [7.1] are equal to $a\ U$:

$$a\ U \oplus 0\ U = 0\ U \oplus a\ U = a\ U$$

And this means, by definition of a neutral element, that the dyad $0\ U$ is one of the law of internal composition called dyadic addition and defined in $\{R,U\}$ by [5.3].

Furthermore, for every concrete scalar entity or dyad $a\ U$ can always form the dyad $(-a)\ U$, because in R there exists the opposite $-a$ of every $a \in R$, by virtue of the additive group structure of the set R of real numbers, and so that it is $a+(-a)=0$; so that, adding the two indicated dyads, applying the definition of addition and operating with the real numbers, we have the chain of equalities:

$$a \ U \oplus (-a) \ U = [a + (-a)] \ U = 0 \ U$$

And so it turns out that $(-a) \ U$ is the scalar dyad opposite to the right of $a \ U$. The commutative property makes it unnecessary to check the condition of neutral element from the left $(-a) \ U \oplus a \ U$, which is also satisfied; and, in sum, for every concrete scalar to $a \ U$ there exists another with the form $(-a) \ U$ such that, added both from the right or from the left, give the neutral element $0 \ U$, which means that the set of dyads scalars $\{R, U\}$, formed with the reals R and the unit U, and endowed with the internal composition law of addition defined by [5.3], has the abelian group structure, because the addition verifies the commutative and associative properties, there is a neutral element and for every dyad there is its opposite[3].

In turn, for the vector concretes $\{R^3, U\}$ and, given the group structure for the vector addition of R^3 or V^3, which are the same space symbolized in two different ways, there are the neutral or null $\overline{0}$ and symmetric or opposite vectors $-\overline{a}$ of every vector $\overline{a}$, so that, by means of a reasoning scheme identical to the previous one, with $\overline{a} + \overline{0} = \overline{0} + \overline{a} = \overline{a}$ and with $\overline{a} + (-\overline{a}) = (-\overline{a}) + \overline{a} = \overline{0}$, it can be concluded that there are null and opposite vector dyads, and that are those indicated by the symbols $\overline{0} \ U$, for the null vector concrete, and $(-\overline{a}) \ U$, for the opposite vector dyad of any other $\overline{a} \ U$, because with [5.4] the following reasonings are spun:

$$\overline{a} \ U \oplus \overline{0} \ U = \overline{0} \ U \oplus \overline{a} \ U = (\overline{a} + \overline{0}) \ U = (\overline{0} + \overline{a}) \ U = \overline{a} \ U$$

$$\overline{a} \ U \oplus (-\overline{a}) \ U = [\overline{a} + (-\overline{a})] \ U = [(-\overline{a}) + \overline{a}] \ U = (-\overline{a}) \ U \oplus \overline{a} \ U = \overline{0} \ U$$

---

[3] An introduction to algebraic structures can be found in the same author's syllabus, «Lesson 37» of Mathematize 1.

Section VIII

DEFINITION OF DYADIC SUBTRATION

The generic operation called subtraction derives from addition, therefore, to be consistent with this algebraic criterion, it must be admitted that the subtraction of dyadic entities must satisfy the same axiom of uniformity of operating with equal units, and this because the definition of concrete subtraction cannot have any other foundation than the dyadic addition, in which all the terms are associated to the same unit, which leads us to the following formulation: **the subtraction of a minuend and a subtrahend, both being dyads scalars of the set $\{R, U\}$ or dyads vectors of the set $\{R^3, U\}$, is said equal to the difference if and only if the subtrahend added to the difference equals the minuend.**

So that addition and subtraction require by the condition of law of internal composition that the addends, in one case, or the minuend and the subtrahend, in the other, refer to the same unit, that is, that they be uniform. Given the abelian group structure of $\{R, U\}$ and $\{R^3, U\}$, we have no problem in defining the subtraction of concrete entities as the sum of the minuend and the opposite of the subtrahend, which always exists, as we have established in the previous section, and thus we will have a result that added to the subtrahend will be equal to the minuend, according to the generic, convenient and usual definition that has been accepted for the subtraction of numerical entities. So we agree to indicate the subtraction of any two scalar dyads from $a\,U$ and b $b\,U$ as the application of $\{R, U\} \times \{R, U\}$ to $\{R, U\}$ defined with the following formula, which assumes, as the addition, that uniform quantities referring to the same unit are subtracted, because it operates as a law of internal composition on the set of the concrete scalars $\{R, U\}$, and written applying the symbolic economy of operations is as follows:

53

$$a\ U-b\ U=[a+(-b)]\ U=(a-b)\ U \tag{8.1}$$

With this definition the concrete or dyadic scalar subtraction is reduced to that of R and, as we pointed out for the addition, the same subtraction sign with the hyphen is used with different meanings, because in [8.1] the minus of $a\ U-b\ U$ refers to the subtraction of scalar dyads and the minus of $(a-b)\ U$ indicates the subtraction of real numbers.

In turn, the subtraction or subtraction of any two vector dyads $\overline{a}\ U$ and $\overline{b}\ U$ is defined as the scalar with the application of $\{R^3, U\} \times \{R^3, U\}$ in $\{R^3, U\}$ such that, like addition, it assumes that uniform quantities referred to the same unit are subtracted, because it operates as a law of internal composition on the set of vector dyads $\{R^3, U\}$, and this according to the definition equation:

$$\overline{a}\ U-\overline{b}\ U=[\overline{a}+(-\overline{b})]\ U=(\overline{a}-\overline{b})\ U \tag{8.2}$$

Here it should also be noted, as we can see in the addition, that the definitions are such that the abbreviation or symbol of the unit behaves for the purposes of formal writing with the distributive property, as if the unit symbol were another algebraic element.

Therefore, the different meanings that correspond to the same sign with which the different operations are indicated, depending on the elements between which it is located, should be noted, so the precise meanings of definitions [8.1] and [8.2] should be understood as the of the exact equations, which differentiate the composition laws and which could be written with «⊖» for scalar or vector dyadic differences, as well as «−» for subtractions of scalars or vectors, in this way:

$$a\ U \ominus b\ U=[a+(-b)]\ U=(a-b)\ U \tag{8.3}$$

$$\overline{a}\ U \ominus \overline{b}\ U=[\overline{a}+(-\overline{b})]\ U=(\overline{a}-\overline{b})\ U \tag{8.4}$$

As we have already indicated, the symbol $-b$ denotes the opposite real number of b, and with $-\overline{b}$ the opposite vector of $\overline{b}$.

The dyadic subtraction can be deduced from the addition and based on the generic subtraction criterion. To do this, consider the following scalar sum:

$$d\ U \oplus s\ U = m\ U \qquad\qquad [8.5]$$

The symbology of addition has simply been adapted to indicate with the letters m a minuend, s a subtrahend and d for the difference that corresponds to them. The usual subtraction criterion, as operation that, given an addition, allows one of the addends to be obtained as a function of the sum and the other by adding, authorizes the establishment of the dyadic difference, distinguished with the sign «$\ominus$», by means of the equation:

$$m\ U \ominus s\ U = d\ U \qquad\qquad [8.6]$$

The definition [5.3] of dyadic addition applied to [8.5], allows us to write:

$$(d+s)\ U = m\ U \qquad\qquad [8.7]$$

The equality criterion of section IV applied to expression [8.7] gives us the relation $(d+s)=m$, and the subtraction in R leads us to write $d=m-s$. So, substituting d in [8.6], we have:

$$m\ U \ominus s\ U = (m-s)\ U \qquad\qquad [8.8]$$

Conclusion [8.8] is identical to definition [8.1], and it means that the dyadic difference between two scalar concretes, called minuend and subtrahend, is a dyad called difference whose primary is the subtraction in R of the primaries and with the same secondary that they.

The reasoning for the subtraction of vector dyads is completely analogous to the previous one with scalars, given the additive and abelian group structure of R^3, which presents the same formal properties for the sum of vectors that are given with real numbers.

Section IX

DEFINITION OF MULTIPLICATION
OF A REAL NUMBER FOR A DYAD

If a quantity $a\ U$ or (a, U) of a certain scalar magnitude, from the set of uniform concretes $\{R, U\}$, is multiplied by a real number p, by definition, we will establish that the result is a dyad such that Its primary or measure is the real product $a \times p$. For now we will use the same multiplication sign for the new operation, but knowing that it is not the product of real numbers, but the multiplication of a scalar by a quantity of magnitude. Then we will detail our own symbology to highlight the difference between these two operations. In analytical terms the definition of this product is:

$$(a\ U) \times p = (a \times p)\ U \qquad [9.1]$$

If in the product $(a\ U) \times p$ its factors are commuted to form the multiplication $p \times (a\ U)$, we must establish by convenient definition that both quantities coincide, so we can axiomatically admit the **commutative property** of the product of a real number times a concrete scalar, expressing it analytically using the expression:

$$(a\ U) \times p = p \times (a\ U) \qquad [9.2]$$

In the interest of mathematical precision, it is worth clarifying that the definition of the product of a scalar by a dyad, defined here analytically with the definition equations [9.1] and [9.2], does not represent more than a law of external composition or application of the Cartesian product $R \times \{R, U\}$ on $\{R, U\}$ on the left and the symmetric on the right of $\{R, U\} \times R$ on $\{R, U\}$.

It should be idle and remember that the same multiplication sign, generally the cross «×» or a blank space, are used to symbolize different laws of composition, depending on the pairs of elements between which they appear. However, let's clarify it: indicating «×» the product of R, designating «•» the product of a scalar by a vector, and indicating with the sign «∘» the product of a real number by a scalar or vector dyadic element, the definitions [9.1] and [9.2] must be interpreted according to the explicit expressions:

$$(a\ U) \circ p = p \circ (a\ U) = (a \times p)\ U = (p \times a)\ U$$

$$(\overline{a}\ U) \circ p = p \circ (\overline{a}\ U) = (\overline{a} \bullet p)\ U = (p \bullet \overline{a})\ U$$

Likewise, it is observed that the previous definitions allow the unit U to be manipulated symbolically as if it were one more algebraic element, such that formally in writing and together with the other elements they appear to be commutative and associative, although the operations on each member are different.

Multiplying the null element of R or zero by any concrete scalar to U of $\{R, U\}$, we will have:

Multiplicando el elemento nulo de R o cero por cualquier concreto escalar $a\ U$ de $\{R, U\}$, se tendrá:

$$0 \circ (a\ U) = (a\ U) \circ 0 = (0 \times a)\ U = (a \times 0)\ U = 0\ U$$

That is, any concrete of $\{R, U\}$ multiplied by the zero scalar, $0 \in R$, the null element of the addition in R, from the right and from the left, is equal to the null element $0\ U \in \{R, U\}$.

In turn, the unit element for the multiplication of R, which is usually symbolized by the number 1, is such that, when compounded with this new external law, it leaves any scalar dyad unchanged, which we can verify without more than taking the generic $a\ U$ and compose it with the unit of R, according to the reasoning that is based on [9.1], [9.2] and on the condition of 1 as a unit element of the product in R, which is such that $a \times 1 = 1 \times a = a$, all of which motivates the following reasoning of analytic dyadic algebra:

$$1 \circ (a\ U) = (a\ U) \circ 1 = (1 \times a)\ U = (a \times 1)\ U = a\ U$$

So, in effect, the unit $1 \in R$ of the multiplicative group R is such that it operates as a unit scalar for the external law «$\circ$» of $R \times \{R, U\}$ in $\{R, U\}$ or $\{R, U\} \times R$ in $\{R, U\}$.

In turn, for the uniform vector dyads of $\{R^3, U\}$, the multiplication by a scalar «$\circ$» must refer to the external law «$\bullet$» of the vector space R^3 or V^3 over R, with the definition having the same form of [9.1] and [9.2], although with the proper meaning of this algebraic structure:

$$(\overline{a}\ U) \circ p = (\overline{a} \bullet p)\ U \qquad\qquad [9.3]$$

$$(\overline{a}\ U) \circ p = p \circ (\overline{a}\ U) \qquad\qquad [9.4]$$

These definition equations represent an external composition law «$\circ$» or application of the Cartesian product $R \times \{R^3, U\}$ in $\{R^3, U\}$ to the left and $\{R^3, U\} \times R$ in $\{R^3, U\}$ by the right, defined as a function of the external law «$\bullet$» of R^3 or V^3 over R. The vector space structure of R^3 guarantees that for the null and unit elements of R we have that it is $0 \bullet \overline{a} = \overline{a} \bullet 0 = \overline{0}$, being $\overline{0}$ the null vector of the vector addition in R^3, and $1 \bullet \overline{a} = \overline{a} \bullet 1 = \overline{a}$, where 1 is the unit element of the multiplication in R, so here too we have the same properties deduced for scalar dyads, according to the following schemes ilatives:

$$0 \circ (\overline{a}\ U) = (\overline{a}\ U) \circ 0 = (0 \bullet \overline{a})\ U = = (\overline{a} \bullet 0)\ U = \overline{0}\ U$$

$$1 \circ (\overline{a}\ U) = (\overline{a}\ U) \circ 1 = (1 \bullet \overline{a})\ U = (\overline{a} \bullet 1)\ U = \overline{a}\ U$$

Once this operation is defined, we are now in a position to complete the dyadic equality criterion in section IV. Let there be two equal dyads $(a_1\ U_1) = (a_2\ U_2)$, the axiom of continuity guarantees that there exists $k \in R$ such that $U_1 = k \circ U_2$. Note that the quantity U_1 is the abbreviated form of the dyad $(1\ U_1)$ and U_2 is the same quantity as $(1\ U_2)$, so we have:

$$U_1 = (1\ U_1) = k \circ U_2 = k \circ (1\ U_2) = (k\ U_2)$$

$$(a_1\ U_1) = a_1 \circ (1\ U_1) = a_1 \circ (k\ U_2) = [(a_1 \times k)\ U_2] = (a_2\ U_2)$$

And so, the criterion of equality of uniform dyads, which requires the equality of the primaries, gives us $a_2 = a_1 \times k$. Therefore, admitting by axiom that all dyadic equality $(a_1 \; U_1) = (a_2 \; U_2)$ requires homogeneity, that is, that equal dyads must refer to the same magnitude, since the axiom of continuity guarantees that there exists k such that $U_1 = k \circ U_2$, it turns out that the dyadic primaries must in turn satisfy that $a_2 = a_1 \times k$.

Let us now use the dyadic notation with an inner comma for greater clarity. **The definition of multiplication by a scalar allows us to write $(a, U) = a \circ (1, U) = a \circ U$, then, $(a, U) = a \circ U$.** We must ask ourselves what happens to a dyad (a, U) when its unit is multiplied by a number p. Have:

$$(a, p \circ U) = a \circ (1, p \circ U) = a \circ (p \circ U) = a \circ (p, U) = (a \times p, U)$$

That is, if the secondary of a dyad (a, U) is multiplied by a number p, the resulting dyad is $(a \times p, U)$, which is the same as that obtained by applying the definition of the product by a scalar. So we can conclude the following property: **in the product of a dyad and a scalar, it is indifferent to multiply its primary or its secondary by said number.**

In other words, when in a dyad only the primary or only the secondary is multiplied by a number, the magnitude quantity is multiplied by that same number. This property will be useful to us later for reasoning with **homogeneous dyadic division** and when we come to formulate the **equivalence classes** of quantities of any magnitude.

Defined the exterior product of a number p by a dyad (a, U) with the expression $p \circ (a, U) = (p \times a, U)$, it is enough to take $(p \times a, U)$ as dividend, p as divisor and (a, U) as a quotient to define the division of a dyad by a number, resulting in a dyadic quotient. It is immediate to observe that **the quotient dyad has as its primary the numerical quotient between the primary of the dividend and the number of the divisor**. In turn, we observe that dividing a dyad by a number gives another uniform dyad as a quotient with the first one of the dividend.

60

Section X

DISTRIBUTIVE PROPERTIES OF THE
DYADIC MULTIPLICATION BY A SCALAR

Given the scalar or real number p of R and the dyad scalars $a\ U$ and $b\ U$ of $\{R, U\}$, let's compose the dyad $p \circ (a\ U \oplus b\ U)$ from the left, the same reasoning would be had for the product from the right. Let us differentiate the signs of each operation. The definition of scalar concrete addition [5.3] will allow to write:

$$p \circ (a\ U \oplus b\ U) = p \circ [(a+b)\ U] \qquad [10.1]$$

The definition of the product of a scalar and a dyad, described by [9.3] and [9.4], allows us to transform the second member of the previous expression:

$$p \circ [(a+b)\ U] = [p \times (a+b)]\ U$$

The distributive property of the product with respect to the sum in R is $p \times (a+b) = (p \times a) + (p \times b)$, which allows the conversion:

$$[p \times (a+b)]\ U = [(p \times a) + (p \times b)]\ U$$

The definition of concrete addition [5.3] favors decomposing the second member into two addends:

$$[(p \times a) + (p \times b)]\ U = (p \times a)\ U \oplus (p \times b)\ U \qquad [10.2]$$

In conclusion, the first member of the equality [10.1] is equal to the second member of [10.2], resulting in the written form of the distributive property of the left product of a scalar with respect to the dyadic addition:

$$p \circ (a\ U \oplus b\ U) = (p \times a)\ U \oplus (p \times b)\ U$$

And by a completely similar reasoning, the right distributive property is also concluded:

$$(a\ U \oplus b\ U) \circ p = (a \times p)\ U \oplus (b \times p)\ U$$

Since in R they are $p \times a = a \times p$ and $p \times b = b \times p$, we verify the equality $p \circ (a\ U \oplus b\ U) = (a\ U \oplus b\ U) \circ p$, in coherence with the commutativity axiom [9.2].

To check that the reciprocal also holds, let's take two scalars p and q of R, and a dyad $a\ U$ of $\{R, U\}$. Let's compose the dyad $(p + q) \circ (a\ U)$. By definition of this external law with [9.1] and [9.2] or their explicit [9.3] and [9.4], we will have on the left and analogously on the right:

$$(p + q) \circ (a\ U) = [(p + q) \times a]\ U \qquad [10.3]$$

The distributive property in R, with $(p + q) \times a = (p \times a) + (q \times a)$, justifies taking the next logical step and writing:

$$[(p + q) \times a]\ U = [(p \times a) + (q \times a)]\ U$$

The definition of addition of concretes [5.3] motivates the passage to the following line of reasoning:

$$[(p \times a) + (q \times a)]\ U = (p \times a)\ U \oplus (q \times a)\ U \qquad [10.4]$$

Therefore, the first member of the expression [10.3] that initiates the reasoning is equal to the second of the last one [10.4], reflecting the reciprocal distributive property on the left:

$$(p + q) \circ (a\ U) = (p \times a)\ U \oplus (q \times a)\ U$$

The same scheme is presented for the reciprocal on the right, with the result:

$$(a\ U) \circ (p + q) = (a \times p)\ U \oplus (a \times q)\ U$$

Since in R they are $p \times a = a \times p$ and $q \times a = a \times q$, we check the equality $(p + q) \circ (a\ U) = (a\ U) \circ (p + q)$, in coherence with the axiom [9.2] of commutativity.

By means of a totally analogous reasoning scheme applied to the definition of concrete subtraction given by the simplified [8.1] or its explicit [8.3], we arrive at the distributive properties of the product by a scalar with respect to the dyadic and real subtractions, from the left and on the right, which can be expressed by the following four equations:

$$p\circ(a\ U\ominus b\ U)=(p\times a)\ U\ominus(p\times b)\ U$$

$$(p-q)\circ(a\ U)=(p\times a)\ U\ominus(q\times a)\ U$$

$$(a\ U\ominus b\ U)\circ p=(a\times p)\ U\ominus(b\times p)\ U$$

$$(a\ U)\circ(p-q)=(a\times p)\ U\ominus(a\times q)\ U$$

Even at the risk of making us heavy, it should be noted again that, for symbolic economy, it is usual to write expressions with the same multiplication, addition and subtraction symbols, to refer indistinctly to the respective composition laws in R and in {R,U} Therefore, depending on the positions of such signs between the composite pairs, the appropriate meanings will have to be attributed to them. With such a principle of symbolic economy, the equations derived earlier would be simplified as follows:

$$p\times(a\ U+b\ U)=(p\times a)\ U+(p\times b)\ U$$

$$(p+q)\times(a\ U)=(p\times a)\ U+(q\times a)\ U$$

$$(a\ U+b\ U)\times p=(a\times p)\ U+(b\times p)\ U$$

$$(a\ U)\times(p+q)=(a\times p)\ U+(a\times q)\ U$$

$$p\times(a\ U-b\ U)=(p\times a)\ U-(p\times b)\ U$$

$$(p-q)\times(a\ U)=(p\times a)\ U-(q\times a)\ U$$

$$(a\ U-b\ U)\times p=(a\times p)\ U-(b\times p)\ U$$

$$(a\ U)\times(p-q)=(a\times p)\ U-(a\times q)\ U$$

For the vector dyads of $\{R^3,U\}$, using an identical logical scheme, although based on the definition of addition of vector concretes [5.2] or its explicit [5.4] and on the distributive property, which guarantees the structure of the space itself R^3 vector, we will have the same formal equations as for the scalar dyads, on the left and on the right, although now referring to vector dyads. With all the explicit operations we will have:

$$p \circ (\overline{a} \ U \oplus \overline{b} \ U) = (p \bullet \overline{a}) \ U \oplus (p \bullet \overline{b}) \ U$$

$$(p + q) \circ (\overline{a} \ U) = (p \bullet \overline{a}) \ U \oplus (q \bullet \overline{a}) \ U$$

$$p \circ (\overline{a} \ U \ominus \overline{b} \ U) = (p \bullet \overline{a}) \ U \ominus (p \bullet \overline{b}) \ U$$

$$(p - q) \circ (\overline{a} \ U) = (p \bullet \overline{a}) \ U \ominus (q \bullet \overline{a}) \ U$$

$$(\overline{a} \ U \oplus \overline{b} \ U) \circ p = (\overline{a} \bullet p) \ U \oplus (\overline{b} \bullet p) \ U$$

$$(\overline{a} \ U) \circ (p + q) = (\overline{a} \bullet p) \ U \oplus (\overline{a} \bullet q) \ U$$

$$(\overline{a} \ U \ominus \overline{b} \ U) \circ p = (\overline{a} \bullet p) \ U \ominus (\overline{b} \bullet p) \ U$$

$$(p - q) \circ (\overline{a} \ U) = (p \bullet \overline{a}) \ U \ominus (q \bullet \overline{a}) \ U$$

And, applying the principle of symbolic economy, the same simplified expressions appear correlative:

$$p \times (\overline{a} \ U + \overline{b} \ U) = (p \times \overline{a}) \ U + (p \times \overline{b}) \ U$$

$$(p + q) \times (\overline{a} \ U) = (p \times \overline{a}) \ U + (q \times \overline{a}) \ U$$

$$p \times (\overline{a} \ U - \overline{b} \ U) = (p \times \overline{a}) \ U - (p \times \overline{b}) \ U$$

$$(p - q) \times (\overline{a} \ U) = (p \times \overline{a}) \ U - (q \times \overline{a}) \ U$$

$$(\overline{a} \ U + \overline{b} \ U) \times p = (\overline{a} \times p) \ U + (\overline{b} \times p) \ U$$

$$(\overline{a} \ U) \times (p + q) = (\overline{a} \times p) \ U + (\overline{a} \times q) \ U$$

$$(\overline{a} \ U - \overline{b} \ U) \times p = (\overline{a} \times p) \ U - (\overline{b} \times p) \ U$$

$$(p - q) \times (\overline{a} \ U) = (p \times \overline{a}) \ U - (q \times \overline{a}) \ U$$

In any case, although the operations of the first and second members do not coincide, it is justified to manipulate all the elements, including the unit U, with the fiction that they are all algebraic terms of R or R^3, even though U really is not, and this on the basis of the definitions and properties of these external laws, not because the illusion of the mere symbolism of the simplified formulas justifies it in any way.

Section XI

DEFINITION OF DIVISION BETWEEN
HOMOGENOUS DYADS

Let us first deal with the quotient between two homogeneous units, and remember that the units are always scalar concretes. For this, if the homogeneous units U_1 and U_2 are of the same magnitude, the continuity axiom [4.3] makes it possible to ensure that the real number k exists such that $U_2 = k\ U_1$. The symbolic definition [3.1] of scalar concrete determines that this expression means the same as this: $(1\ U_2) = k \times (1\ U_1)$. Note that this multiplication is actually the dyadic symbolized by «∘», given by $(1\ U_2) = k \circ (1\ U_1)$. The common concept of division allows us to consider in the abstract that $(1\ U_2)$ is associated with a dividend, that $(1\ U_1)$ is a divisor and that k is a quotient, and thus it can be considered that the division between $(1\ U_2)$ and $(1\ U_1)$, which nothing prevents symbolizing with the form of a quotient with a double bar $(1\ U_2) /\!/ (1\ U_1)$, or what would mean the same $U_2 /\!/ U_1$, will be equal to a quotient k, which is a real number. From this we can conclude that the ratio or division between two homogeneous units must always be equal to a real number. If the units were the same, we would have that k is the unit of R, that is, the quotient of every unit between itself will be the real unit, which justifies the way of operating by simplifying the symbols of the units that appear in the numerator and denominator of physical equations. We thus observe that it is in the axiom of continuity [4.3] that the germ of the definition of division of homogeneous units is found.

To practice, let's reason with symbolic economy and now take two homogeneous scalar concretes a1 $a_1\ U_1$ and $a_2\ U_2$, we know that there exists k such that $U_2 = k\ U_1$, so that the concrete $a_2\ U_2$, given the definition [3.1] of scalar dyad, is you can write like this:

$$a_2 \ U_2 = a_2 \times (1 \ U_2) = a_2 \times (k \ U_1) = (a_2 \times k) \ U_1$$

Multiplying the equality by a_1 and, operating with the product [9.1] and [9.2] by a scalar and the laws of R, we will have:

$$a_1 \times (a_2 \ U_2) = a_1 \times [(a_2 \times k) \ U_1] = (a_1 \times a_2 \times k) \ U_1 = (a_2 \times k) \ (a_1 \ U_1)$$

Now multiplying by a_1^{-1}, knowing that $a_1 \times a_1^{-1} = a_1^{-1} \times a_1 = 1$, because these entities are elements of R, we will have:

$$(a_2 \ U_2) = (a_1^{-1} \times a_2 \times k) \ (a_1 \ U_1)$$

Assuming that $(a_2 \ U_2)$ is a dividend and $(a_1 \ U_1)$ a divisor, the quotient between the two, which can be symbolized with the common notation $(a_2 \ U_2)/(a_1 \ U_1)$, will be given by the real number that results from the operation of the second member of [11.1]:

$$\frac{a_2 \ U_2}{a_1 \ U_1} = a_1^{-1} \times a_2 \times k = \frac{a_2}{a_1} \times k \qquad [11.1]$$

This means that the quotient between any two homogeneous scalar dyads is the real number given by the last member of the definition equation [11.1].

For vector concretes, we must note that vector algebra is such that multiplication by a scalar relates collinear vectors, so that the concrete division will only be possible when the dividend and divisor primaries are in turn collinear. So that the vector concretes $\overline{a}_1 \ U_1$ and $\overline{a}_2 \ U_2$ are now, such that the vectors $\overline{a}_1$ and $\overline{a}_2$ are collinear and that the units U_1 and U_2 are homogeneous. As in the previous case of scalar concretes, homogeneity assumes that there exists k such that $U_2 = k \ U_1$. In turn, vector algebra ensures that there exists a scalar λ such that $\overline{a}_2 = \lambda \times \overline{a}_1$. Operating with vector algebra and using the definition [9.3] and [9.4] of a concrete product by a scalar, we can write with full justification the following reasoning:

$$\overline{a}_2 \ U_2 = (\lambda \times \overline{a}_1) \ (k \ U_1) = [(\lambda \times \overline{a}_1) \times k] \ U_1 =$$

$$=[(\lambda\times k)\times\overline{a}_1]\,U_1=(\lambda\times k)\times(\overline{a}_1\,U_1)$$

Looking at the first and last members, according to the usual division criteria, we can assume that $(\overline{a}_2\,U_2)$ is a dividend, that $(\overline{a}_1\,U_1)$ is a divisor and that $(\lambda\times k)$ is a quotient. With this, we will arrive at the division formulation between collinear and homogeneous vector concretes, which with the usual symbology will be established by the following definition equation:

$$\frac{\overline{a}_2\,U_2}{\overline{a}_1\,U_1} = \lambda\times k \qquad\qquad [11.2]$$

So, by definition, the quotient of two collinear vector dyads is the scalar $(\lambda\times k)\in R$.

This section has been developed and reasoned with intention, taking into account the principle of symbolic economy, without explicitly differentiating the homonymous operations of multiplication and division, for which, strictly speaking, divisions [11.1] and [11.2] should be written with their own dyadic quotient sign, for which at the beginning we had chosen the double bar, with which these equations should explicitly understand as follows:

$$\frac{a_2\,U_2}{a_1\,U_1} = \frac{a_2}{a_1}\times k$$

$$\frac{\overline{a}_2\,U_2}{a_1\,U_1} = \lambda\times k$$

These expressions justify the simplification of the homogeneous units in numerator and denominator, so that the dyadic ratio of two homogeneous quantities will always result in a determined real number. On the other hand, it is possible to conceive of two other dyadic divisions, which we will also indicate with a double line, although they are different composition laws: those that correspond to a dyadic, scalar or vector dividend, divided by a divisor that is a real number. It is enough to observe the previous equations to deduce these other two, which relate all the elements of the division of a scalar dyad by a real number and the division of a vector dyad by a real number, according to the following two equations:

$$\frac{a_2 \, U_2}{\dfrac{a_2}{a_1} \times k} = a_1 \, U_1 \qquad\qquad [11.3]$$

$$\frac{\overline{a_2} \, U_2}{\lambda \times k} = \overline{a_1} \, U_1 \qquad\qquad [11.4]$$

Therefore, the division of a scalar dyad by a real number produces another scalar dyad, but not just any one, but the one established by equation [11.3]. In turn, the division of a vector dyad by a real number will give rise to another collinear vector dyad that verifies equation [11.4].

Continuing with the completion of the dyadic equality criterion, to indicate that two quantities of the same magnitude $(a_1 \, U_1)$ and $(a_2 \, U_2)$ are equal, we will write $(a_1 \, U_1) = (a_2 \, U_2)$. The parentheses are superfluous, but we specify them to mark the dyadic pairs well. By the scalar multiplication defined above we can write $(a_2 \, U_2) = 1 \circ (a_2 \, U_2)$, with which $(a_1 \, U_1) = 1 \circ (a_2 \, U_2)$. And in this equality we can define $(a_1 \, U_1)$ as a dividend, $(a_2 \, U_2)$ as a divisor and 1 as a quotient, with respect to the multiplication «∘», which we can notice $(a_1 \, U_1) /\!/ (a_2 \, U_2) = 1$. Since we have called this operation homogeneous dyadic division, we will have that the dyadic quotient of two equal dyads, representative of the same quantity of a certain magnitude, is the unit of real numbers. In analytical terms, the following important property of equality results, which is the basis of physical equations:

$$\left(a_1 \, U_1\right) = \left(a_2 \, U_2\right) \;\Rightarrow\; \frac{\left(a_1 \, U_1\right)}{\left(a_2 \, U_2\right)} = 1$$

Section XII

DEFINITION OF MULTIPLICATION
DYADIC OF SCALAR DYADS

Just as the addition of dyads is defined as an internal law, we will see that geometry teaches us that multiplication must be conceived as an external law. To support this conception, we will begin by referring to the geometric experiment of areas and volumes, described in figure 1, which reveals how the multiplication of lengths defined by metric geometry gives rise to two new magnitudes, the surface and the volume, according to let the multiplied factors be two lengths or three, respectively [4]. We observe that, if we take two lengths $a\, U_1$ and $b\, U_2$, where U_1 and U_2 are any two units of length, and whose measures a and b are integers, that is, a and $b \in Z$, with both lengths forming an abstract rectangle without base scale $a\, U_1$ and height $b\, U_2$, the magnitude that we call the area or surface of the rectangle thus formed would be expressed as a concrete scalar equal to $a \times b$ times the area of an elementary rectangle with a base unit U_1 and a height unit U_2, which is would indicate with the dyadic form $(a \times b)\,(U_1 \times U_2)$. This geometric operation is called length multiplication, and we immediately observe that it does not correspond to the classical algebraic notion of as many times the multiplicand as indicated by the multiplier, which reducing to an addition would require homogeneity in both factors, that is, they should be referred to the same unit, and in turn the product would also be given in that same unit. By contrast, geometric multiplication allows lengths to be expressed in different units,

---

[4]   If the reader is not familiar with the geometric multiplication of segments, or what is the same, of lengths, you can consult the detailed development of this topic in «Lesson 32» of Mathematize 1 and in «Lesson 3» of Mathematize 3, both publications of the same author's agenda.

Geometric experiment of the areas

Given two lengths expressed in the same unit U, if an **abstract rectangle without scale** is formed with its numerical parts, it is observed that, dividing it into ideal squares of unit side, the number of these is equal to the product of the measures of the lengths given with respect to unity. This observation of the geometry allows defining the product of two lengths $a\ U$ and $b\ U$ or two dyad quantities with the same unit, interpreting it as an area that is symbolized:

$$a\ U \times b\ U = (a \times b)\ (U \times U) = (a \times b)\ U^2$$

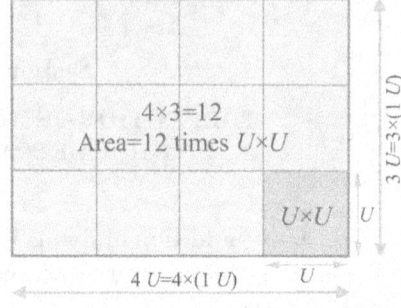

On the left, the case in which the lengths or dyads are not expressed in the same unit as $a\ U_1$ and $b\ U_2$, in the abstract rectangle built with them it is observed that their product can be associated with the quantity called area, which is measured by means of rectangles equal to the symbolized area unit $U_1 \times U_2$, justifying the same definition of product:

$$a\ U_1 \times b\ U_2 = (a \times b)\ (U_1 \times U_2)$$

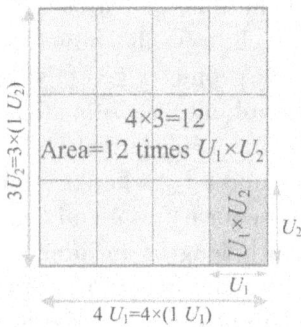

On the right the product of two lengths with fractional measure $(3/5)\ U_1 \times (2/3)\ U_2$. Dividing one of the dimensions into five equal segments and the other into three, results in a set of equal rectangles whose sides measure 1/5 of U_1 and 1/3 of U_2, the number of these equal elements that make up the unit is equal to $5 \times 3 = 15$, which coincides with the product of the denominators, and the number of equal elements that fit in the assumed fractional measure is $3 \times 2 = 6$, which coincides with the product of the numerators; the fractional area will be 3×2 elements of the 5×3 total rectangles, which is the fraction $(2 \times 3)/(3 \times 5)$, which is equal to the product $(3/5) \times (2/3) = 6/15$, so here also the form of the definition of the dyad multiplication is fulfilled.

$$(3/5)\ U_1 \times (2/3)\ U_2 =$$
$$= (6/15)\ U_1 \times U_2$$

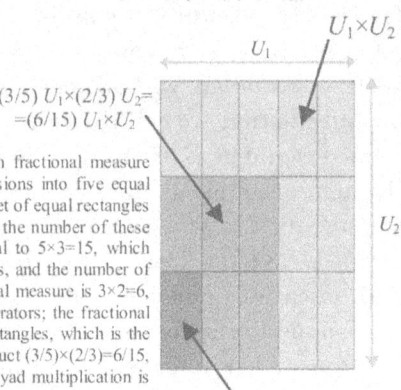

$(1/5)\ U_1 \times (1/3)\ U_2$

Figure 1

provided that the unit in which the product is measured is an abstract rectangle with dimensions precisely equal to the units in which the multiplied factors or segments are expressed.

If the measures of the segments were given, instead of with integers, with rational numbers, the dyads $(p/q)\ U_1$ and $(r/s)\ U_2$, it is also observed that the area of the abstract rectangle formed with both measures also It can be expressed as the scalar concrete with a numerical part equal to the product of the initial rationals, according to the definition of multiplication of fractions, which corresponds to the rational number that has the product of the numerators as the numerator and the product of the denominators as the denominator , and this as a multiplier of the area of the unit rectangle whose base and height are the given units U_1 and U_2, which would be expressed analytically with the written form $[(p \times r)/(q \times s)] \, (U_1 \times U_2)$.

In sum, given any two segments or lengths $a \; U_1$ and $b \; U_2$, forming an abstract rectangle with them and taking as a unit of surface the unit rectangle with dimensions U_1 and U_2, which are the units of the multiplied lengths, whose area will be written By definition, in the form of the product $U_1 \times U_2$, it turns out that the area of the rectangle is measured with the scalar dyad $(a \times b) \, (U_1 \times U_2)$.

Well, this geometric fact serves as the basis for defining the multiplication of lengths, and by axiomatic generalization, it also conceptualizes the **multiplication of quantities of scalar magnitudes** or **dyads** $a \; U_1$ and $b \; U_2$, no matter what the units are associated with them, from according to the analytic expression:

$$a \; U_1 * b \; U_2 = (a \times b) \, (U_1 * U_2) \qquad [12.1]$$

The definition equation above symbolizes an external composition law of the Cartesian product $\{R, U_1\} \times \{R, U_2\}$ in $\{R, U_1 * U_2\}$. So that the multiplication of dyads does not operate on the same set, as was had with the addition, which is an internal law, but the three sets involved in the multiplication, in general, can be different. If the units of the multiplied concretes coincide, designating them with the letter U, the generating external law of multiplication would apply the Cartesian product $\{R, U\} \times \{R, U\}$ in $\{R, U * U\}$. The product $U * U$, for symbolic convenience, is by definition symbolized by the power form U^2, so that the set $\{R, U * U\}$ can also be written $\{R, U^2\}$. And here we see clearly how the notation of square meters m^2, or seconds

71

squared s^2, or any other unit squared U^2 arises; with this, the unknown that eminent mathematical philosophers of the stature of Fourier or Maxwell wondered, as we explained in the exordium, about what could be the meaning or motivation of multiplication and of the powers of physical units or quantities of magnitudes. In turn, the division between quantities of any scalar magnitudes and, therefore, between non-homogeneous units, will also be described and justified later, as soon as it is conceived as an operation derived from multiplication.

As with the previous definitions on laws of composition, it should be noted here that the product of $a\ U_1 * b\ U_2$ means the newly defined dyadic multiplication, that the product of $(a \times b)$ points to the multiplication of the real numbers, and that the asterisk of $(U_1 * U_2)$ refers again to the multiplication of dyads, in this case as a particular case that multiplies the two units, with the meaning of $(U_1 * U_2) = (1\ U_1) * (1\ U_2)$.

If instead of two segments or lengths we think of composing three, $a\ U_1$, $b\ U_2$ and $c\ U_3$, where U_1, U_2 and U_3 are three units of length, which can be different, the geometric multiplication of segments is defined as the composition of a abstract straight parallelepiped with the dimensions of the three segments. We observe that a certain volume thus results, which can be measured as a function of the volume of the unit parallelepipeds of dimensions U_1, U_2 and U_3, which can be indicated with the form of the product $U_1 \times U_2 \times U_3$, with the meaning of the new unit of volume magnitude created in this way; well, the total volume will be described by the real product of the numerical part of the multiplied lengths, which is $a \times b \times c$, or the number of times that the total volume comprises the abstract or numerically unspecified unit volume $U_1 \times U_2 \times U_3$. A numerical example with lengths is shown in figure 2 to illustrate the reasoning. With which, all that has been said can be included in an analytical formulation, which defines the geometric multiplication of three quantities of lengths or segments and, generalized in the abstract, of any three quantities of magnitudes or any specific scalars,

Experimental significance of the product geometric of three lengths, and by abstraction of any three magnitudes

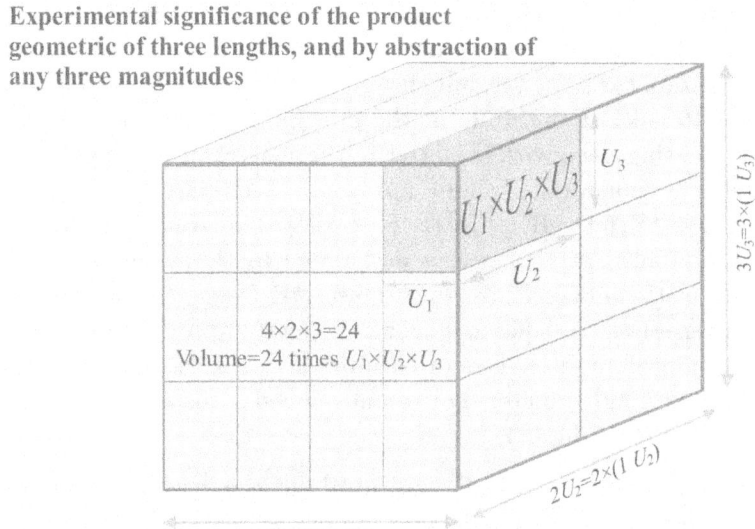

Given three lengths $4\ U_1$, $2\ U_2$ and $3\ U_3$, an abstract straight parallelepiped without scale can be formed with them and ideally decomposed by delimiting the corresponding symbolic length on each edge. Thus, they result in a series of parallelepipeds with the same ideal unit measurements, which is why they are congruent and equal. The new magnitude that results from composing three lengths is called volume, and the fact that the number of elementary parallelepipeds is equal to 24 makes it possible to refer to the amount of volume indicating that one of these elements measures 24 times, which nothing prevents symbolizing with the Notation similar to the algebraic $U_1 \times U_2 \times U_3$, writing this result 24 $(U_1 \times U_2 \times U_3)$. With this, the operation of composing three lengths consisting of forming a straight parallelepiped with them can be called multiplication of the initial dyads given by three lengths, symbolized $(4\ U_1) \times (2\ U_2) \times (3\ U_3) = (4 \times 2 \times 3)\ (U_1 \times U_2 \times U_3)$, resulting in that the numerical part is given by $4 \times 2 \times 3 = 24$. So, in general, it can be defined that multiplying dyads is obtaining another dyad whose numerical element is the usual product of the numerical parts of the factors and whose dimension is expressed as the product of the units of the factors. As the unit elements are composed in the same way regardless of the order in which the factor units are composed, the commutative and associative properties of concrete multiplication must be axiomatized.

Figure 2

according to the definition equation following, explicit for dyadic algebra:

$$a\ U_1 * b\ U_2 * c\ U_3 = (a \times b \times c)\ (U_1 * U_2 * U_3) \qquad [12.2]$$

Equation [12.2] it is the analytical form of the generative external composition law the Cartesian product set

$\{R,U_1\}\times\{R,U_2\}\times\{R,U_3\}$ in $\{R,U_1*U_2*U_3\}$. So that the multiplication of three dyads, as when it operates on two, does not act on the same set, but the sets related by it, in general, can be different. If the units of the multiplied concretes coincide, designating them with the letter U, **the generating external law of the new magnitude** would apply the Cartesian product set $\{R,U\}\times\{R,U\}\times\{R,U\}$ in $\{R,U*U*U\}$. The product $U*U*U$, for symbolic convenience, is designated by definition with the form of the power U^3, so that the set $\{R,U*U*U\}$ can also be written $\{R,U^3\}$. And we once again clearly appreciate how the notation of the new generated units arises with mathematical coherence, cubic meters m^3, seconds cubed s^3, or any other unit raised to the cube U^3.

To explicitly describe dyadic multiplication without symbolic economy, let us differentiate this operation with the asterisk sign «$*$» in the definition equations [12.1] and [12.2]. We avoid the character «$\otimes$» so as not to confuse it with the tensor product. Such definitions provide the logical motivation to generalize the definition of the product of scalar concretes with any number n of factors and units U_1, U_2, up to U_n; by applying the multiple Cartesian product $\{R,U_1\}\times\{R,U_2\}\times$ ... $\times\{R,U_n\}$ in $\{R,U_1*U_2*$ ... $*U_n\}$, so that, given the dyads $a_1\,U_1$, $a_2\,U_2$, up to $a_n\,U_n$, the dyadic product will be defined by the generic equation:

$$a_1\,U_1*a_2\,U_2*\ ...\ *a_n\,U_n=$$

$$=(a_1\times a_2\times\ ...\ \times a_n)\,(U_1*U_2*\ ...\ *U_n) \qquad [12.3]$$

The multiplication of the first member «$*$» is the dyadic, as well as the one that appears between the units U_i of the second member; while the multiplication marked with the cross sign «$\times$» between the numbers a_i symbolizes the product of the real numbers:

$$a_1\,U*a_2\,U*\ ...\ *a_n\,U=(a_1\times a_2\times\ ...\ \times a_n)\,(U*U*\ ...\ *U)=$$

$$=(a_1\times a_2\times\ ...\ \times a_n)\,U^n\ \text{con}\ U^n=U*U*\ ...\ *U\ \text{y}\ n\ \text{factores}$$

Section XIII

COMMUTATIVE AND ASSOCIATIVE PROPERTIES
OF THE MULTIPLICATION OF SCALAR DYADS

Taking into account that the abstract rectangle that integrates the unit of the multiplication of two dyads has to geometrically present the same quantity of area whether it is taken as base U_1 and height U_2, or if the base is equal to U_2 and the height to U_1, because both rectangles will be geometrically congruent and, therefore, of equal area, it is justified **to axiomatize the commutativity of the multiplication of units whatever they are:**

$$U_1 * U_2 = U_2 * U_1 \qquad [13.1]$$

And this unavoidable axiom results in the product of any two scalar dyads being commutative. Indeed, given the dyads $a\ U_1$ and $b\ U_2$, equation [12.1] for the definition of the explicit product is:

$$a\ U_1 * b\ U_2 = (a \times b)\ (U_1 * U_2) \qquad [13.2]$$

The commutative axiom [13.1] of units and the commutative property of multiplication in R of real numbers allow us to write:

$$(a \times b)\ (U_1 * U_2) = (b \times a)\ (U_2 * U_1)$$

The same equation [12.1] for the definition of multiplication justifies the following logical step:

$$(b \times a)\ (U_2 * U_1) = b\ U_2 * a\ U_1 \qquad [13.3]$$

Resulting in conclusion that the first member of [13.2] is equal to the second member of [13.3], whose meaning is that of the commutative property of the multiplication of diads, which will be written as follows:

$$a\ U_1 * b\ U_2 = b\ U_2 * a\ U_1$$

If instead of two factors there were three, it would be necessary to consider that the abstract unitary right parallelepiped formed with the units U_1, U_2 and U_3 should include the same amount of volume regardless of how its edges are ordered, because all of them will be geometrically congruent and equals; so that here too we are obliged to axiomatize that the dyad product of three units is commutative, which will be written:

$$U_1 * U_2 * U_3 = U_1 * U_3 * U_2 = U_2 * U_1 * U_3 =$$
$$= U_2 * U_3 * U_1 = U_3 * U_1 * U_2 = U_3 * U_2 * U_1$$

This means that the units can be multiplied in any order, because the dyadic product will always represent the same quantity of the compound magnitude. This axiom, associated with the commutative property of the multiplication of real numbers for any number of factors, allows us to conclude, with the same ease as for the product of two dyads, that in the case of three the commutativity of the dyadic product can also be affirmed , regardless of the order of the multiplied dyads. And, generalized in the abstract for any number of factors, this geometric foundation authorizes us to **postulate that the dyadic product is commutative whatever the multiplied dyadic factors are.**

As usual, understanding the parentheses as a priority command in the operations that group, with the same geometric foundations indicated, **the multiplication of units must be axiomatized that it is associative**, which means that the units can be associated in any way in the multiplication without varying your product:

$$U_1 * (U_2 * U_3) = (U_1 * U_2) * U_3 = U_1 * U_2 * U_3 \qquad [13.4]$$

So the general associative property is immediate. Indeed, let the product $a\ U_1 * (b\ U_2 * c\ U_3)$, by the definition described in [12.3], we will have:

$$a\ U_1 * (b\ U_2 * c\ U_3) =$$
$$= a\ U_1 * [(b \times c)\ (U_2 * U_3)] =$$
$$= [a \times (b \times c)]\ [U_1 * (U_2 * U_3)] \qquad [13.5]$$

76

The associative property of the product in R is $a \times (b \times c) = (a \times b) \times c$, with which, substituting in [13.5] and taking into account [13.4]:

$$[a \times (b \times c)] \, [U_1 * (U_2 * U_3)] = [(a \times b) \times c] \, [(U_1 * U_2) * U_3]$$

The definition [12.3] of the multiplication of scalar dyad entities allows us to write:

$$[(a \times b) \times c] \, [(U_1 * U_2) * U_3] =$$

$$= [(a \times b) \, (U_1 * U_2)] * c \; U_3 =$$

$$= (a \; U_1 * b \; U_2) * c \; U_3 \qquad\qquad [13.6]$$

Since the first member of [13.5] is equal to the last of [13.6], the associative property of the dyadic multiplication of scalar dyads finally results:

$$a \; U_1 * (b \; U_2 * c \; U_3) = (a \; U_1 * b \; U_2) * c \; U_3$$

Therefore, scalar dyads behave with the associative form because of the definition of dyadic multiplication and its properties, not because the validity of traditional symbolic logic, whose lack of prior foundation is patently and radically unacceptable, must be admitted outright. since it clearly consists of an invalid scheme of reasoning.

Section XIV

INEXISTENCE OF UNIT OR REVERSE ELEMENTS
FOR THE MULTIPLICATION OF SCALAR DYADS

Given a set of scalar dyads associated to the unit U, which we have symbolically indicated with the notation $\{R,U\}$, let 1 be the multiplicative unit in R, whose existence is guaranteed by the body structure of the set of real numbers; so it will always be possible to form the dyad 1 U of $\{R, U\}$. It could be intuited that the dyad 1 U would have to be the unit element of the multiplication of homogeneous dyads. To check this, let's take any other element a U of $\{R,U\}$, being any number of R. The product between the two will give a U^2, because the dyadic multiplication a $U*1$ U, by the definition equation [12.1] and the condition of unit element in R, they allow to write the following:

$$a\ U*1\ U=(a\times 1)\ (U*U)=a\ U^2$$

The commutative property of dyadic multiplication, followed by the same two previous foundations, lead us to string together this reasoning:

$$a\ U*1\ U=1\ U*a\ U=(1\times a)\ (U*U)=a\ U^2$$

Therefore, in any case 1 U is such that multiplied dyadically by any dyad a U gives another dyad a U^2, with the same primary a, but with different secondary, U in one case and U^2 in the other. Observing that U^2 is a unit of another magnitude different from the magnitude that corresponds to the unit U, the dyads a U and a U^2 are different and it can be stated that 1 U is not the neutral element or unit of the concrete multiplication defined on $\{R,U\}$. Furthermore, as this would happen with any element of $\{R,U\}$, because multiplying it by a U would always result in another different measurable magnitude in the unit U^2, it can be concluded that it is not possible to find any element of $\{R,U\}$

79

that behaves like a typical unit element. We insist, this is because concrete or dyadic multiplication is an external generating law that applies the Cartesian product $\{R,U\}\times\{R,U\}$ in $\{R,U^2\}$, and it turns out that the elements of $\{R,U^2\}$ they are different from those of $\{R,U\}$, because their unit parts are units of different magnitudes, so both sets are different and this prevents the algebraic behavior of the concept of unit element from being verified.

With the inverse element something similar happens. For all non-zero concrete $a\ U$, that is, with $a\neq0$, given the field structure of R, there exists $a^{-1}\ U$, a^{-1} being the inverse of a for the product of the reals. One could suspect that the $a^{-1}\ U$ dyad is the inverse of $a\ U$, but this dyad is such that multiplied dyadically with the other produces the $a\ U*a^{-1}\ U$ dyad.

Thus, considering definition [12.1], knowing that in R it is $a\times a^{-1}=1$, and taking into account the commutative property in the second chain of equalities, we easily have the reasoning scheme of the following two lines:

$$a\ U*a^{-1}\ U=(a\times a^{-1})\ (U*U)=1\ U^2$$

$$a\ U*a^{-1}\ U=a^{-1}\ U*a\ U=(a^{-1}\times a)\ (U*U)=1\ U^2$$

As $1\ U$ is a dyad of different magnitude than $1\ U^2$, they cannot be equal and this result allows us to ensure that $a^{-1}\ U$ is not the inverse for the multiplication of dyad $a\ U$, because U^2 is a unit of certain magnitude always different from the magnitude associated with U.

And this happens not only for the dyad $a^{-1}\ U$, but for any other, since multiplying it with $a\ U$ will produce in any case a quantity of another magnitude, to which the unit $U^2\neq U$, so the secondaries will always be different and the criterion of dyadic equality can never be applied.

The foregoing is reinforced by the fact that, as for this composition law there is no unitary element, because it is an external generating law, neither can there be an inverse element in the strict sense of ordinary algebra, leaving possible innovations safe, because no one can be multiplied. dyad by another so that an element that does not exist results.

This same result is quickly reached as follows: since the set $\{R, U\}$ is different from $\{R, U^2\}$, for the dyadic multiplication or application of $\{R, U\} \times \{R, U\}$ in $\{R, U^2\}$, the unit and inverse elements of any dyad of $\{R, U\}$ cannot be found in $\{R, U\}$, because the product of any element of $\{R, U\}$ by another element of the same set it produces elements from another set, that is, the factors belong to a different magnitude than their multiplication, so the criterion of dyadic equality cannot be applied to them. In particular, the inverse elements U^{-1} of the unit $U = (1, U)$ cannot be found. So unit notations with negative exponents, such as m^{-1}, kg^{-1}, or s^{-1} are absurd and non-existent, unless they are given the proper and consistent meaning as divisors of the dyadic division to be defined later. As will be duly verified, these symbols, which cannot be rationally associated with any quantity of magnitude, must have as their only coherent algebraic meaning that of mere indications that the quantities indicated with negative exponents are part of the divisor of a certain unit composed in the form of quotient, resulting from the composition laws of dyadic algebra.

If this is the case for the multiplication of homogeneous dyads, all the more so for the heterogeneous products or applications of $\{R, U_1\} \times \{R, U_2\}$ in $\{R, U_1 * U_2\}$, since in these the three related sets are different, so they are associated with measurements of different magnitudes between which the equality relationship cannot be established.

So that the multiplication of physical measurements or scalar dyads can never satisfy the algebraic conditions of existence of unit and inverse elements, departing in this aspect from isomorphism with the field of real numbers.

It could be judged that an unnecessary detour has been taken to demonstrate the obvious non-existence of unitary and inverse elements with respect to dyadic multiplications, but we have wanted to underpin this fact to the maximum, which due to its negative nature always presents greater evidentiary resistance, to refute its current presumed existence, since currently the International System of Units admits by mere conventionalism that the set of physical quantities has an abelian multiplicative group structure in which every magnitud can be expressed as a function of the integer powers of a certain number of magnitudes called base.

The algebra of magnitudes established here shows us that there cannot be such an abelian multiplicative group, because the operations defined are external laws and, therefore, lacking unitary and inverse elements, two qualities that cannot be lacking in every group. We are, therefore, before a crass error of the International System of Units, caused by the omission of definition of the multiplicative laws of composition for physical magnitudes. Omission that is saved with full physical and mathematical coherence in this First Algebra of Magnitudes.

Without prejudice to the above, in the annex included at the end of the text, a theory is developed, among the many possible ones, on the logical meaning of dyadic algebra that seems to correspond to the meaning of unitary and inverse magnitudes.

Section XV

DISTRIBUTIVE OWNERSHIP OF THE MULTIPLICATION
ON THE ADDITION OF THE DYADIC ALGEBRA

Let's check if the distributive behavior of dyadic multiplication with respect to the addition of measurements holds. To do this, let's take the dyads $a\ U_1$, $b\ U_2$ and $c\ U_2$. The second and third dyad with the same unit have been chosen so that they are uniform and can be added together. Let us form the compound dyad $a\ U_1*(b\ U_2 \oplus c\ U_2)$. Equation [5.3] for the definition of addition supports the first logical step:

$$a\ U_1*(b\ U_2 \oplus c\ U_2)=a\ U_1*[(b+c)\ U_2)] \qquad [15.1]$$

The equation [12.1] for the definition of the dyadic product allows us to write the equality:

$$a\ U_1*[(b+c)\ U_2)]=[a\times(b+c)]\ (U_1*U_2)$$

The distributive property of the product with respect to the multiplication in the field R of the reals is $a\times(b+c)=(a\times b)+(a\times c)$ and allows us to write:

$$[a\times(b+c)]\ (U_1*U_2)=[(a\times b)+(a\times c)]\ (U_1*U_2)$$

The definition [5.3] of addition of dyads, makes it easier for us to advance again in the reasoning, by simply doubling the sum of the second member:

$$[(a\times b)+(a\times c)]\ (U_1*U_2)=(a\times b)\ (U_1*U_2)\oplus (a\times c)\ (U_1*U_2)$$

The definition [12.1] of the product leads us directly to the expression:

$$(a\times b)\ (U_1*U_2)\oplus (a\times c)\ (U_1*U_2)=$$
$$=(a\ U_1*b\ U_2)\oplus (a\ U_1*c\ U_2) \qquad [15.2]$$

The conclusion arises from the equality between the first member of equation [15.1] and the second of [15.2]:

$$a\ U_1*(b\ U_2 \oplus c\ U_2)=(a\ U_1*b\ U_2)\oplus(a\ U_1*c\ U_2)$$

This formula reproduces the well-known distributive form, which here reflects this property of the product with respect to the sum of scalar dyads and for equal units it is reduced to:

$$a\ U*(b\ U \oplus c\ U)=(a\ U*b\ U)\oplus(a\ U*c\ U)$$

In any case, the unit symbols behave formally like any other algebraic element; but, as for the other properties analyzed, this is not an immediate consequence of the symbolic operation, but due to the definitions and properties of dyadic algebra.

As a consequence of everything explained so far in the previous sections, any set of scalar measurements referring to the same unit $\{R,U\}$, endowed with the internal laws that we have called dyadic addition and multiplication, which respond to the definition equations [5.3] and [12.1], would verify all the conditions that configure the algebraic structure of the commutative field, if it were not for the fact that dyadic multiplication is an external law instead of an internal one, making the existence of the unit and inverse elements impossible. However, the algebraic structure of scalar dyadic elements is isomorphic with that of the field of real numbers. On the other hand, we have verified the fact that every set of homogeneous scalar $\{R,U\}$ or vector $\{R^3,U\}$ concretes, endowed with the internal composition law of addition, defined in section V, and with the law external of the multiplication by a scalar, defined in section IX, satisfies all the conditions of a vector space over the field R of the real numbers[5].

---

[5]   The vector space structure can be found in «Lesson 32» of Mathematize 1, on the fundamentals of algebraic structures, and more extensively in «Lesson 2» of Mathematize 2, which studies vector spaces.

Section XVI

DEFINITION OF DYADIC DIVISION
BETWEEN SCALAR MEASUREMENTS

In section XI we define the division between dyads homogeneous scalars as a function of dyadic multiplication by a scalar. In this section we will define the division based on the multiplication of dyads defined by the definition equation [12.1]. With it, the dyadic multiplication with two factors establishes the relationship between the multiplicand, the multiplier and the product, by means of an abstract rectangle in which the base is the multiplicand, the height the multiplier and the area of the rectangle the product.

Well, changing the symbology and identifying said area with a dividend, one of its dimensions with a divisor and the other with the resulting quotient, we will have the notion of division as an operation such that the quotient multiplied by the divisor equals the dividend, and all this through the dyadic algebra of measurements.

Let's start by shaping the division between any two units U_1 and U_2. Nothing prevents us from establishing as a symbol of the quotient between the two a notation similar to that of abstract algebraic elements, for example, separating them with two inclined or horizontal bars $U_1 /\!/ U_2$. We must pay attention to the fact that the unit U_1, dividend, will be associated here with the area of the abstract rectangle of dimensions U_2, divisor, and $U_1 /\!/ U_2$, quotient. Therefore, the units related by the dyadic product must satisfy the following equation:

$$(U_1 /\!/ U_2) * U_2 = U_1 \qquad [16.1]$$

It is vital to observe that [16.1] justifies the simplification rule of the factors U_2, as it would happen with the algebra of R, but not because it applies, but because the dyadic definition of the product by means of abstract rectangles creates a specific algebra that thus determines it .

Under these conditions, the writing of the division between the concretes or dyads $a\ U_1$ and $b\ U_2$, with $b \neq 0$, must have the symbolic form of this definition equation:

$$\frac{a\,U_1}{b\,U_2} = \frac{a}{b}\frac{U_1}{U_2} \qquad\qquad [16.2]$$

The epistemic equation [16.2] will be fully justified when verifying that it satisfies the condition that the dyadic product of the quotient and the divisor is equal to the dividend. To do this, let us write this dyadic product and operate with the properties of R and taking into account equation [16.1] for the definition of the quotient of units, resulting in:

$$\left(\frac{a}{b}\frac{U_1}{U_2}\right) * \left(b\,U_2\right) = \left(\frac{a}{b} \times b\right)\left(\frac{U_1}{U_2} * U_2\right) = a\,U_1$$

Therefore, definition [16.2] is motivated and sufficiently grounded, because it describes with the general criterion of division the quotient between any two scalar dyads or concretes.

It is possible to deduce the definition [16.2] without more than attending to the generic concept of division. To do this, just imagine an abstract rectangle whose surface is identified with a dyadic dividend $a\ U_1$, one of its dimensions with $b\ U_2$ and the other with the concrete quotient $c\ (U_1 /\!/ U_2)$. The unit associated with c must be identified with the dyadic quotient of units $U_1 /\!/ U_2$, because the unit rectangle must have the unit U_1 by area and by dimensions U_2 and $U_1 /\!/ U_2$, as seen in [16.1]. In the same way, the three indicated dyads cannot be independent, but must satisfy the division condition, that is, the quotient multiplied by the divisor must give the dividend; or, in other words, the dyadic product of the two dimensions of the abstract rectangle must be equal to its surface; and it will be written analytically like this:

$$a\,U_1 = b\,U_2 * c\frac{U_1}{U_2} \qquad [16.3]$$

The equality [16.3] can be interpreted as a dyadic division, for which it is enough to consider the factor $c\,(U_1 /\!/ U_2)$ as the quotient between the total surface of the abstract rectangle $a\;U_1$ and the other of its two dimensions $b\;U_2$. And this analytically can be described in this way:

$$c\frac{U_1}{U_2} = \frac{a\,U_1}{b\,U_2}$$

The geometry of the abstract rectangle is such that $a=b\times c$, so $c=a/b$ with the algebra of R. So, substituting $c=a/b$ in the first member of the last equality, we will finally have this other:

$$c\frac{U_1}{U_2} = \frac{a\,U_1}{b\,U_2} = \frac{a}{b}\frac{U_1}{U_2}$$

And the same equality of the definition [16.2] is already observed between the second and third terms, to establish the dyadic division between the dyads $a\;U_1$ and $b\;U_2$, which was postulated there and has been deduced here by means of the preceding reasoning. So it can be concluded that the quotient of two dyads is equal to a concrete whose primary is the ordinary quotient of the primaries of the factors and whose secondary is the dyadic division of the units of the dividend and the divisor. Expressed analytically:

$$\frac{a\,U_1}{b\,U_2} = \frac{a}{b}\frac{U_1}{U_2} \qquad [16.4]$$

We check in this way, as for the rest of the operations previously analyzed, that the symbols of the units behave ideally like the other elements of R, but this consequence is not due to the traditional symbolic logic, and we insist, it would be a crass error and it is inadmissible to consider it this way, because we have irrefutably justified that this formal behavior is due to the concept of dyadic multiplication by means of abstract rectangles.

On the other hand, let us note that the division of scalar concretes analyzed in this chapter has been symbolized with the double bar, a different operation from the quotient of homogeneous dyads in section XI, which we have agreed to represent with the same sign. And it is that the diversity of algebraic laws is such that, although symbolic exhaustiveness is sought for didactic clarity, it is inevitable and even at times convenient to resort to a certain degree to the principle of symbolic economy, if one does not want to fall into a kind of confused batahola operational.

To better visualize the operation of the dyadic division, let's analyze the case of the magnitude that is known by the name of density. The analysis begins with the observation that bodies have two proper magnitudes, the volume they occupy and the mass that corresponds to that volume. The geometry of the case would be the one described in figure 3. We can start by defining an abstract rectangle such that its area is identified with the amount of mass of the considered body, which we will assume equal to M kg, and one of its dimensions with the amount of volume that occupies that same mass, indicated by V m^3. It is not incongruous that a volume is represented by a length, because in dyadic or physical algebra the quantity of any magnitude can be indistinctly similar, by definition, to a segment, an area, a volume or a hypervolume. In turn, in this case, these two quantities, mass and volume, are related to each other through the dyadic product and through another magnitude represented by the second dimension of the abstract rectangle. Let us symbolize the quantity of this third magnitude with d U_d and to understand ourselves let's call it density. The three quantities thus related,

Dyadic analysis of the composite magnitude called DENSITY

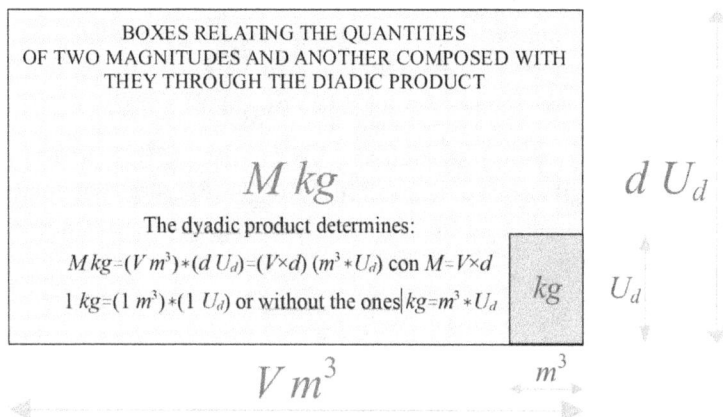

Given a body of mass $M\,kg$ and volume $V\,m^3$, the dyadic product allows us to relate these quantities to that of a third magnitude derived from the first ones and called density, such that $M\,kg$ corresponds to the area of the rectangle with dimensions equal to the others two, $V\,m^3$ and $d\,U_d$. Thus, the density measure is given by the dyadic quotient $M\,kg//V\,m^3$. In turn, in the unit rectangle the three units of the related quantities must be such that the unit of density U_d is the dyadic quotient between a kg and a m^3, that is, $1\,U_d=(1\,kg)//(1\,m^3)$ or written for short $U_d=kg//m^3$.

Figure 3

mass, volume and density, it is clear that they are not independent, but must satisfy the condition imposed by the dyadic product, that is, the following concrete equation:

$$M\,kg=(V\,m^3)*(d\ U_d) \qquad\qquad [16.5]$$

In the unit rectangle of dimensions $1\,m^3$ and $1\,U_d$, whose area must be identified with the quantity of mass to $1\,kg$, we will have:

$$1\,kg=(1\,m^3)*(1\ U_d),\ \text{abbreviated } kg=m^3*U_d \qquad [16.6]$$

The definition of the dyadic product [12.1] transforms the expression [16.5] into this one:

$$M\,kg=(V\times d)\ (m^3*U_d) \qquad\qquad [16.7]$$

Equation [16.7] means that the quantity of mass M kg or area of the rectangle is equal to $V{\times}d$ times the area of the unit rectangle of dimensions 1 m^3 and 1 U_d, which is symbolized by the dyadic product $m^3 * U_d$ and that has to be equal to the unit of mass or kg, given the equation of the product operation [16.6].

The dyadic product is such that $M = V{\times}d$, multiplication that is that of R, with the justification of the geometric experiment of the areas exposed in section XII. Expressing it as a quotient in R, we will have:

$$d = \frac{M}{V} \qquad\qquad [16.8]$$

In turn, the general division criterion that arises from multiplication, applied to the dyadic products [16.5] and [16.6], allows them to be symbolized in this other way:

$$d\,U_d = \frac{M\,kg}{V\,m^3} \;\; ; \;\; U_d = \frac{kg}{m^3}$$

Taking into account [16.8] and the second of the last two equations, substituting d and U_d in the first member of the first of these, it results:

$$\frac{M\,kg}{V\,m^3} = \frac{M}{V}\;\frac{kg}{m^3} \qquad\qquad [16.9]$$

It is observed that [16.9] indicates the same result as [16.4] or generic dyadic quotient, in the present assumption applied to the case of density and its other two magnitudes related by the dyadic product. The reading of [16.9] should be understood as follows: the dyadic quotient between a quantity of mass M kg and a quantity of volume V m^3 is a quantity of a compound magnitude called density, which is equal to the physical or concrete dyad whose primary is the real quotient of the primaries M and V, and whose secondary is the dyadic quotient of the secondaries, that is, in this case the unit of mass or kilogram kg divided by the unit of volume or cubic meter m^3. Even at the risk of being repetitive, the concept of «quantity» is emphasized so that the dyadic nature of the entities that are composed is not forgotten.

The physical meaning of density can be assessed simply by multiplying it by the unit of volume to determine its corresponding mass, according to the following reasoning from physical algebra:

$$\left(1\ m^3\right)*\left(d\ \frac{kg}{m^3}\right)=\left(1\times d\right)\left(m^3*\frac{kg}{m^3}\right)$$

The compound unit of the second member indicates the dyadic product of m^3 by the dyadic quotient $kg/\!/m^3$, which is U_d, so it is the unit rectangle whose abstract surface is one kg; so that the symbols of m^3 that appear both as a multiplier and as a divisor can be simplified as in R, but not by the properties of R, but by the definition of a dyadic product with abstract rectangles, as defined in section XII . Under these conditions, it is concluded that the first member must be equal to $d\ kg$, with which it is observed that the mass of the unit of volume called m^3 is precisely $d\ kg$; and this must be the meaning of density: mass of each unit of volume of material bodies.

Another significant case and analogous to that of density is the composite magnitud called velocity. His analysis starts from the material observation that every mobile takes a certain time to travel each specific distance. So suppose that the distance $L\ m$, which is a length measured in meters, is covered in an interval $t\ s$, time expressed in seconds. Nothing prevents us from assembling the abstract rectangle in figure 4, which has an area $L\ m$ and such that one of its sides is $t\ s$, so that the other dimension will be univocally established by both measurements. It is not incongruous that the area represents the length $L\ m$, although geometrically this does not seem to make sense, because we are operating in the abstract with the dyadic algebra of magnitudes.

Dyadic analysis of the compound magnitude called VELOCITY

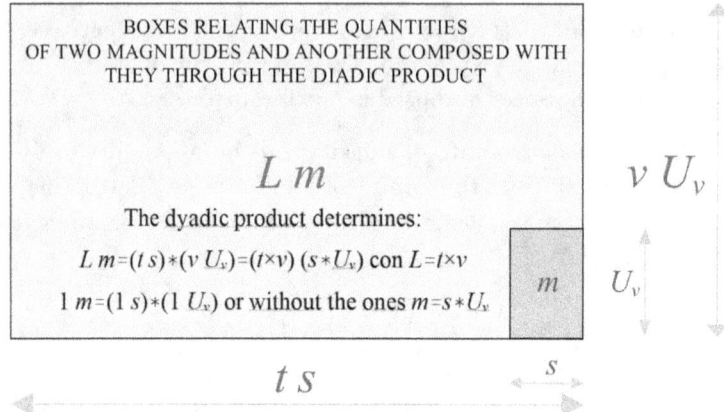

BOXES RELATING THE QUANTITIES
OF TWO MAGNITUDES AND ANOTHER COMPOSED WITH
THEY THROUGH THE DIADIC PRODUCT

$$L\,m$$

The dyadic product determines:

$L\,m=(t\,s)*(v\,U_x)=(t{\times}v)\,(s*U_x)$ con $L=t{\times}v$

$1\,m=(1\,s)*(1\,U_x)$ or without the ones $m=s*U_x$

$v\,U_v$

m U_v

$t\,S$ s

Given a mobile that travels the distance $L\,m$ in the time of $t\,s$, the dyadic product allows us to relate these quantities to that of a third magnitude derived from the first ones and called velocity, such that $L\,m$ corresponds to the area of the rectangle of equal dimensions to the other two, $t\,s$ and $v\,U_v$. Thus, the measure of velocity is given by the dyadic quotient $L\,m/\!/t\,s$. In turn, in the unit rectangle the three units of the related magnitudes must be such that the unit of speed U_y is the dyadic quotient between one m and one s, that is, $1\,U_y=(1\,m)/\!/(1\,s)$ or abbreviated $U_y=m/\!/s$.

Figure 4

The argument for the velocity magnitude is completely analogous to that seen above for density; however, it is developed again step by step for greater clarity.

In this assumption, the three related quantities are length, time and velocity, which are not independent, but must satisfy the condition imposed by the dyadic product, that is:

$$L\,m=(t\,s)*(v\,U_v) \tag{16.10}$$

In the unit rectangle of dimensions $1\,s$ and $1\,U_v$, whose area must be identified with the amount of length equal to $1\,m$, we will have:

$$1\,m=(1\,s)*(1\,U_v),\ \text{abbreviated}\ m=s*U_v \tag{16.11}$$

The definition of dyadic product [12.1] transforms the expression [16.10] into this one:

$$L\, m = (t \times v)\, (s * U_v) \qquad [16.12]$$

Equation [16.12] means that the amount of length $L\, m$ o area of the rectangle is equal to $t \times v$ times the area of the unit rectangle of dimensions $1\ s$ and $1\ U_v$, which is symbolized by the dyadic product $s * U_v$ and must be equal to the unit of length om, given the equation of the product operation [16.11].

The dyadic product is such that $L = t \times v$, multiplication that is that of R, with the justification of the geometric experiment of the areas exposed in section XII. Expressing it as a quotient in R, we will have:

$$v = \frac{L}{t} \qquad [16.13]$$

In turn, the general division criterion that arises from multiplication, applied to the dyadic products [16.10] and [16.11], allows them to be symbolized in this other way:

$$v U_v = \frac{L\, m}{t\, s} \ ; \ U_v = \frac{m}{s}$$

Taking into account [16.13] and the second of the last two equations, substituting v and Uv in the first member of the first one, it results:

$$\frac{L\, m}{t\, s} = \frac{L}{t}\, \frac{m}{s} \qquad [16.14]$$

It is observed that [16.14] indicates the same result as [16.4] or generic dyadic quotient, in the present assumption applied to the case of speed and its other two magnitudes related by the dyadic product. The reading of [16.14] should be understood as follows: the dyadic quotient between a quantity of length $L\, m$ and a quantity of time $t\, s$ is the quantity of a compound magnitude

93

called velocity, which is equal to dyad whose primary is the real quotient of the primaries L and v, and whose secondary is the dyadic quotient of the secondaries, that is, in this case the unit of length or meter m divided by the unit of time or second s.

The physical meaning of velocity can be assessed simply by multiplying it by the unit of time to determine the corresponding length, according to the following reasoning from dyadic algebra:

$$\left(1s\right)*\left(v\,\frac{m}{s}\right) = \left(1\times v\right)\left(s*\frac{m}{s}\right)$$

The compound unit of the second member indicates the dyadic product of a second s by the dyadic quotient $m/\!/s$, which is U_v, so it is the unit rectangle whose abstract surface is one meter; so that the symbols of s that appear both as a multiplier and as a divisor can be simplified as in R, but not by the properties of R, but by the definition of a dyadic product with abstract rectangles, as defined in section XII . Under these conditions, it is concluded that the first member must be equal to $v\ m$, with which it is observed that the length traveled in the unit of time called second s is precisely $v\ m$; and this must be the meaning of speed: length or distance traveled in each unit of time.

Any other magnitude derived from two others by means of dyadic division will show an analysis completely analogous to those of density and speed, without more than taking into account the units that correspond to the dividend and the divisor, which will determine the compound unit in which it should be measured the derived magnitude or quotient.

Finally, we must now give the meaning in dyadic algebra of negative exponents, which is nothing more than a simple notation to write the division. Let's take definition [14.4], it is obvious that nothing prevents us from writing it with the product notation, following the steps of ordinary algebra, according to the following notation:

$$\frac{a\,U_1}{b\,U_2} = \frac{a}{b}\frac{U_1}{U_2} = \frac{a}{b}\left(U_1 * U_2^{-1}\right)$$

But here it should be noted that the inverse notation U_2^{-1} does not mean that there exists an entity such that, when multiplied by U_2, it gives the unit of the real numbers, because we have shown in section XIV that such a thing does not exist. There we concluded that U_2^{-1} cannot be a quantity of the same magnitude as U_2, so the dyadic notation $U_1 * U_2^{-1}$ must be considered equivalent to the dyadic quotient $U_1/\!/U_2$, solely for symbolic purposes.

Therefore, in dyadic algebra of magnitudes, unlike what occurs in arithmetic algebraic structures, inverse notations with units or physical quantities have their own meaning and do not correspond to the phantasm of inverse elements isolated from internal operations, but to divisors of some ratio that cannot be missing. Thus, in general, an expression like $U_1 * U_2^{-n}$ does not mean the product of U_1 by the figurative inverse element of U_2^{n} (see the following section on exponentiation), but rather represents the dyadic quotient $U_1/\!/U_2^{n}$, resulting from the operation of dividing the dividend U_1 by the power U_2^{n}. We insist, $U_1 * U_2^{-n}$ does not denote the product of the quantity U_1 by the inverse quantity U_2^{-n}, which has no physical meaning. In affine geometry it symbolizes a rectangle with abstract surface U_1 and sides U_2^{n} and $U_1/\!/U_2^{n}$. In turn, with algebraic operation nomenclature, the expression indicated by $U_1/\!/U_2^{n} = U_1 * U_2^{-n}$ intervenes in a generating external composition law that applies the Cartesian product $\{R, U_1\} \times \{R, U_2^{n}\}$ in the dyadic set $\{R, U_1/\!/U_2^{n}\}$, which is a dyadic division. As already indicated, the specific dyadic inverses for the generating external composition laws, specific to magnitudes, are developed with greater algebraic precision in the annex at the end of the text.

Section XVII

DEFINITION OF POWER AND
RADICATION OF SCALAR DYADS

The definition equation [12.3] of the dyadic product allows us to define potentiation without controversy. Let us consider an element $a\ U$ of the set $\{R, U\}$ and let n be any natural number of N, we will call the n power of $a\ U$ the scalar dyad $(a\ U)^n$ defined by the following equation:

$$(a\ U)^n = (a^n)\ (U^n) \qquad [17.1]$$

It is required that n be natural, because the definition of the concrete multiplication law operates on whole units, not on fractions of units, a concept that would be meaningless by the definition of unit itself.

Let's analyze the meaning of equation [17.1] that defines the power of a scalar dyad, because it is not strange to find opinions that judge it as obvious, since these understandings suppose that it obeys the most elementary algebra of real numbers. Well, such a prejudice would mean not having understood anything about the algebra of magnitudes, let's see why: equation [17.1] does not relate real numbers but scalar physical dyads; if its elements were considered real numbers, it would mean that the power n of the product $a\ U$ would be equal to the product of the powers of the factors $(a^n) \times (U^n)$; however, since $a\ U$ is not a product of real numbers, but a quantity of some magnitude, with the meaning of the quantity of it equal to times that of the unit U, its power n, given by [17.1], means the measure a^n of another magnitude in the unit indicated by $U^n = U * U * \ldots * U$, with n factors, and this product is not the real but the dyadic of scalar measurements that arises from the abstraction of the multiplication of the geometric segments. Thus, the definition equation [17.1] is not obvious from the algebra of R, but is a consequence motivated by two causes:

the first, due to the geometric fact derived from the definition of segment multiplication and its subsequent abstraction generic; the second, because when developing the algebra of the quantities of magnitudes we lead the symbology along the path of formal similarity with the common of real numbers, so that the resulting notation is isomorphic with it. But, of course, dyadic algebra, like any mathematical or scientific entity, cannot be considered valid or obvious without having adequately defined it in advance, as we are doing here or by means of any other scheme of appropriate definitions. Hence the scruples shown by eminent authors such as Planck or Maxwell, among many others, in relation to operations with magnitudes and the meanings of the expressions constructed with them, as we outlined in the exordium, concerns that can only be overcome by means of the definition of an epistemic algebra that supports operations with magnitudes, since the surprising scientific gap in this matter is not admissible, and solving it is the object of this humble and well-intentioned work.

Once the empowerment is defined, the formulation of the filing is immediate. To do this, the first step must be to fix the natural root of any unit U, which can be established analytically with the following definition equation:

$$\sqrt[n]{U} = U_n \text{ such that } U_n * U_n * \ldots * U_n = U \text{ with } n \text{ factors}$$

This definition equation allows to define analytically the natural dyadic root n of any concrete to U:

$$\sqrt[n]{a\ U} = \sqrt[n]{a}\ U_n \qquad\qquad [17.2]$$

Definition [17.2] could also be struck off as obvious on the wrong basis of assimilating it to an algebraic expression of R, which can be refuted with the same reasons given for potentiation.

Section XVIII

DEFINITION OF DYADIC LOGARITHMATION
SCALAR AND THE LEGENDARY CALCULATION RULE

Equation [17.1] for the definition of the empowerment of measurements, in combination with the dyadic multiplication and division of the operations [12.3] and [16.2] allow the notion of logarithmation to be defined in an isomorphic way with R. It is enough to consider that in [17.1] the number n represents the logarithm at the base $(a\ U)$ of the given scalar concrete and equal to $[(a^n)\ (U^n)]$.

Thus, the dyadic logarithm is established as follows: given a dyad $(a\ U^n)$ called antilogarithm, and another $(b\ U)$ called base, it will be said that the real number n is the dyadic logarithm in the considered base of the indicated antilogarithm if and only if the condition that $(b\ U)^n = (a\ U^n)$ is verified, which will be written with the following form:

$$\mathscr{Log}_{(b\ U)}(a\ U^n) = n \qquad\qquad [18.1]$$

The intention of this definition is none other than to maintain the isomorphism with the definition of logarithm in R, hence, according to the criterion of dyadic equality, it must be $b^n = a$, which means that in R it is also n the logarithm base b of a, that is:

$$log_b\ (a) = n$$

Remember that, for the logarithmic definition to be coherent and reproduce a one-to-one and continuous function between every logarithm n and its corresponding antilogarithm associated to, the base b must be positive and different from unity ($b > 0$ y $b \neq 1$). Remember also the equivalent notations:

$$b^n = a \Leftrightarrow log_b\ (a) = n;\ b^{log_b\ (a)} = a;\ antilog_b\ (n) = a;\ antilog_b\ (n) = b^n$$

Let's check if the dyadic logarithm satisfies the important properties that are given in R, insofar as the logarithm of a product is equal to the sum of the logarithms, and that of a quotient is the difference of logarithms. To do this, take a base $(b\ U)$ and two dyadic antilogarithms $(a_1\ U^m)$ and $(a_2\ U^n)$. It is clear that m and n are, by definition, the dyadic logarithms of these two antilogarithms, so they are verified:

$$(b\ U)^m = (a_1\ U^m) \ \text{y} \ (b\ U)^m = (a_2\ U^n) \qquad [18.2]$$

By multiplying these two equations dyadically, we arrive at the formulation:

$$(b\ U)^m * (b\ U)^n = (a_1\ U^m) * (a_2\ U^n)$$

The definition [12.3] of the scalar dyadic product allows transforming the first member of this formula into this other:

$$(b\ U)^{m+n} = (a_1\ U^m) * (a_2\ U^n)$$

Reading this equation according to the definition of dyadic logarithm [18.1], we have that $m+n$ is the base logarithm $(b\ U)$ of the dyadic product $(a_1\ U^m) * (a_2\ U^n)$; therefore, with m being the logarithm of the first factor and n being the logarithm of the second, we have that the logarithm of a dyadic product is the sum of the logarithms of the factors.

In a totally analogous way, simply dividing equations [18.2] member by member, it is concluded that the logarithm of a dyadic quotient is the difference of the logarithms, by virtue of the following reasoning:

$$\frac{\left(b\,U^m\right)}{\left(b\,U^n\right)} = \left(b\ U^{m-n}\right) = \frac{\left(a_1\,U^m\right)}{\left(a_2\,U^n\right)}$$

Indeed, it is observed that $m-n$ is the logarithm of the dyadic quotient and, in turn, $m-n$ is the difference of the logarithms of the numerator and denominator.

It has already been established that dyadic composition laws are based on the algebra of geometric segments. Hence one of the most spectacular applications of the fascinating relationship between segment addition, logarithmation, and arithmetic operations cannot be omitted. This is the **slide rule**, which facilitated the development of technology for centuries. Suffice it to say that bridges like the Golden Gate and so many great engineering works or that NASA's Apollo missions were possible thanks to this ingenious calculation instrument, until modern electronics removed it from circulation, not without some loss of training in mental calculation and mathematical foundations, undermining above all the quality of education, since the slide rule infused those values in those who understood it.

Well, **the slide rule is based on dyadic algebra and the properties of logarithms**, which make it possible to convert numerical multiplication into a geometric sum of segments and numerical division into a geometric subtraction of segments, in addition to other operations such as the numerical empowerment, which is reduced to the geometric addition of segments through the double logarithm. In any case, the principle of the slide rule is always the same: additive geometric algebra of segments or lengths. Let's see below the foundations of this principle.

A slide rule is a device that consists of three elements: a fixed part or ruler; another mobile rule that moves over the other with a rectilinear movement; and a cursor, which serves to improve the accuracy of the readings and indicate the vertical lines that relate the readings on the ruler and the mobile ruler. The fact of being able to slide the mobile ruler over the ruler is clear that it allows the geometric addition and subtraction of segments to be easily reproduced.

To do this, it is enough to arrange a certain quantity of length or segment S_1 on the ruler, juxtapose another segment S_2, this one located on the mobile ruler, and make the final end of S_1 coincide with the initial of S_2. In this way, the sum segment will be given by the dyadic addition $S_1 \oplus S_2$. In turn, the subtraction of segments is reproduced with the mobile ruler placing the end of S_2 on the end of S_1, resulting in the geometric difference $S_1 \ominus S_2$. The

SLIDE RULE
Linear scales for design unit U in ruler and mobile ruler
Configurations for arithmetic addition and subtraction

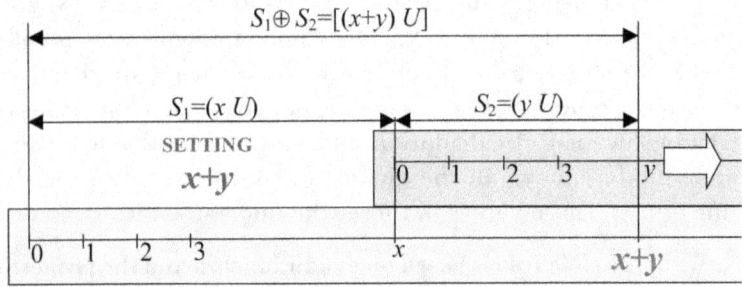

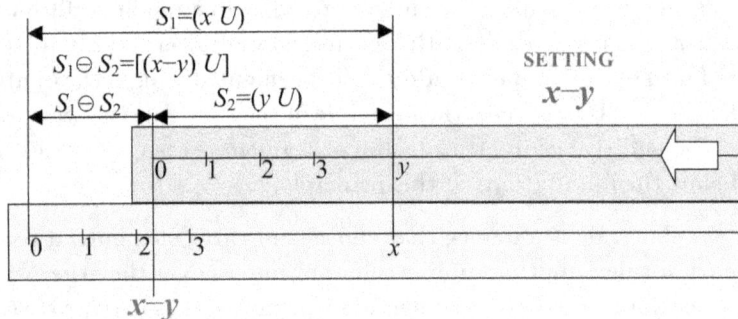

With linear scales in the design length unit U, the distances to the origin are lengths or dyads $(x\,U)$ or $(y\,U)$, so segment addition and subtraction serves to materialize arithmetic addition and subtraction.

Figure 5

schematic in figure 5 clarifies the above and what follows. If both the ruler and the mobile ruler are graduated in a certain unit of length U and proportionally, forming linear scales in ruler and mobile ruler, the slide rule will allow to reproduce the addition and subtraction of numbers, representative of the measurements of the segments in the unit of length U set by the design.

102

And this is where the great contribution of logarithms to mathematical calculations comes: if the ruler and the mobile ruler, instead of linear graduations, are arranged with a logarithmic scale at any base b and always with reference to the design length unit U, adding segments will reproduce arithmetic multiplication and subtracting segments will materialize arithmetic division. Let us see why with the help of figure 6: given two dyads $(M\ U)$, supposed by multiplying, and the multiplier $(m\ U)$, let's look on the rule for the number that indicates the measure M, indicative of the multiplicand. This is equivalent to establishing the segment S_1 such that its measure in the design unit U is the logarithm in the base b of M, that is, $log_b\ M$. Let us place the number m on the grid, which will indicate the multiplier or measure of segment S_2, which is $log_b\ m$. **It should be noted that the measures of the logarithms are not labeled in the ruler and the mobile ruler on the grid, but the antilogarithms M and m themselves**, as is usual in logarithmic scales, so that the multiplicand and multiplier can be read directly and easily. This is the reason why on the logarithmic scale the graduation starts at one, which is the antilogarithm base b of zero, since $b^0 = 1$. So, if on a linear scale the graduation with the unit of length U one by one is 0, 1, 2, 3, etc., on a logarithmic scale with base b, the correlative graduation from unit to unit is marked with the antilogarithms of the previous sequence, thus resulting in the series b^0, b^1, b^2, b^3, etc.

Under these conditions, the geometric addition of the segments S_1 and S_2, symbolized $S_1 \oplus S_2$, will have as its measure the sum of the logarithms:

$$log_b\ M + log_b\ m = log_b\ (M \times m)$$

Therefore, looking in the rule for the end of the segment $S_1 \oplus S_2$, which must be in the vertical of m, the result of the arithmetic product $M \times m$ will be read.

The configuration for the division is the inverse of the previous one and is based on the geometric subtraction of segments with logarithmic scale. On the rule we look for the dividend D and the

SLIDE RULE
Logarithmic scales for design unit U in ruler and mobile ruler
Settings for arithmetic multiplication and division

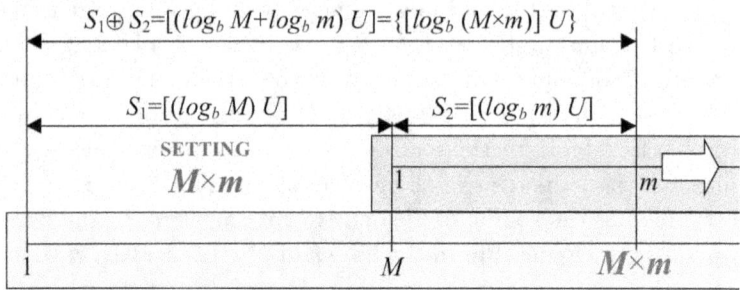

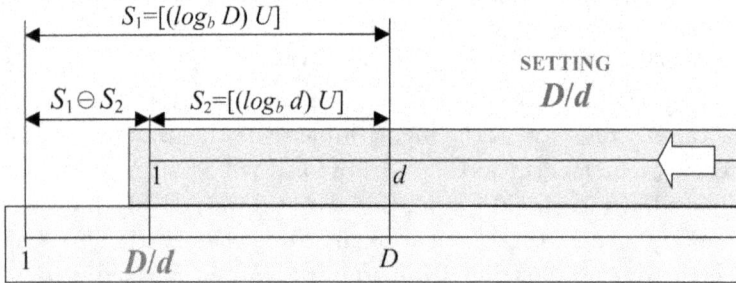

$$S_1 \ominus S_2 = [(log_b D - log_b d)\ U] = \{[(log_b\ (D/d)]\ U\}$$

With logarithmic scales the geometric sum of segments reproduces the arithmetic multiplication and the difference of segments materializes the arithmetic division.

Figure 6

divisor d is placed on the mobile ruler. With this configuration, the difference segment $S_1 \ominus S_2$ is given by the dyadic expression:

$$S_1 \ominus S_2 = [(log_b D - log_b d)\ U] = \{[(log_b\ (D/d)]\ U\}$$

Therefore, the measure of the segment $S_1 \ominus S_2$ in unit U is $log_b\ (D/d)$, resulting in that the arithmetic quotient between D and d is indicated in the rule by the vertical of the origin of the mobile ruler, marked with one. Configuration that obviously coincides with the multiplication of the quotient D/d by d to obtain D, and which could be considered the geometric proof of the division.

The last operation to be analyzed is potentiation, with the help of figure 7. On a doubly logarithmic scale, which is symbolized LL, the reference point of the rule must be the antilogarithm of the antilogarithm base b of zero, which is b. With this arrangement, take the arithmetic power $B^p = x$ and apply the logarithm in any base b, resulting in the equality indicated below:

$$log_b \ B^p = p \times log_b \ B = log_b \ x$$

SLIDE RULE
Logarithmic scale L on the mobile ruler for design unit U
and double logarithmic scale LL on the ruler
Settings for arithmetic empowerment

SETTING
$$B^p$$

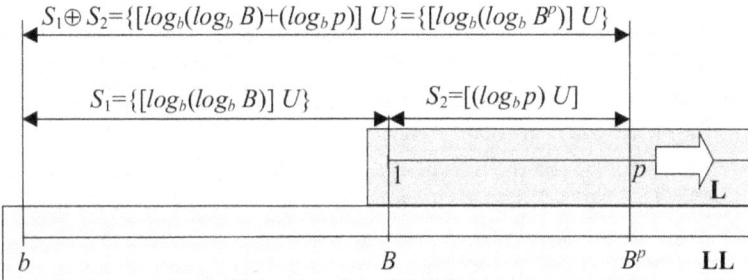

With scales doubly logarithmic LL on the ruler and logarithmic L on the ruler, the configuration of the power B^p results as that which would correspond to the multiplication $B \times p$, but here with the meaning of power, instead of product.

Figure 7

105

In the previous identity, let's calculate the logarithm base b again, we will have:

$$log_b(p \times log_b B) = log_b p + log_b(log_b B) = log_b(log_b x)$$

Let the segments $S_1 = \{[log_b(log_b B)] \ U\}$ and $S_2 = [(log_b p) \ U]$. The geometric addition of these two segments is represented by the analytical form and the reasoning presented in the following logical sequence:

$$S_1 \oplus S_2 = \{[log_b(log_b B)] \ U\} \oplus [(log_b p) \ U] = \{[log_b(log_b B) + log_b p] \ U\} =$$
$$= \{[log_b(log_b x)] \ U\} = \{[log_b(log_b B^p)] \ U\}$$

In conclusion, the resulting configuration indicates that, taking the base B of the power on the double-logarithmic scale rule LL, moving the scale on logarithmic scale L until its origin is located on B and reading the quantity p, the power $x = B^p$ will lie on the rule LL in the vertical of p.

The slide rule became possible after the invention of logarithms in 1614 by the Scottish mathematician John Napier. In 1622 the englishman William Aughtred, the first to use the Greek letter π to symbolize the constant quotient between the length of every circumference and its diameter, was the one who knew how to apply the properties of logarithms to operate with numbers and is considered the inventor of the slide rule. Until 1972, when the first electronic pocket calculators appeared, the slide rule was an emblematic and essential instrument for technicians of all fields, it was widely used in Europe and the US. Its manufacture stopped in 1975 due to the advancement of modern computer equipment available to everyone.

To operate with the slide rule, it is necessary to understand the mathematical foundations on which it is based, as well as skill in mental calculation, in order to be able to establish the integer and decimal parts of the results for the different operations and configurations. Hence, it is an instrument of great pedagogical potential. Although its practical benefits have been left behind by advances in electronics, it is no less true that its close relationship with Mathematics and, especially, with additive dyadic algebra, make it a first-rate teaching tool. And this without forgetting that it is an iconic instrument and a very powerful assistant for technicians and scientists until the seventies of the 20th century, with whose contribution the vigorous social development of recent modernity was possible.

Section XIX

DEFINITION OF SCALAR PRODUCTS
AND VECTOR OF VECTOR DYADS [6]

In Physics, vector magnitudes use mathematical vectors and products between vectors called scalar product and vector product. It is usual to indicate the scalar with a mathematical point «·» and the vector with the same cross «×» used as a symbol of the various multiplications, by the principle of symbolic economy; although here with the meaning of this kind of product between vectors. We know that this should not be misleading, because, depending on which elements are multiplied, the meaning of the multiplication sign will be established, even if it is the same for different operations. Thus, if the operators are real numbers, the cross will indicate the multiplication in R; if scalar dyads are multiplied, it will refer to the product of these entities; if a real number and a vector are composed, the multiplication indicated by «×» will be the external law of the vector space R^3 over R; if the factors are vectors, this same cross will indicate the vector product of vectors; or, if vector concretes are multiplied, the product will be the concrete vector, which we are going to define here. And something similar can be indicated about the sign of the scalar product with a point «·», which can adopt different meanings depending on the context of the factors. However, in what follows we will distinguish each specific symbol,

---

[6] The scalar and vector products of vectors of R^3 or V^3 or E^3, which are three more or less synonymous and isomorphic notations, as well as their properties, can be found in «Lesson 5» of Mathematize 2. Roughly speaking we have both Spaces of the points and geometric vectors, when they are associated with each other, the affine space is shaped. If the scalar product or interior connection of vectors is also defined, the Euclidean space results, all in three dimensions; and, by simple abstract generalization, they are defined for any dimension n.

for greater pedagogical clarity. The mathematical point «·» will indicate the operation scalar product of vectors, the angle «∧» will indicate the vector product of vectors, and for the dyadic homonyms of quantities of vector magnitudes we will reserve the circle with a point «⊙» for the scalar dyadic product and the circle with an asterisk «⊛» for the vector dyadic product.

There is no better way to define both the scalar product and the vector product of vector magnitudes than in terms of their counterparts of mathematical vectors. Let's start with the dot product of two vector dyads $\overline{a}\ U_1$ and $\overline{b}\ U_2$ of the dyadic or concrete vector spaces $\{R^3, U_1\}$ and $\{R^3, U_2\}$. We distinguish here the first elements $\overline{a}$ and $\overline{b}$ the pairs as mathematical vectors of R^3, because for scalars neither the scalar product nor the vector make sense. We have to define the dot product of said vector concretes with the following definition equation:

$$\overline{a}\ U_1 \odot \overline{b}\ U_2 = (\overline{a} \cdot \overline{b})\,(U_1 * U_2) \qquad [19.1]$$

That is, by definition, the scalar product of two vector dyads measures a scalar magnitude with the scalar concrete of the set $\{R, U_1 * U_2\}$ such that its primary is the scalar product of the vector primaries of the factors and the unit is the product dyadic of the factor units.

Regarding the vector dyadic product, we will have the following definition equation in an analogous way:

$$\overline{a}\ U_1 \circledast \overline{b}\ U_2 = (\overline{a} \wedge \overline{b})\,(U_1 * U_2) \qquad [19.2]$$

In this case, by definition, the vector product of two vector dyads measures a vector magnitude with the dyad of the set $\{R^3, U_1 * U_2\}$ such that its primary is another vector equal to the vector product of the vectors that integrate the primaries of the given concrete or dyads, and whose unit is the dyadic product of the factor units.

The definition formulas [19.1] and [19.2] will facilitate the correct interpretation of the physical equations involving these composition laws, such as the work magnitude of the dyad force, in the case of the scalar product, or the moment magnitude related to the dyad force, for the vector product.

The scalar product of vectors of R^3 is commutative, associative with respect to the scalars of R and distributive with respect to vector addition, properties of the vector space R^3 that are reproduced in the vector dyadic algebra, as can be easily verified, as is done in what follows.

The commutative property of the scalar product is derived from the definition equation [19.1]. The commutativity of the scalar product of vectors is such that we have $\overline{a \cdot b} = \overline{b \cdot a}$. The axiomatic commutative property [13.1] of the multiplication of units $U_1 * U_2 = U_2 * U_1$, exposed at the beginning of section XIII, produces as a consequence the following chain of equalities:

$$\overline{a}\ U_1 \odot \overline{b}\ U_2 = (\overline{a \cdot b})\,(U_1 * U_2) = (\overline{b \cdot a})\,(U_2 * U_1) = \overline{b}\ U_2 \odot \overline{a}\ U_1$$

The associative property in relation to the product by a scalar p of R on the left, analogously on the right, is established by the reasoning that follows, based on the associative behavior of the corresponding operations with scalars and vectors, thus as in the definition [19.1] of the dot product of vector dyads:

$$p \circ (\overline{a}\ U_1 \odot \overline{b}\ U_2) = p \circ [(\overline{a \cdot b})\,(U_1 * U_2)] \qquad [19.3]$$

The definitions [9.1] and [9.2] of multiplication of a scalar by a scalar dyad, or their explicit [9.3] and [9.4], justify that:

$$p \circ [(\overline{a \cdot b})\,(U_1 * U_2)] = [p \times (\overline{a \cdot b})]\,(U_1 * U_2)$$

For vectors, we have that $p \times (\overline{a \cdot b}) = (p \bullet \overline{a}) \cdot \overline{b}$, which allows us to write:

$$[p \times (\overline{a \cdot b})]\,(U_1 * U_2) = [(p \bullet \overline{a}) \cdot \overline{b})]\,(U_1 * U_2) \qquad [19.4]$$

The definition [19.1] of the dot product of dyads or vector leads us to:

$$[(p \bullet \overline{a}) \cdot \overline{b})]\,(U_1 * U_2) = (p \bullet \overline{a})\ U_1 \odot \overline{b}\ U_2$$

111

And again the definitions [9.3] and [9.4] authorize us to write the following equality:

$$(p \bullet \overline{a})\ U_1 \odot \overline{b}\ U_2 = [p \circ (\overline{a}\ U_1)] \odot \overline{b}\ U_2 \qquad [19.5]$$

Therefore, the first member of [19.3] and the second of [19.5] are equal, we have the first associative property of the scalar dyadic product, which is described by the following expression:

$$p \circ (\overline{a}\ U_1 \odot \overline{b}\ U_2) = [p \circ (\overline{a}\ U_1)] \odot \overline{b}\ U_2$$

In turn, for vectors, we have that $p \times (\overline{a} \cdot \overline{b}) = \overline{a} \cdot (p \bullet \overline{b})$, which allows us to write the first member of [19.4] like this:

$$[p \times (\overline{a} \cdot \overline{b})]\ (U_1 * U_2) = [\overline{a} \cdot (p \bullet \overline{b})]\ (U_1 * U_2)$$

The definition [19.1] of the scalar product of vector dyads leads us to:

$$[\overline{a} \cdot (p \bullet \overline{b})]\ (U_1 * U_2) = \overline{a}\ U_1 \odot (p \bullet \overline{b})\ U_2$$

Again the definition [9.3] and [9.4] leads us to write the following equality:

$$\overline{a}\ U_1 \odot (p \bullet \overline{b})\ U_2 = \overline{a}\ U_1 \odot [p \circ (\overline{b}\ U_2)] \qquad [19.6]$$

And, the first member of [19.3] and the second of [19.6] being equal, we have the second associative property of the scalar product:

$$p \circ (\overline{a}\ U_1 \odot \overline{b}\ U_2) = \overline{a}\ U_1 \odot [p \circ (\overline{b}\ U_2)]$$

The distributive property of the dyadic scalar product with respect to the addition of vector dyads is also derived from the corresponding property of the vectors of R^3. Reasoning is described from the left, similarly it would be spun from the right. The definition [5.4] of addition of vector dyads serves to give the first step of the reasoning:

$$\overline{a}\ U_1 \odot (\overline{b}\ U_2 \oplus \overline{c}\ U_2) = \overline{a}\ U_1 \odot [(\overline{b} + \overline{c})\ U_2] \qquad [19.7]$$

Note that the principle of uniformity of addition requires that the addends refer to the same unit U_2.

The definition [19.1] of the dot product of vector dyads allows us to write the second member of [19.7] with the form:

$$\overline{a}\ U_1 \odot [(\overline{b} + \overline{c})\ U_2] = [\overline{a} \cdot (\overline{b} + \overline{c})]\ (U_1 * U_2)$$

The distributive property of the scalar product of vectors of $\mathbf{R}^3$ is $\overline{a} \cdot (\overline{b} + \overline{c}) = (\overline{a} \cdot \overline{b}) + (\overline{a} \cdot \overline{c})$, which leads us to:

$$[\overline{a} \cdot (\overline{b} + \overline{c})]\ (U_1 * U_2) = [(\overline{a} \cdot \overline{b}) + (\overline{a} \cdot \overline{c})]\ (U_1 * U_2)$$

The definition [5.3] of addition of scalar dyads justifies the equality:

$$[(\overline{a} \cdot \overline{b}) + (\overline{a} \cdot \overline{c})]\ (U_1 * U_2) = [(\overline{a} \cdot \overline{b})\ (U_1 * U_2)] \oplus [(\overline{a} \cdot \overline{c})\ (U_1 * U_2)]$$

And by virtue of the definition [19.1] of the dot product of vector dyads we arrive at:

$$[(\overline{a} \cdot \overline{b})\ (U_1 * U_2)] \oplus [(\overline{a} \cdot \overline{c})\ (U_1 * U_2)] =$$
$$= [(\overline{a}\ U_1) \odot (\overline{b}\ U_2)] \oplus [(\overline{a}\ U_1) \odot (\overline{c}\ U_2)] \qquad [19.8]$$

As the first member of [19.7] and the second of [19.8] are equal, the distributive property of the scalar product of vector dyads with respect to their dyadic addition remains explicit, with the analytic form:

$$\overline{a}\ U_1 \odot (\overline{b}\ U_2 \oplus \overline{c}\ U_2) = [(\overline{a}\ U_1) \odot (\overline{b}\ U_2)] \oplus [(\overline{a}\ U_1) \odot (\overline{c}\ U_2)]$$

In turn, for the vector product of dyads we must use the properties of the vector multiplication of vectors of $\mathbf{R}^3$, which we know is not commutative, but anticommutative or antisymmetric, which is not associative, and which does verify the distributive property. The anticommutativity of the vector product of vectors is written $\overline{a} \wedge \overline{b} = - \overline{b} \wedge \overline{a}$, together with the definition [19.2] and by the commutative axiom [13.1] for the multiplication of units $U_1 * U_2 = U_2 * U_1$, we have the following logical reasoning of dyadic algebra or concrete in relation to this lack of symmetry:

$$\overline{a}\ U_1 \circledast \overline{b}\ U_2 = (\overline{a} \wedge \overline{b})\ (U_1 * U_2) = -(\overline{b} \wedge \overline{a})\ (U_2 * U_1) = -\overline{b}\ U_2 \circledast \overline{a}\ U_1$$

We take the opportunity to develop another associative form, which is also verified with the vector product and which is verified here in a similar way, which is the associative property in relation to the product by two scalars p and q of R on the left, it is analogous on the right. It is established by the sequence set out below, based on the associative behavior of the corresponding operations with scalars and vectors, as well as on definitions [9.3] and [9.4] of the product by scalars of vector concretes:

$$(p \circ \overline{a} \ U_1) \circledast (q \circ \overline{b} \ U_2) = [(p \bullet \overline{a}) \ U_1] * [(q \bullet \overline{b}) \ U_2] \qquad [19.9]$$

The definition [19.2] of the vector dyadic product of vector dyads justifies that:

$$[(p \bullet \overline{a}) \ U_1] * [(q \bullet \overline{b}) \ U_2] = [(p \bullet \overline{a}) \wedge (q \bullet \overline{b})] \ (U_1 * U_2)$$

In vector algebra we have $(p \bullet \overline{a}) \wedge (q \bullet \overline{b}) = (p \times q) \bullet (\overline{a} \wedge \overline{b})$, with the meaning to be given to the multiplication signs, depending on the elements that make up: in $(p \bullet \overline{a})$ and $(q \bullet \overline{b})$ it is about the multiplication of scalars by vectors, in $(p \times q)$ it indicates the multiplication of scalars of R and in $\overline{a} \wedge \overline{b}$ it describes the vector product of vectors. In these conditions, we have:

$$[(p \bullet \overline{a}) \wedge (q \bullet \overline{b})] \ (U_1 * U_2) = [(p \times q) \bullet (\overline{a} \wedge \overline{b})] \ (U_1 * U_2)$$

The composition law defined by [9.3] and [9.4] or multiplication of a scalar by a vector dyad leads us to:

$$[(p \times q) \bullet (\overline{a} \wedge \overline{b})] \ (U_1 * U_2) = (p \times q) \circ [(\overline{a} \wedge \overline{b}) \ (U_1 * U_2)]$$

The definition [19.2] of a dyadic vector product of vector dyads allows us to write:

$$(p \times q) \circ [(\overline{a} \wedge \overline{b}) \ (U_1 * U_2)] =$$
$$= (p \times q) \circ [(\overline{a} \ U_1) \circledast (\overline{b} \ U_2)] \qquad [19.20]$$

And, the first member of [19.19] and the second of [19.20] being equal, we have the analytical form of the investigated associative property:

$$(p \circ \overline{a} \ U_1) \circledast (q \circ \overline{b} \ U_2) = (p \times q) \circ [(\overline{a} \ U_1) \circledast (\overline{b} \ U_2)] \qquad [19.21]$$

As regards the distributive property of the vector product of vector dyads, it can be deduced with similar ease by the following reasoning, that we start with the definition [5.4] of the addition of uniform vector dyads applied to the first member of [19.22]:

$$\overline{a}\ U_1 \circledast(\overline{b}\ U_2 \oplus \overline{c}\ U_2) = \overline{a}\ U_1 \circledast[(\overline{b} + \overline{c})\ U_2] \qquad [19.22]$$

The definition [19.2] of the vector product itself legitimizes us to write:

$$\overline{a}\ U_1 \circledast[(\overline{b} + \overline{c})\ U_2] = [\overline{a} \wedge (\overline{b} + \overline{c})]\ (U_1 * U_2)$$

The distributive property of the vector product of vectors of R^3 is $\overline{a} \wedge (\overline{b} + \overline{c}) = (\overline{a} \wedge \overline{b}) + (\overline{a} \wedge \overline{c})$, with which:

$$[\overline{a} \wedge (\overline{b} + \overline{c})]\ (U_1 * U_2) = [(\overline{a} \wedge \overline{b}) + (\overline{a} \wedge \overline{c})]\ (U_1 * U_2)$$

Definition [5.4] of addition of vector dyads leads us to the following line of reasoning:

$$[(\overline{a} \wedge \overline{b}) + (\overline{a} \wedge \overline{c})]\ (U_1 * U_2) = [(\overline{a} \wedge \overline{b})\ (U_1 * U_2)] \oplus [(\overline{a} \wedge \overline{c})\ (U_1 * U_2)]$$

With the definition [19.2] of the vector product of vector dyads we arrive at:

$$[(\overline{a} \wedge \overline{b})\ (U_1 * U_2)] \oplus [(\overline{a} \wedge \overline{c})\ (U_1 * U_2)] =$$
$$= [(\overline{a}\ U_1) \circledast (\overline{b}\ U_2)] \oplus [(\overline{a}\ U_1) \circledast (\overline{c}\ U_2)] \qquad [19.23]$$

And, resulting that the first member of [19.22] is equal to the second of [19.23], we arrive at the analytic form of the distributive property:

$$\overline{a}\ U_1 \circledast(\overline{b}\ U_2 \oplus \overline{c}\ U_2) =$$
$$= [(\overline{a}\ U_1) \circledast (\overline{b}\ U_2)] \oplus [(\overline{a}\ U_1) \circledast (\overline{c}\ U_2)] \qquad [19.24]$$

Note, as already pointed out before, that in the distributive properties dyads have been added referring to the same unit U_2, because we have established that the dyadic addition requires that the addends be uniform.

In this section, the principle of symbolic economy has not been deliberately applied in an absolute way, in order to highlight the various composition laws that intervene in the properties that relate them. However, in scientific or educational practice it is usual to indicate all additive operations with the same signs of addition, just as all multiplicative laws are identified with the same multiplication graph, and the same can be said about subtractions or divisions. However, the spell that this causes leads to reason under the effect of the symbolic illusion that this simplification produces, since the elements with which it operates appear as if they were entities of R, and the custom associated with the properties of the real numbers induces unconscious reasoning with the error of subliminally admitting this fantasy, causing that, although correct conclusions are reached, in reality the logical argument is not valid, because the true algebraic nature of the related composition laws is ignored by the equations.

When the principle of symbolic economy is applied, it is necessary to make an effort to observe that the same signs used for different operations represent different composition laws depending on the nature of the elements between which they are located. Thus, when written between scalars, they refer to the addition, multiplication, subtraction, or quotient of R; when they are found between vectors, they will indicate the vector addition or subtraction, the scalar product or the vector product in R^3; and, situated between dyadic entities, they will refer to dyadic operations of $\{R, U\}$ or $\{R^3, U\}$. Otherwise, the algebraic reasoning would not have been truly understood, although the conclusion may appear to be symbolically correct.

Section XX

PRODUCT DEFINITION BETWEEN
SCALAR AND VECTOR DYADS

Another composition law that must be defined, because it appears constantly in physical equations, is the multiplication of scalar dyads by other vectors. To do this, let be the scalar dyad $a\ U_1$ of $\{R, U_1\}$ and the vector $\overline{b}\ U_2$ of $\{R^3, U_2\}$. Let us use the sign «⊚» for this composition law. We have to establish the definition of this product with the external law of the vector space R^3 over R and dyadic multiplication, so there is no better formulation than with the following two definition equations:

$$a\ U_1 ⊚ \overline{b}\ U_2 = (a \bullet \overline{b})\ (U_1 * U_2) \qquad [20.1]$$

$$\overline{b}\ U_2 ⊚ a\ U_1 = (\overline{b} \bullet a)\ (U_2 * U_1) \qquad [20.2]$$

In [20.1] and [20.2] the sign «⊚» of the first member symbolizes the composition law that we are defining, the product of a scalar dyad by another vector, the sign «•» placed in the factor $(a \bullet \overline{b})$ of the second member indicates the external law of R^3 over R, application of $R \times R^3$ in R^3, or product of a scalar by a vector, and the multiplications of the terms $(U_1 * U_2)$ and $(U_2 * U_1)$ indicate the dyadic product of two scalar dyadic, defined in section XII. Although it is usual to use the same sign «×» for all these multiplicative laws.

Due to the algebraic structure of the vector space R^3, as well as the commutative axiom [13.1] of the multiplication of concrete units, it can be written:

$$a\ U_1 ⊚ \overline{b}\ U_2 = (a \bullet \overline{b})\ (U_1 * U_2) = (\overline{b} \bullet a)\ (U_2 * U_1) \qquad [20.3]$$

The definition [20.2] of this new composition law leads to identity:

117

$$(\overline{b} \bullet a)\,(U_2 * U_1) = \overline{b}\,U_2 \circledcirc a\,U_1 \qquad\qquad [20.4]$$

And, the first member of [20.3] and the second of [20.4] are equal, the commutative property of the operation defined by [20.1] and [20.2] is concluded:

$$a\,U_1 \circledcirc \overline{b}\,U_2 = \overline{b}\,U_2 \circledcirc a\,U_1$$

This new operation contains many possibilities. For example, the real number a could indicate the scalar product of two vectors $a = \overline{a}_1 \cdot \overline{a}_2$, and this would be the case of the scalar product of the vector dyadic entities $\overline{a}_1\,A_1$ and $\overline{a}_2\,A_2$, where the unit U_1 would be given by the compound $U_1 = A_1 * A_2$. Under these conditions, equation [20.1] would become:

$$[(\overline{a}_1\,A_1)\circledcirc(\overline{a}_2\,A_2)]\circledcirc \overline{b}\,U_2 = [(\overline{a}_1 \cdot \overline{a}_2)\bullet \overline{b}]\,(A_1 * A_2 * U_2) \quad [20.5]$$

All its own vector properties can be applied to the product $(\overline{a}_1 \cdot \overline{a}_2)\bullet \overline{b}$ in R^3.

For its part, the vector dyad $\overline{b}\,U_2$ can be such that it is given by a vector dyadic product $\overline{b}\,U_2 = \overline{b}_1\,B_1 \circledast \overline{b}_2\,B_2$, which would have the form for [20.1]:

$$a\,U_1 \circledcirc (\overline{b}_1\,B_1 \circledast \overline{b}_2\,B_2) = [a \bullet (\overline{b}_1 \wedge \overline{b}_2)]\,(U_1 * B_1 * B_2) \qquad [20.6]$$

All the proper properties of this space can be applied to the cross product $a \bullet (\overline{b}_1 \wedge \overline{b}_2)$ in R^3.

In both hypotheses of [20.5] and [20.6] the compound units have been written taking into account their associative property [13.4].

The simple verification is left to the reader that the dyadic composition law defined in this section verifies all associative and distributive forms. In turn, definitions [20.1] and [20.2] allow to deduce the corresponding divisions of this product without more than taking the second members as a dividend and one of the factors of the first members as a divisor, resulting in the scalar quotient between a vector dividend and a vector divisor, and the vector quotient between a vector dividend and a scalar divisor.

118

Section XXI

DEFINITION OF DYADIC ENTITIES
IMAGINARIES AND THEIR LAWS OF COMPOSITION

We have used the algebraic structures of R and R^3 to define the entities and composition laws of scalar and vector dyads. However, we still have a very interesting algebraic structure, the C field of complex or imaginary numbers[7]. Recall that imaginary numbers are defined as pairs of real numbers x and y related to the number $i = \sqrt{-1}$ and symbolized with the form $z = x + i \times y$. So we agree to define on the basis of them the imaginary concrete entities as those in which the primary is an element $z \in C$ and in whose secondary a unit is arranged, which must necessarily be scalar, given the axiom established for this purpose in the section III. We will symbolize this new entity with the notation $z\,U$, and the set of homogeneous imaginary concretes will be written $\{C, U\}$.

Once the imaginary dyad and its sets $\{C, U\}$ have been created, the definitions of its internal and external laws of composition must be undertaken. In the first place, following the usual order, one must begin with addition: let the imaginary dyads $z_1\,U$ and $z_2\,U$ be, which must be homogeneous, given the axiom of uniformity, which is also required in this case, because we know that imaginary numbers operate as vectors in the R^2 plane. So we define the additive internal law «⊕» or application of the Cartesian product $\{C, U\} \times \{C, U\}$ in $\{C, U\}$, by means of this definition equation:

$$z_1\,U \oplus z_2\,U = (z_1 + z_2)\,U$$

---

[7] In «Lesson 42» of Mathematize 1 the algebraic structure of the field C of imaginary numbers with their composition laws is analyzed.

The addition of the first member corresponds to the sum of imaginary dyads defined here and differs from the additives of scalar or vector dyadics, although they are all marked with the same sign «⊕»; while the addition of the term $z_1 + z_2$ is the sum of C, which also differs from the addition of R and is commutative and associative, given the abelian additive group structure of C, which entails the commutative and associative properties of the addition of imaginary dyads, which we have already succinctly exposed, so as not to result excessively repetitive in the reasoning:

$$z_1 \ U \oplus z_2 \ U = (z_1 + z_2) \ U = (z_2 + z_1) \ U = z_2 \ U \oplus z_1 \ U$$

$$z_1 \ U \oplus (z_2 \ U \oplus z_3 \ U) = z_1 \ U \oplus (z_2 + z_3) \ U = [z_1 + (z_2 + z_3)] \ U =$$

$$= [(z_1 + z_2) + z_3] \ U = (z_1 + z_2) \ U \oplus z_3 \ U = (z_1 \ U \oplus z_2 \ U) \oplus z_3 \ U$$

In an analogous way we can and do define the multiplicative external generating law «*», or application of the Cartesian product set $\{C, U_1\} \times \{C, U_2\}$ in $\{C, U_1 * U_2\}$, with the defining equation:

$$z_1 \ U_1 * z_2 \ U_2 = (z_1 \times z_2) \ (U_1 * U_2)$$

The multiplication of the first member is the dyadic of imaginary dyads that is defined in the equation itself and differs from the dyadic with real entities, although it is indicated with the same sign «*», while the product of the term $z_1 \times z_2$ is the multiplication of C, not R, and is commutative and associative, given the commutative field structure of C. All this, together with the commutative axiom [13.1] and the associative [13.4] of the product of units, leads us to the properties commutative and associative of the dyadic product «⊛» of imaginary measurements:

$$z_1 \ U_1 * z_2 \ U_2 = (z_1 \times z_2) \ (U_1 * U_2) = (z_2 \times z_1) \ (U_2 * U_1) = z_2 \ U_2 * z_1 \ U_1$$

$$z_1 \ U_1 * (z_2 \ U_2 * z_3 \ U_3) = z_1 \ U_1 * (z_2 \times z_3) \ (U_2 * U_3) =$$

$$= [z_1 \times (z_2 \times z_3)] \ (U_1 * U_2 * U_3) = [(z_1 \times z_2) \times z_3] \ (U_1 * U_2 * U_3) =$$

$$= (z_1 \times z_2) \ (U_1 * U_2) * z_3 \ U_3 = (z_1 \ U_1 * z_2 \ U_2) * z_3 \ U_3$$

In a totally isomorphic way with the reasoning followed for $\{R, U\}$, since C has the same algebraic structure as R, each of these two numerical sets with its own additive and multiplicative internal laws, and even in analogy with $\{R^3, U\}$, since C also behaves as a vector space over R, the rest of the properties are deduced and the other composition laws are defined for $\{C, U\}$. With this and remembering that the additive part has to refer to the same unit U, given the axiom of uniformity, we will have the distributive property of the product with respect to the addition of imaginary dyads:

$$z_1 \, U_1 * (z_2 \, U \oplus z_3 \, U) = z_1 \, U_1 * (z_2 + z_3) \, U =$$

$$= [z_1 \times (z_2 + z_3)] \, (U_1 * U) = [(z_1 \times z_2) + (z_1 \times z_3)] \, (U_1 * U) =$$

$$= (z_1 \times z_2) \, (U_1 * U) + (z_1 \times z_3) \, (U_1 * U) = (z_1 \, U_1 * z_2 \, U) \oplus (z_1 \, U_1 * z_3 \, U)$$

The definition of multiplication by a scalar p of C, which in turn could be singularly an element of R, because $R \subset C$, must be associated with a map of the Cartesian product $C \times \{C, U\}$ in $\{C, U\}$ on the left, and $\{C, U\} \times C$ on $\{C, U\}$ on the right, with the definition equations:

$$p \circ (z \, U) = (p \times z) \, U$$

$$(z \, U) \circ p = (z \times p) \, U$$

Since in C we have the commutative property $p \times z = z \times p$, it is immediate that the imaginary concrete multiplication by an imaginary or real scalar is commutative:

$$p \circ (z \, U) = (z \, U) \circ p$$

The properties of C guarantee the associative behavior of this dyadic multiplication by two scalars p and q of C, or singularly of R, since $R \subset C$:

$$(p \times q) \circ (z \, U) = [(p \times q) \times z] \, U = [p \times (q \times z)] \, U = p \circ [(q \times z) \, U]$$

Various distributive properties such as the following are also verified for this law:

$$p \circ (z_1 \, U \oplus z_2 \, U) = p \circ [(z_1 + z_2) \, U] = [p \times (z_1 + z_2)] \, U =$$

$$[(p \times z_1) + (p \times z_2)] \ U = (p \times z_1) \ U \oplus (p \times z_2) \ U =$$

$$= [p \circ (z_1 \ U)] \oplus [p \circ (z_2 \ U)]$$

$$(p + q) \circ (z \ U) = [(p + q) \times z] \ U = [(p \times z) + (q \times z)] \ U =$$

$$= (p \times z) \ U \oplus (q \times z) \ U = [p \circ (z \ U)] \oplus [q \circ (z \ U)]$$

In sum, we also observe here how the daily algebraic structures of the primaries are accommodated to those already established for the mathematical entities of R, R^3 or C, so that the secondary or dimensional part respond independently to the generalization of the algebra of the geometric segments, with the **affinity postulate**, derived from the possibility of establishing one-to-one correspondence between the quantities of length of the geometric segments and the quantities of any other magnitude, thereby justifying operations with magnitudes affine to length based on the geometric algebra.

In turn, the symbolic forms of the usual rules are maintained with all the terms, including the symbols of the units, as a consequence of the definitions and properties of the multiple laws of composition that configure the various daily algebraic structures.

Section XXII

EFFECTS OF THE PRINCIPLE
OF SYMBOLIC ECONOMY

Throughout the development of this first physical algebra we have observed how the geometric experiment of the multiplication of segments or lengths is generalized in the abstract, giving rise to the generic algebra of dyads, as mathematical representatives of the quantities of physical magnitudes. We have repeatedly warned about the hypnotic effect that can be produced by availing itself of the symbolic economy, understood as the simplification of signs for the different operations of the same species, such as the additive ones, all denoted with the typical cross «+», the multiplicative ones indicated generically , for example, with the cross «×», subtraction with the hyphen «–» or divisions with the slash «/». To break this spell and warn for pedagogical purposes about how easy it is to be fascinated by it and to believe that what really remains in the dark is understood, we have made an effort of symbolic detail, to make explicit the maximum number of distinguishable composition laws, as well as the relationships that arise between them; although, given the large number of these, as the symbolism is limited and it would not be useful to take such differentiation to the absolute extreme, it is inevitable and even convenient that some share common signs, which is not an obstacle for the phenomenon to be explained with sufficient clarity didactic.

For a better overview, the symbols of the operations involved in dyadic algebra can be detailed, represented by the sign $\mathscr{D}$, unlike the structures of R, C and R^3 or any other. Such operations are those of the usual structures and those specifically defined for specific entities. In this way the following synoptic diagram results:

Take, for example, the expression [19.21], which is the result of previous reasoning with dyadic algebra:

Type of dyadic composition law / Section of the *First Algebra of Magnitudes*		Ordinary number algebra (see note)		Dyadic algebra or physical	With the principle of symbolic economy
		In R y C	In R³	In $\mathscr{D}$	
Scalar magnitudes and vector	Addition (V)	+	+	⊕	+
	Subtraction (VIII)	−	−	⊖	−
	Multiplication by a number (IX)	×	•	○	×
	Homogeneous division (XI)	/ ÷		// ≑	/ ÷
Scalar magnitudes	Heterogeneous multiplication (XII)	×		✶	×
	Heterogeneous division (XVI)			// ≑	/ ÷
Vector magnitudes	Scalar product (XIX)		•	⊙	•
	Vector product (XIX)		∧	⊛	×
	Product of mixed magnitudes (XX)			◎	×

(Note) The symbols of the operations in R, C and R³ obviously refer to the addition, subtraction, multiplication and division of these algebraic structures, not to the dyadic or concrete ones that are defined in the sections of the first column.

$$(p\circ\overline{a}\ U_1)\circledast(q\circ\overline{b}\ U_2)=(p\times q)\circ[(\overline{a}\ U_1)\circledast(\overline{b}\ U_2)]$$

The sign «∘» indicates the multiplicative operation of a scalar of R by a vector dyad; on the other hand, «⊛» represents the composition law called the vector product of vector dyads; and, finally, with the symbol «×» the multiplication of real numbers is named. The principle of symbolic economy consists in denoting all these laws of composition of the same multiplicative species with the same character, for example, the cross «×». And with that the traditional notation results:

124

$$(p \times \overline{a}\ U_1) \times (q \times \overline{b}\ U_2) = (p \times q) \times [(\overline{a}\ U_1) \times (\overline{b}\ U_2)]$$

Observing this last expression, unless one has algebraic expertise, it is difficult to escape the illusion caused by the constant sign of the cross «×» and it is easy to believe that the property that describes equality is evident by the laws of R^3. However, this is not the case, because what it relates are physical dyads, and the full meaning of equality is given by the algebraic reasoning that has led to the deduction of [19.21], by virtue of the different composition laws that relate the own equation and specifically defined between the spaces $\{R^3, U_1\}$, $\{R^3, U_2\}$ and $\{R^3, U_1 * U_2\}$.

The same can be said about any other expression in dyadic algebra, such as the distributive property described in [19.24]:

$$\overline{a}\ U_1 \circledast (\overline{b}\ U_2 \oplus \overline{c}\ U_2) = [(\overline{a}\ U_1) \circledast (\overline{b}\ U_2)] \oplus [(\overline{a}\ U_1) \circledast (\overline{c}\ U_2)]$$

If the additive signs «⊕», which refers to the sum of vector dyads, and the multiplicative «⊛», which symbolizes the dyadic vector product of vector concretes, are replaced by the usual «+» and «×», we will have simplified or implicit equality:

$$\overline{a}\ U_1 \times (\overline{b}\ U_2 + \overline{c}\ U_2) = [(\overline{a}\ U_1) \times (\overline{b}\ U_2)] + [(\overline{a}\ U_1) \times (\overline{c}\ U_2)]$$

The observation of the equation, unless one knows how to distinguish each operation according to the elements that it relates, seduces the intellect almost without remedy, making it believe that the distributive property is immediately fulfilled; however, this is not the case, because we are dealing with an equality of physical algebra, for which an exhaustive reasoning has had to be followed, such as that presented instead for equation [19.24].

So the isomorphic appearance of physical algebra with the structures of R or R^3, which is appreciated when the principle of symbolic economy is applied, is not much less presumed, as in practice it has been assumed, not without a good deal of uncertainty and negligence, suggested in their writings by the best authors; although other less rigorous minds, enraptured by the symbology in the manner described above, might believe in this imposture and take the dyadic properties that require specific proof for granted. Which, not because it is common, ceases to be erroneous and clearly arbitrary, violating the most elementary logic of knowledge and forgetting or disregarding the algebraic obligation to define the laws of composition between entities destined to represent quantities of related scientific magnitudes through physical equations.

Therefore, the principle of symbolic economy would lead directly to the conclusion that to operate with quantities of magnitudes, the isomorphic appearance with the operations of R and R^3 could be relied upon, admitting as evident the following formal rule:

To compose concrete entities or physical dyads, it is enough to specify in the equations the abbreviations or symbols of the units that intervene and operate with them according to the algebraic laws of real numbers and geometric vectors, and considering that they multiply the accompanying measurements, that they can be manipulated in the same way; which is simply pretending that the units are common algebraic elements.

However, we have already sufficiently justified that, although this rule may have practical utility, it should not be admitted on the sole basis of mere symbolic logic and its appearance isomorphic with the structures of R and R^3, but is a logical consequence of physical algebra defined and developed through the various laws of composition duly configured.

Fortunately, the stated rule is correct, as physical algebra shows, but tradition has arrived at it by invalid reasoning, which is inadmissible and further justifies the need for algebra of magnitudes, if only because Physics must settle on good logical foundations, so it is not understood that to date it has remained hidden in the imposture described by an operational rule without prior proof, which we can now understand fully justified and was

126

previously a mere whim of the most vulgar intuition[8]. But there is another reason, if possible, more transcendent to appreciate the unappealable need for dyadic or physical algebra, and that is that it reveals something of the utmost importance, which is nothing but how composite units are built, which are all the result of the abstract generalization of the geometric multiplication of segments or lengths; so, as in this abstraction a good dose of arbitrariness inseparable from the equations of definition is observed, this should move us to the greatest prudence when trying to assess the essence of any of these derived magnitudes, because in principle they have no other character than that of mathematical entities originated by composition laws defined by means of an illusion that generalizes the geometric multiplication of segments, artificially assimilating any quantity of any magnitude to a quantity of length or abstract segment[9].

On the other hand, it should be noted that the rule described has exceptions, like any intellectual simplification, because we have observed that the multiplication of dyads, defined in section XII, given its status as a law of external composition, does not allow the existence of unitary elements and inverse, which distances the structure of scalar dyads from that of the body of reals. This alerts us to the meaning to be given to the negative exponents that often appear in physical equations, because they cannot have any other meaning than to symbolize the divisors or denominators in which powers of units appear in this way; but they cannot at all indicate the inverse of any unit, because such an entity does not exist. Therefore, the rule established here cannot in any way replace the algebra of magnitudes and should

---

[8]   In «Lesson 1» of Mathematize 3, a mathematical method of logic is developed and it is described how an invalid reasoning could lead to a correct conclusion, without assuming that such logical schemes are admissible.

[9]   A somewhat more elementary exposition, but equally valid for understanding the algebra of magnitudes, can be found on the same author's syllabus, «Lesson 3» of Mathematize 3.

only be understood as an aid that speeds up operational practice, but without giving it greater significance, because something else could only induce serious errors in the analysis of magnitudes and physical equations.

A more vulgar reason about the need for dyadic algebra, although no less conclusive, is that to repudiate it would be something like legitimizing that the arithmetic multiplication table also remains subject to the discretion of each one, and on this it does not seem that any normal brain can have any doubt.

Furthermore, this First Algebra of Magnitudes clearly points out the possibility of constructing other more abstract and complex ones, linked to algebraic structures of dimensions greater than three, or even to other non-Euclidean metrics, which will undoubtedly contribute to the development of innovative models. of Theoretical Physics.

Meanwhile, with the composition laws defined in this monograph, all the operations that can be found in the physical equations can be based, so that, although not all possible cases have been analyzed, a task that would be difficult and even unfeasible, with analysis used could solve any imaginable assumption.

Section XXIII

CLASSES OF MAGNITUDES

We have dedicated sections I and II to the concepts of quantity, magnitude and measurement, which refer to three fundamental entities of Physics, and we have agreed that the measurements are represented by dyadic entities and their specific algebra, which allows defining some units starting from others, so that, by composing some of them, the others will be deduced according to the theory to which they belong. The magnitudess from which others are established by means of the algebra of magnitudes and **generating external laws** of sections XII to XVII are called **primary** or **simple**. Those that are expressed in terms of the former will be said to be **secondary** or **compound**. Various authors tend to name the secondary **derived** magnitudes. In turn, we will also call **fundamental** the magnitudes that are taken as the basis of a certain dimensional system, to compose them algebraically as appropriate. Finally, we will end with the introduction of a new concept of magnitude, derived from the «dysmetric» variant, which we will define at the end of this section. In relation to this criterion, we will establish the innovative concepts of **rigid** and **flexible** magnitudes.

The truth is that there is no unanimity in ordinary nomenclature, although this seems irrelevant to us, since the essential thing in each specific case will be to establish what the independent magnitudes are for abstract algebra, because these non-composite magnitudes will be the ones that can shed the most light on the physical meaning of the phenomena analyzed.

Physics is an experimental science, but its experiments, described using common language, would only serve to accumulate historical knowledge with little prospective value: it would be useless to know the trajectory of a particle if there were no way to understand the laws of its motion. However, if scientific facts are written in mathematical language, taking advantage of the entities of this abstract science to couple natural observations into them, something wonderful happens: numbers, vectors, functions and other mathematical instruments allow observations to be organized in such a way that invariant relationships emerge between the different established variables, which allow us to determine some as a function of others. These variables from the mathematical point of view are for Physics or other sciences measurements of magnitudes, in the style of linear, superficial or volumetric geometry.

It is observed in nature that there are phenomena such as time or distance that lack direction, although they can be taken in one direction or the opposite; These can be represented by scalar measurements of the type $\{R, U\}$, whose real part can be positive or negative, for which a convention on the meaning adopted as positive will suffice, and it will be possible to operate according to dyadic algebra. Other magnitudes such as mass are always positive, which constitutes a particular case of the previous ones. In turn, other phenomena such as a speed or a force, in addition to their size or quantity, are not indifferent to the direction or the sense within it, so these physical facts fit very well as vectors and will have to be referred to a adequate reference system in which its components can be determined to operate with them according to vector algebra; in the dyadic field they can be represented by the set $\{R^3, U\}$, if their scope were the three-dimensional Euclidean space, or in general $\{R^n, U\}$ for a space of dimension n. In this way, we verify that the great classification of the physical quantities that reflect certain natural properties has to be established between these two: scalars and vector; thus, once this conceptualization is admitted, it turns out that Physics is absorbed by the properties and compositions of numbers and vectors, members of their corresponding dyadic entities, with which the magic of subsuming natural facts in the appropriate laws will have been achieved pre-existing abstract mathematics, taking advantage of the general truth content of these entities to inflate them with physical meanings. Therefore, the first physical operation relevant to the measurement is to identify a magnitude and base its scalar or vector nature, to establish the type of algebra that is going to be implemented when operating with the measurements, and also in accordance with the essential dyadic algebra.

It should not be overlooked that vector algebra can refer to that which is proper to the field of complex numbers, with its specific laws of composition, because it is already known that some natural phenomena are subsumed by the algebraic structure conferred on such imaginary numbers; It seems incredible, but it is a fact that a mathematical abstraction such as complex algebra, born long before alternating current, serves to explain electrical phenomena of this nature; although sometimes, it must also be recognized that Physics stimulates the development of mathematical structures, which does not change the fact that Physics is included in Mathematics in any case. So we could metaphorically indicate that doing Physics is nothing more than reducing mathematical abstraction to a concrete form that is justified by the relevant essays, which implies that Physics can be considered as the Mathematics of experiment.

Examples of scalar magnitudes are length, area, volume, time, mass, density, temperature, work, energy, power, intensity of electric current in linear conductors, voltage, among many others. On the other hand, the displacement of a mobile, the speed, the acceleration, the force and all those that are characterized by a quantity or module, a direction and a sense of action are vector magnitudes.

The magnitudes that show their indivisible elements are called **discrete** and the element that determines them can be taken as a fundamental unit; for example, registered vehicles, inhabitants of a country, packaging of an industry, and the like are examples of discrete magnitudes. On the contrary, physical quantities such as length, weight, temperature or energy do not have this discrete characteristic and any quantity, no matter how small, is divisible into other smaller ones, such magnitudes are called **continuous** and there are no better ones referring to them than the real numbers, given their symbolized continuity on the real line. Continuous magnitudes are such that they leave no choice but to establish the pattern by means of experimental physical references that implicitly indicate a certain unit quantity that is not determinable or numerically expressible of the magnitude measured and indicated by a specific and arbitrary abstract symbol, being mathematically represented by a set dyadic of the type $\{R, U\}$ or $\{R^n, U\}$, depending on whether they are scalars or vector, respectively. For their part, discrete magnitudes, which do show natural unit elements, can be mathematically described with dyads formed with the set N of natural numbers, denoted $\{N, U\}$, if the measurements could only be positive, or that of the integers Z, referred to the concrete set $\{Z, U\}$, if the measurements could be positive and negative.

Entering the dyadic universe, we cannot escape an important and unobjectionable observation that substantially expands the mathematical field to represent physical phenomena. Every dyad is composed of two elements, a mathematical primary, number, vector or tensor, and a physical secondary indicative of a material reality that is taken as a standard unit of some magnitude, the meter, the kilogram or the second, for example. Then, all measurement is established by that pair of elements closely related to each other, which we have come to call a concrete number, a physical dyad or simply measurement. Since the beginning of time, the tendency towards the «arithmetization» of Physics has paid attention to the measurements of the quantities of magnitudes, tacitly assuming that the standard units should always have the same invariable quantity of the magnitude implicit in them. It is precisely on this invariance that the International System of Units is based. Thus, it has always been assumed that the standard meter must contain the same amount of length in any position in space and under any circumstance. And the same assumption is attributed to the standard kilogram and the second standard. We could distinguish this hypothesis with the qualifier of **isometry**. However, nothing prevents us from formulating the opposite and more general variant of imagining that the quantity of each implicit magnitude in physical patterns can vary from one point to another in space, for various reasons that are not of interest at the moment in this phase of abstract formulation and logic of the mathematical tool of dyadic algebra. We can name this new variant with the term **«dysmetry»** and it consists of the following:

Let us take any standard unit U of some magnitude, such as the meter or the kilogram. For this, it would be enough to prepare a straight rod of a certain extension indicative of a length or a material body formed by a certain grouping of matter, which we will call a weight. These material elements can present the same quantity of their associated magnitude at all points in space-time, isometric hypothesis, or on the contrary, the same physical bodies can vary in their quantity of magnitude depending on the spatial position, which is a «dysmetric» axiom, with which the two existing logical possibilities are completed and which is more generic than the first hypothesis, which is contained in it as a particular case.

Let O be a point in space and let U_O symbolize the quantity of length or mass that the rod or weight in question contains implicit in that point. Let P be any point in space and let U_P be the implicit quantity of length or mass of the same material elements, which are the selected rod and weight, so that the dyadic ratio $U_P/\!/U_O$ can be established. As we are dealing with homogeneous magnitudes, this ratio corresponds to the division in section XI, which we know always results in a real number and, therefore, dimensionless. Let us designate this number with the symbol δ_P. We will call this ratio the «dysmetric» density of the quantity in question at the point P in relation to the point O. Obviously, we have the relation $U_P = \delta_P \circ U_O$, where the operation «$\circ$» is the multiplication of a measurement by a scalar of section IX, an expression that we could also have established from the beginning without more than considering that U_P will have to be a quantity equal to δ_P times the quantity of the same implicit magnitude in

133

U_O, defining the real number δ_P as the «dysmetric» density of the magnitude considered in P with respect to O.

When the «dysmetric» density δ_P is constant, we are in an isometric space. On the other hand, when it varies in any way, we will find a «dysmetric» space. In this way, «dysmetry» was born as a more comprehensive and powerful mathematical tool than current isometry to represent natural phenomena in diverse and variable settings.

The second part of this work develops in more detail the mathematization of «dysmetry», as well as some consequences and applications for generalizing physical laws with the «dysmetric» formulation, demonstrating that we are faced with a fruitful mathematical tool that allows us to develop a new physics. Here, we limit ourselves to introducing the innovative concepts of **flexible magnitudes** and **«dysmetric» spaces** resulting from the possible variation of the various measurements that the same patterns can yield at each point in space-time.

XXIV

EQUALITY, IDENTITY, EQUATION AND PHYSICAL LAW

In Physics, the experiments aim to determine **invariant relations** expressed with dyadic algebra between quantities of different magnitudes and possible **constants**[10] that relate them and it is usual to apply the principle of symbolic economy with operations of the same kind. For example, Newton's second law states that the ratio between the amount of force applied to a body and the amount of acceleration it gives it is constant, and this constant is precisely what is called the mass of inertia of the body. This law or invariant relationship is symbolized by the abbreviated dyadic equation $\overline{F} = m \odot \overline{a}$, the dyadic product between a scalar and a vector dyadic, or economically $\overline{F} = m \times \overline{a}$, where $\overline{F}$ and $\overline{a}$ are vectors, which we have been distinguishing with the upper dash, although any other could serve symbology, and m is a positive real number, and all the factors accompanied by their inseparable units.

Equations that equate vector dyads like $\overline{F} = m \odot \overline{a}$ should actually be written explicitly $\overline{F} \; U_F = m \; U_m \odot \overline{a} \; U_a$, where U_F is the unit of force, U_m the unit of mass, and U_a the unit of acceleration. Due to homogeneity, these units cannot be independent and the relation $U_F = U_m * U_a$ must exist between them, economically $U_F = U_m \times U_a$, since otherwise the dyads of the first and second member could not be equal, according to the equality criterion in section IV. In turn, any vector dyadic equation can always be reduced to scalar forms, using figure 8: let $\overline{e}_F$ be the versor or unit vector in the direction of $\overline{F}$, it will be linked to the vector dyad

---

[10]   We will not question here the existence of physical constants, in the ordinary sense of the term, although the second part of this work explains the reasons why this notion should be revised in «dysmetric» spaces.

135

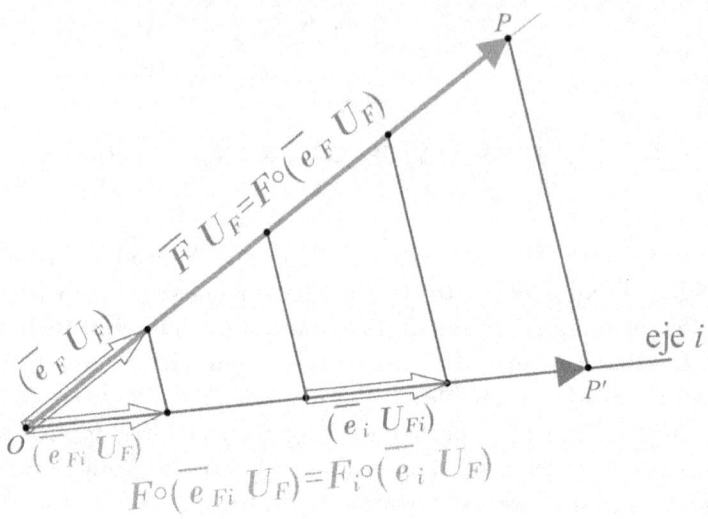

$$F \circ (\overline{e}_{Fi}\, U_F) = F_i \circ (\overline{e}_i\, U_F)$$

Geometrically, if OP', a parallel and generally oblique projection of OP on any axis i, is measured by the unit U_{Fi}, the dyad $F\, U_{Fi}$ results. If the same segment OP' is measured with the unit U_F, the dyad $F_i\, U_F$ results. Then, so that OP' is referred to the same unit U_F of OP, it is enough to express the measure through the projection F_i of OP.

Figure 8

$(\overline{e}_F\, U_F)$ and thus we will have the equality $\overline{F}U_F = F \circ (\overline{e}_F\, U_F)$, where F is the component of on $\overline{e}_F$ and the module of $\overline{F}$ represents an abstract length equal to the unit of force U_F. Let $\mathbb{R}^3$ be any basis, whose versores are denoted $\overline{e}_1$, $\overline{e}_2$ and $\overline{e}_3$, abbreviated $\{\overline{e}_i\}$, with i = 1, 2, 3. We admit the geometric fiction that $\overline{e}_i$ has as a module the abstract unit of length representing U_F, that is to say, that the modules of $\overline{e}_F$ and $\overline{e}_i$ are equal and we pretend that they represent the unit U_F. We define $\overline{e}_{Fi}$ as the projection or component on the i axis of the versor $\overline{e}_F$, its module will represent the quantity U_{Fi}, or the fake projection of U_F on the same axis i, because U_F is not a length. Let us designate by F_i the components of $\overline{F}$ on each axis i. Thales' Theorem guarantees the following: given a segment divided into lengths equal to a certain unit of length, projecting the divisions in parallel onto another line, the same number of projected units result as in the origin segment, so that the measure of the projected segment in projected units is the same as the original segment measurement in the first few units[11]. Applying this geometric property to the vector $\overline{F}$ and its projection F_i on an axis i, we will have that F_i will be represented by the scalar concrete $F\ U_{Fi}$; we want to find out how much F_i has to be worth for the projection to be represented by the dyad $F_i\ U_F$. The vectors $\overline{e}_{Fi}$ and $\overline{e}_i$ are collinear, so there will be a scalar k_i such that $\overline{e}_{Fi}=k_i{\bullet}\overline{e}_i$, with which, the relationship between U_{Fi} and U_F must be admitted to be the same, with $U_{Fi}=k_i{\circ}U_F$, a result which also we would arrive by means of the axiom of continuity of section IV. Under these conditions, you will have:

$$F_i\ U_F=F\ U_{Fi}=F{\circ}(k_i{\circ}U_F)=(k_i{\times}F)\ U_F$$

In accordance with the equality criterion in section IV, it is concluded that for all i, it is:

$$F_i=k_i{\times}F$$

Therefore, the measure F_i of the segment OP' in the unit U_F in relation to the measure F of OP with the same unit is in the same ratio k_i as the collinear vectors $\overline{e}_i$ and $\overline{e}_{Fi}$, projection of $\overline{e}_{Fi}$. And so it is justified that to determine the projection on any axis of every vector dyad $\overline{F}U_F$ it is sufficient to find the component $F_i=k_i{\times}F$ of $\overline{F}_i$, resulting in the projected dyad $\overline{F}_iU_F$.

And so is the important forgotten property that we could state like this: **given a vector measurement, its projection on any axis is in the same dyadic ratio as the versor of the dyad projected on said axis between the versor of this same axis.**

Applying this property to the acceleration vector, the dyadic vector equation of Newton's second law $\overline{F}\ U_F=m\ U_m{\circ}\overline{a}\ U_a$ is

---

[11] The statement, the meaning and the geometric deduction of Thales' Theorem can be found in «Lesson 26» of Mathematize 1.

divided into its three scalar components, referred to the same units U_F, U_m and U_a:

$$F_i\ U_F = m\ U_m * a_i\ U_a \text{ con } i = 1, 2, 3$$

We insist, the preceding confusing reasoning of dyadic algebra shows that the criterion for assigning units to a vector formula is far from obvious, as it is negligently shown in educational settings. This is manifested by the fact that the segment OP measures F times U_F and the segment OP' measures F times U_{Fi}, as a consequence of Thales' Theorem. The same measure F, but different units U_F and U_{Fi}. Hence, to express OP' in the same U_F unit as OP, the coefficient k_i, given by trigonometry, must be taken into account.

Leaving aside such uncertainties, we axiomatize the previous criterion and thus every vector physical equation can be transformed into its corresponding scalars, one for each reference axis, and with the same units, so from here on we will limit ourselves to analyzing only equations of scalar dyadic nature. And thus, the form of equality of the equations implicitly assumes that the dyads of the first and second member must be homogeneous, in accordance with the criterion established by [4.1], with which it will be generally assumed that $a_1\ U_1 = a_2\ U_2$. Multiplying with the operation of section IX by the real number a_1^{-1} inverse of a_1, the following dyadic expression immediately results:

$$a_1^{-1} \circ (a_1\ U_1) = a_1^{-1} \circ (a_2\ U_2)$$

According to the algebra of R, with definition [3.1] and with the multiplication of a dyad by a scalar, an operation defined in section IX, we will have:

$$U_1 = (a_1^{-1} \times a_2)\ U_2$$

Which can also be written with the fractional notation of real numbers:

$$U_1 = \frac{a_2}{a_1}\ U_2$$

Finally, by virtue of the definition of the division of homogeneous units, analyzed in section XI, we will have as a conclusion:

$$\frac{a_2}{a_1} = \frac{U_1}{U_2} \qquad [24.1]$$

Equation [24.1] would seem obvious, if it were admitted without more than with the symbols of the units, it would operate as if they were elements of R; but they are not, so the second member does not indicate an arithmetic but a dyadic quotient between units of homogeneous magnitudes, the one in section XI, so formula [24.1] requires, as we have justified here, to apply a dyadic algebra as the one developed in this monograph, so that only after having justified it with the foundation of the laws of composition of scalar dyads, applied to the equality of homogeneous dyads, does it acquire the important meaning that has been attributed to it since Fourier, although taking into account here the dyadic quotient of the second member counts: **the arithmetic quotient of the measurements of the same quantity of a certain magnitude expressed with homogeneous units is equal to the inverse dyadic ratio between the units.**

Precisely such a ratio between the units is the quotient that dimensional analysis schemes forget to justify, because of the absolute absence of a physical algebra, since they limit themselves to operating with the unit symbols as if they were elements of R without dealing with why they compose them in that way and admitting without more what subjective intuition may dictate in this regard; hence, based on this prejudice, the various theories of dimensional analysis start with equation [24.1] «arithmetized», interpreting the quotient between units as a numerical quotient; while we have previously traveled a long and rigorous algebraic path to base that same formula and give testimony of its unequivocal meaning, rejecting any arbitrary operational form, which we have tried to save in this work with the generic postulation of the algebra of magnitudes and specifically [24.1] in its section XI, precisely defining the quotient between homogeneous units.

The equations of Physics such as the «arithmetized» classical vector $\overline{F}=m\times\overline{a}$ or in their complete dyadic formulation $\overline{F}\ U_F=m\ U_m\odot\overline{a}\ U_a$ are called **universal laws**, because they have the characteristic that they are mathematical symbolic forms that subsume empirical observation, therefore, In order to have an unequivocal meaning, its elements, in this case $\overline{F}$, m and $\overline{a}$, must be established through appropriate epistemic definitions, in order to build an exact knowledge of the observations, well founded, and methodically and rationally worked out.

On the other hand, other equations, such as the one that establishes the speed as a function of the distance traveled in a certain time, or in mathematical terms the derivative of the position vector with respect to time, are the only consequence of the arbitrariness of physical thought to compose magnitudes. In this case, the velocity would be a magnitude derived from the magnitudes of length and time, so an epistemic definition of the velocity is not required, but only a definition equation as a function of these primary magnitudes. Hence, these types of physics equations must be called **definition equations**.

Both the universal laws and the equations of definition are equalities between dyadic entities, so when operating on them it will be necessary to adhere to the guidelines of the algebra of these elements. Thus, for example, for Newton's second law in abbreviated notation $\overline{F}=m\odot\overline{a}$, $\overline{F}$ it is indicated by a vector dyad of $\{R^3,N\}$, where N is the symbol of the unit of force called newton, assuming that it operates in the System International; for its part, the mass m will indicate another scalar dyad of the type $\{R,kg\}$, with the unit of mass called the standard kilogram; and the acceleration will belong to the dyadic set $\{R^3,m/\!/s^2\}$, referred to the compound unit called standard meter divided by the second pattern raised to the dyadic square, all of this dyadically. In turn, the universal law $\overline{F}=m\odot\overline{a}$ will be divided into three equations, one per coordinate, which will have the form

$F_i = m * a_i$, with $i = 1, 2, 3$, and so that F_i, m and a_i will be elements respectively of $\{R, N\}$, $\{R, kg\}$ and $\{R, m//s^2\}$, so it must always be borne in mind that such equations relate concrete entities and that they must be operated with through the algebra of magnitudes. Note that the operation «◎» is the multiplication of section XX, the «*» is the one defined in section XII and the one indicated by «//» is the division of section XVI, all of them **generating laws**.

However, despite the clarity of what has been said, it is notorious and striking that all texts from any scientific field, even the most reputable ones, absolutely forget this evidence and develop their expositions and theories omitting any reference to the laws of composition of the physical units, with which scientific equations are presented as if they related real numbers or vectors; something totally erroneous and inappropriate, because we have already sufficiently justified up to now that the basic mathematical elements of the physical sciences are the dyadic entities, requiring a specific algebra, which cannot be left to the subjective discretion of each one.

So it is necessary and healthy to save that unanimous vice with which texts undertake operations with units without any preamble or algebraic motivation, trusting that readers or students are enlightened by a kind of epiphany that guides them along the correct path of operations with magnitudes.

We hope with this monographic work to shed light on this matter and contribute to unearthing the forgotten pillar of science, laying the foundations for an algebra of magnitudes that provides objective criteria to judge and manipulate these entities according to their true physical nature. In particular, in relation to the physical equations object of this section, their generic form will be represented by an equality like [4.1] of homogeneous scalar concrete entities $a_1 U_1 = a_2 U_2$ or by an equality like [4.2] for vector entities $\overline{a}_1 U_1 = \overline{a}_2 U_2$. The uniformity axiom [4.3] makes it possible to ensure in both cases that there exists $k \in R$ such that $U_2 = k \circ U_1$. Substituting, we will have:

$$a_1 \ U_1 = a_2 \ U_2 = a_2 \ (k \circ U_1)$$

$$\overline{a}_1 \ U_1 = \overline{a}_2 \ U_2 = \overline{a}_2 \ (k \circ U_1)$$

Where «○» is the multiplication of section IX. The properties of this operation allow us to write these scalar and vector equations in the following way:

$$a_1 \ U_1 = a_2 \ U_2 = (a_2 \times k) \ U_1$$

$$\overline{a}_1 \ U_1 = \overline{a}_2 \ U_2 = (\overline{a}_2 \bullet k) \ U_1$$

The operation «×» is the multiplication of R and «•» is the multiplication of a real number by a vector. Once both members are uniform, the equality criterion in section IV establishes the identity of the primaries:

$$a_1 = a_2 \times k = k \times a_2$$

$$\overline{a}_1 = \overline{a}_2 \bullet k = k \bullet \overline{a}_2$$

In summary, in terms of the algebra of magnitudes, every physical equation is represented by the equality of two dyadic entities, scalar or vector, which must be homogeneous without the need for them to be uniform, and the equality of dyads is split into relations between its primaries and secondaries, according to the following scalar and vector reasoning schemes:

$$a_1 \ U_1 = a_2 \ U_2 \Rightarrow \text{Si} \ U_2 = k \circ U_1 \Rightarrow a_1 = k \times a_2 \qquad [24.2]$$

$$\overline{a}_1 \ U_1 = \overline{a}_2 \ U_2 \Rightarrow \text{Si} \ U_2 = k \circ U_1 \Rightarrow \overline{a}_1 = k \bullet \overline{a}_2 \qquad [24.3]$$

In the particular case that the physical equations identify uniform dyads, the relationships between their primaries and secondaries will correspond to the particular case $k = 1$.

The decomposition of dyadic equalities, described by [24.2] and [24.3], will be called the **doubling theorem**. We will verify that it is of utmost importance for the analysis of any scientific equation, as will be seen in section XXVI on physical constants, where some highly relevant universal laws are valued and unfolded.

In order to practice with the algebra of magnitudes and observe its functionality, let's proceed to deduce some physical equations of importance and some complexity. Let's start with the **continuity equation for hydrodynamics.** Let's imagine a fluid in stationary motion. Its state will be represented by a dyadic vector field with primary defined by an application of R^3 in R^3, independent of time and symbolized $\overline{v} = v(\underline{x,y,z})$, so that each point $P(x,y,z)$ of the fluid present a velocity $\overline{v}$ as a function of the coordinates (x,y,z) of the point, with its unit in the secondary.

Suppose that the density ρ of the fluid is variable, it will be represented by a dyadic scalar field with the primary described by the function $\rho = \rho(x,y,z,t)$, which represents an application of R^4 in R, with its unit of measured in the secondary. Let us admit that both functions $\overline{v}$ and ρ are differentiable.

Figure 9 represents in ordinary geometric space an elementary parallelepiped from a generic point, taking the increments of the variables Δx, Δy and Δz. Let the versors $\overline{i}$, $\overline{j}$ and $\overline{k}$ be an orthonormal basis of R^3 and the components of $\overline{v}$ in this basis $v_1(x,y,z)$, $v_2(x,y,z)$ and $v_3(x,y,z)$. Let us observe how the mass of the fluid inside said parallelepiped varies at any given instant t.

To fix ideas we will use the units of the International System. The quantity of fluid per unit of time that crosses the face of the parallelepiped that passes through P and is normal to the versor $\overline{i}$ will be described by the following product of four dyadic entities:

$$[\rho(x,y,z,t) \; kg/\!/m^3] * [v_1 \; (x,y,z) \; m/\!/s] * [\Delta y \; m] * [\Delta z \; m] \quad [24.4]$$

According to the definition of multiplication in section XII, its properties and the division in section XVI, the quantity [24.4] becomes the dyad:

$$[\rho(x,y,z,t) \times v_1 \; (x,y,z) \times \Delta y \times \Delta z] \; kg/\!/s \qquad [24.5]$$

Similarly, the amount of fluid that passes through the face of the parallelepiped parallel to the previous one per unit of time and that passes through the point $P + \Delta P$ of coordinates $(x + \Delta x, y + \Delta y, z + \Delta z)$ will be given by:

143

Analysis of the variation of a vector field through an elementary parallelepiped at any point *P*

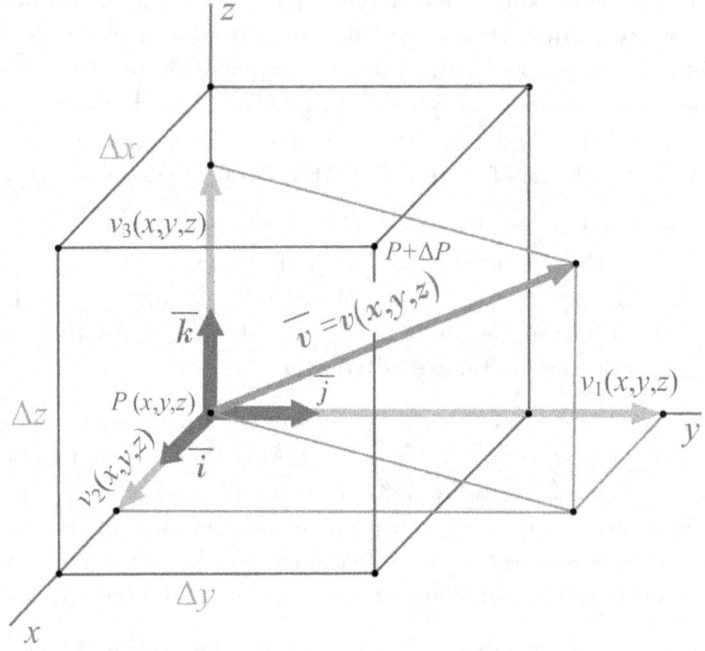

Figure 9

$$[\rho(x+\Delta x,y,z,t)\times v_1\,(x+\Delta x,y,z)\times\Delta y\times\Delta z]\;kg/\!/s \qquad [24.6]$$

Finding the difference between [24.6] and [24.5], dividing and multiplying it by Δx, so that it will not vary, and by definition of a partial derivative, in the limit when Δx tends to zero, we arrive at the variation of the amount of fluid that crosses both parallel faces of the parallelepiped at the point $P(x,y,z)$, which will be represented by the expression:

$$\left[\frac{\partial\left[\rho(x,y,z,t)\times v_1(x,y,z)\right]}{\partial x}\times\Delta x\times\Delta y\times\Delta z\right]\frac{kg}{s} \qquad [24.7]$$

144

Similarly, the amount of fluid that passes through the two faces parallel to each other and normal to the versor $\overline{j}$ will be given by:

$$\left[\frac{\partial\left[\rho(x,y,z,t)\times v_2(x,y,z)\right]}{\partial y}\times \Delta x \times \Delta y \times \Delta z\right]\frac{kg}{s} \quad [24.8]$$

And, finally, the amount of fluid that passes through the two normal faces to the versor $\overline{k}$ will be:

$$\left[\frac{\partial\left[\rho(x,y,z,t)\times v_3(x,y,z)\right]}{\partial z}\times \Delta x \times \Delta y \times \Delta z\right]\frac{kg}{s} \quad [24.9]$$

The dyadic sum of the uniform quantities [24.7], [24.8] and [24.9] will indicate the quantity of fluid entering or leaving the elementary parallelepiped per unit time at the point $P(x,y,z)$.

Dividing this total quantity by the dyad [$(\Delta x \times \Delta y \times \Delta z)$ m^3], according to the definition of division in section XVI, we will have said quantity per unit of geometric volume in P indicated by the following concrete:

$$\left[\frac{\partial\left[\rho(x,y,z,t)\times v_1(x,y,z)\right]}{\partial x}+\frac{\partial\left[\rho(x,y,z,t)\times v_2(x,y,z)\right]}{\partial y}+\frac{\partial\left[\rho(x,y,z,t)\times v_3(x,y,z)\right]}{\partial z}\right]\frac{kg}{s*m^3} \quad [24.10]$$

In mathematical field theory, the div divergence of a vector field $\overline{A}=A_1(x,y,z)\times\overline{i}+A_2(x,y,z)\times\overline{j}+A_3(x,y,z)\times\overline{k}$ is defined with the expression:

$$div\ \overline{A}=\frac{\partial A_1(x,y,z)}{\partial x}+\frac{\partial A_2(x,y,z)}{\partial y}+\frac{\partial A_3(x,y,z)}{\partial z}$$

With this notation, applied to the vector field product $\rho(x,y,z,t)\times\overline{v}(x,y,z)$, the dyad [24.10], assuming the coordinates x,

145

y, z and t, for simplicity, and recalling since the analysis is being carried out for a certain instant t, the variation of the amount of fluid in the parallelepiped with the following dyad can be written in a synthetic way:

$$[div\ (\rho \times \overline{v})]\ kg /\!/ (s*m^3) \qquad [24.11]$$

So far we have only taken into account the variation in mass due to the instantaneous movement of the fluid, but the density being generally variable with time, this effect must also be introduced. To do this, we must observe that the change in mass due to the change in density in the elementary parallelepiped at point P between the instants t and t y $t + \Delta t$ will have the dyadic form:

$$[[\rho(x,y,z,t+\Delta t) - \rho(x,y,z,t)] \times (\Delta x \times \Delta y \times \Delta z)]\ kg$$

Dividing by the measurement $[\Delta t\ s]$ and taking the limit when ?t tends to zero, we arrive at the dyad expressed with the partial derivative with respect to t for the change in mass in the parallelepiped per unit time:

$$\left[\frac{\partial [\rho(x,y,z,t)]}{\partial t} \times \Delta x \times \Delta y \times \Delta z \right] \frac{kg}{s}$$

Dividing this quantity by the dyad $[(\Delta x \times \Delta y \times \Delta z)\ m^3]$ with the operation in section XVI, we arrive at the variation in mass per unit of time and volume:

$$\left[\frac{\partial [\rho(x,y,z,t)]}{\partial t} \right] \frac{kg}{s*m^3} \qquad [24.12]$$

Adding [24.11] and [24.12] we arrive at the variation in total unit mass, which we write in abbreviated form, regardless of the coordinates, which are understood, by the following dyadic quantity:

$$\left[div\ (\rho \times \overline{v}) + \frac{\partial \rho}{\partial t} \right] \frac{kg}{s*m^3} \qquad [24.13]$$

146

Imagining two scalar fields that represent the contribution and loss of fluid at each point $P(x,y,z)$ per unit of time and volume, which we can represent by $\varphi(x,y,z)$ for the sources and $\sigma(x,y,z)$ for drains or sinks, grouping them into a single scalar field function represented by the form $\psi(x,y,z) = \varphi(x,y,z) - \sigma(x,y,z)$, where the minus sign must have the meaning of decrease in mass, equaling [24.13], we arrive at the physical equation known as **hydrodynamic continuity**:

$$\left[div \left(\rho \times \overline{v} \right) \right] \frac{kg}{s*m^3} = \left[\psi - \frac{\partial \rho}{\partial t} \right] \frac{kg}{s*m^3} \qquad [24.14]$$

Note that, according to the equality criterion in section IV, it has been assumed that the first and second members of [24.14] are uniform, although it would suffice if they were homogeneous, as we will analyze next. Under these conditions, the doubling theorem [24.2] with $k=1$ will allow us to write the algebraic equality without the secondaries:

$$div \left(\rho \times \overline{v} \right) = \psi - \frac{\partial \rho}{\partial t}$$

In general, if the units of length, mass and time were U_{L1}, U_{m1} and U_{t1} for the first member, and U_{L2}, U_{m2} and U_{t2} for the second, equation [24.4] will be formulated as follows:

$$\left[div \left(\rho \times \overline{v} \right) \right] \frac{U_{m1}}{U_{t1}*U_{L1}^3} = \left[\psi - \frac{\partial \rho}{\partial t} \right] \frac{U_{m2}}{U_{t2}*U_{L2}^3} \qquad [24.15]$$

To simplify, designating by U_1 and U_2 the compound units of the first and second members, respectively, and assuming $U_2 = k \circ U_1$, the doubling theorem [24.2] leads us to the generic algebraic equation:

$$div \left(\rho \times \overline{v} \right) = k \times \left[\psi - \frac{\partial \rho}{\partial t} \right] \qquad [24.16]$$

Formula [24.15] is the universal dyadic expression of the hydrodynamic continuity equation and [24.16] is its form of unfolded ordinary algebra.

In the particular case that the fluid was incompressible, the density would be constant in space and in time, the corresponding partial derivative would cancel out and we would have:

$$\rho \times div\ \overline{v} = k \times \psi$$

If there were also no sources or drains or sinks, we would have at all points ψ=0, with which the continuity equation would have the form:

$$div\ \overline{v} = 0$$

Needless to remember again that, if the units of the first and second members of the concrete equation were uniform, it is enough to take k=1 in the doubling.

Let's look at a second example of a physical equation that refers to the phenomenon of **heat conduction**. Let us consider a dyadic vector field defined by an application of R^3 in R^3, independent of time and symbolized $\overline{Q} = Q(x,y,z)$, so that at each point $P(x,y,z)$ there is a given heat transfer by the vector, which indicates the **heat current** in units of heat per unit of time and unit of area in the normal direction, which will be a function of the coordinates (x,y,z) of the point. Suppose that the temperature T of the fluid is variable, it will be represented by a dyadic scalar field described by a function of R^4 in R denoted $T = T(x,y,z,t)$ and its unit of measurement. Let us admit that both functions $\overline{Q}$ and T are differentiable.

Let's look at the same figure 6, in which we will now have to substitute $\overline{v}$ for $\overline{Q}$, which are silent or indifferent symbols for the purposes of analysis. The figure represents an elementary parallelepiped in ordinary geometric space from a generic point, taking the increments of the variables Δx, Δy and Δz. Fixing ideas, we will use the units of the International System for the different magnitudes that intervene in the phenomenon.

Under these conditions, the change in heat due to the field $\overline{Q}$ in the elementary parallelepiped and per unit volume will be described by the same equation [24.11], without more than substituting the field $\rho \times \overline{v}$ for it $\overline{Q}$, associating the corresponding units with it, placing in the numerator the calorie as a unit of heat, or any other, which leads us to the following concrete:

$$[div \ \overline{Q}] \ cal /\!/(s*m^3) \qquad [24.17]$$

Next, it is necessary to assess the contribution to heat by the variation of the thermal field at a fixed point and with respect to time t. Thus, being c with its units the specific heat of the body, the variation of heat in the elemental parallelepiped due to the change in temperature with time will be determined by the product of the following dyadic entities:

$$\left[\rho \frac{kg}{m^3}\right] * [\Delta x \ m] * [\Delta y \ m] * [\Delta z \ m] * \left[c \frac{cal}{kg*K}\right] \times \left[\frac{\partial T}{\partial t} \frac{K}{s}\right]$$

Dividing the previous quantity between the dyad that defines the volume of the parallelepiped $[(\Delta x \times \Delta y \times \Delta z) \ m^3]$ and operating with the dyadic algebra, we arrive at the expression of the heat variation at point P due to the thermal variation:

$$\left[\rho \times c \times \frac{\partial T}{\partial t}\right] \frac{cal}{s*m^3} \qquad [24.18]$$

Adding the uniform quantities [24.17] and [24.18], we arrive at the total change in heat at point P per unit of time and volume with the dyad:

$$\left[div \ \overline{Q} + \rho \times c \times \frac{\partial T}{\partial t}\right] \frac{cal}{s*m^3}$$

Assuming that there are no heat sources or sinks, it is enough to set the previous quantity to zero to obtain the conduction equation:

$$\left[div \ \overline{Q} + \rho \times c \times \frac{\partial T}{\partial t}\right] \frac{cal}{s*m^3} = 0 \qquad [24.19]$$

149

The law of heat conduction axiomatizes the experience that the relationship between heat current and the change in temperature is given by:

$$\left[\overline{Q}\ \frac{cal}{s*m^2}\right] = -\left[\lambda\ \frac{cal}{s*m*K}\right] * \left[grad\ T\ \frac{K}{m}\right] \qquad [24.20]$$

Where λ is called the **coefficient of thermal conductivity** and is specific to each substance. Its secondary is expressed by the fundamental units indicated in [24.20]. In turn, the operator *grad* or «nabla» ∇ of a scalar field T is the gradient vector field, which in the theory of mathematical field analysis is defined on an orthonormal basis $(\overline{i},\overline{j},\overline{k})$ by the equation:

$$grad\ T = \nabla T = \frac{\partial T}{\partial x} \times \overline{i} + \frac{\partial T}{\partial y} \times \overline{j} + \frac{\partial T}{\partial z} \times \overline{k}$$

Substituting [24.20] in [24.19], we have the following dyadic equation:

$$\left[div\left(-\lambda \times grad\ T\right) + \rho \times c \times \frac{\partial T}{\partial t}\right] \frac{cal}{s*m^3} = 0$$

In mathematical field theory, the **Laplacian** operator is defined, which is represented by the same letter delta Δ of the increments for variables and means the divergence of the gradient vector field. With which, the previous equation is symbolically transformed into the well-known fundamental equation of heat conduction:

$$\left[-\lambda \times \Delta T + \rho \times c \times \frac{\partial T}{\partial t}\right] \frac{cal}{s*m^3} = 0$$

In the foregoing we have developed two significant physical equations, which illustrate the deduction procedure to be followed to analyze physical phenomena based on the algebra of magnitudes, through the operations defined in this monograph for the dyadic entities of Physics.

The two phenomena analyzed have no relation to each other, because they refer to very different material realities; however, the mathematical components of the primaries of the magnitudes described are very similar and in some respects appear identical or at least isomorphic. On the contrary, the different nature between both cases is evident in the units of the secondary ones, which are different because they refer to different magnitudes. This shows that the mathematical apparatus of the primary ones can acquire different meanings depending on the context of the fact under investigation, and hence the importance of not losing sight of the secondary ones and their units, which determine the intervening magnitudes in each phenomenon, otherwise it would be impossible to fully understand the physical meaning of the derived equations.

Thus, the importance of operating with dyads instead of ordinary algebraic entities has become evident, since otherwise, apart from the objective incorrectness that this implies, the equations lose much of their significance, becoming mere very elegant symbologies mathematical in nature, though physically abstruse.

On the contrary, operations with dyads allow logical steps to be established without the slightest confusion, as can be seen in the dyadic division that gives rise to the quantity [24.10], which unequivocally deduces a quotient between measurements, both with their two primary elements and secondary, being established unambiguously by means of that composition law. With all this, the convenient **generator procedure** is revealed to systematically determine scientific formulations, operating in the first place not only with simple common algebraic entities, but with quantities of magnitudes, to later equate those that are homogeneous and that must be identified with each other, by virtue of its own observed nature, to **create a physical equation**.

We observe in this process the decisive intervention of the **generating external laws of composition**, without which Physics would be emptied of content. And this is how the algebra of magnitudes decisively helps the best understanding and exact formulation of the analyzed natural phenomena, while at the same time establishing the essential composition laws so that physical language is complete and univocal, safe from subjective, intuitive interpretations or arbitrary.

Let us finally clarify that we have established the **criterion of dyadic equality** with the general condition that equal dyads represent the same quantity of magnitude. This definition opens up three possibilities: first, that the dyads of both members are homogeneous, in which case we will speak of **identity**; second, that the dyad of one of the members is generated by those of the other through generative operations, in which case equality will represent a **physical equation**; and third, that a physical equation has been established experimentally, which gives rise to an equality that we will call a **physical law**.

For example, all equality of homogeneous quantities such as $150\ cm = 1.50\ m$ are identities. Equalities that relate dyads through generating operations, such as the definition of surface $3\ m * 2\ m = 6\ m^2$ or speed $6\ m /\!/ 2\ s = 3\ m /\!/ s$, are physical equations. And, finally, generating empirical relations such as Newton's second law, $2\ kg * 6\ m /\!/ s^2 = 12\ N$, are physical laws.

Therefore, the definitions of additive operations are identities (sections V to XI); the definitions of the generating multiplicative operations are equations (sections XII to XX); and formulations such as *Newton's second law* are physical laws. However, for convenience in practice we use the terms equality, identity and equation as synonyms, a simplification that lacks physical-mathematical relevance.

Section XXV

DEFINITION OF DIMENSIONS
OF THE PHYSICAL MAGNITUDES

Lord Kelvin used to say: «There is something extremely interesting in the fact that we can establish a metric system based on a unit of length and a unit of time. There is nothing new in it, since it has been known since Newton's time, but it retains all its interest and relevance».

Although Kelvin reduced the dimensional basis to two magnitudes, length and time, to which today mass has been incorporated, his reflection is nonetheless valid. A rational or coherent system of units must be such that it includes the minimum number of fundamental magnitudes from which all the others are derived. Hence, three systems of units have traditionally been formed: Cegesimal System, International System and Technical or Terrestrial System. The **Cegesimal System**, known by the acronym CGS, initials of the units centimeter, gram and second, which are adopted as primary units of the fundamental magnitudes of length, mass and time; It should be noted that the gram, symbolized g, is one thousandth of the standard kilogram. For its part, the **International System** or MKS, by the initials of meter, kilogram and second, which are the primary units adopted by this system, is recommended by the International Committee of Weights and Measures and was implemented in Spain by law in 1967 Finally, the **Terrestrial or Technical System** differs from the previous ones in that it does not use mass as a fundamental magnitude, but rather force, which is the action that interests technical applications, which is why it is often used in the field engineering; the other two fundamental magnitudes correspond to length and time, with their primary units the meter and the second, and the kilopond is established as the primary unit of force, defined as the force with which the

Earth attracts a mass of one kilogram in a point of latitude equal to 45 degrees at sea level, where the acceleration of gravity is 980.665 centimeters per second squared, abbreviated 980.665 $cm/\!/s^2$ «arithmetizing» 980.665 cm/s^2.

The above allows to approach the concept of dimension and dimensional equation of a universal law or a definition equation. To specify, let us assume as the International system of units, in which the fundamental magnitudes are length, mass and time, which will be denoted for dimensional purposes with the initials L, M and T. Suppose we want to find the form symbolic that represents the composition of the magnitudes of a surface without quantities or units, only specifying the magnitudes concerned; the dimensional shape of a surface will be symbolized $[S]$, the brackets to refer to the dimensional equation and the S to indicate the proper magnitude of a surface; It is clear that every unit of surface, by the definition of dyadic multiplication, is the geometric product of two lengths, which can be written, making abstraction of the units and paying attention only to the magnitudes, with $[S]=L*L=L^2$, and this is, by definition, the form of the equation dimensional of the magnitude called surface, described by the square of two symbolic lengths. The square brackets indicate the meaning that the equation that follows does not address specific values of magnitudes, but is limited to composing magnitudes in the abstract of the quantities indicated in the second member. For the example of the area, this means that this magnitud derives from the length and is equivalent to a length multiplied geometrically by another length, a relationship that the respective units of area and length must respect. Remember that geometric multiplication has nothing to do with arithmetic, for the reasons widely explained in this work. Therefore, it should not be forgotten that this product refers to that of dyadic entities, although it commonly coincides in nomenclature with the multiplication of real numbers, with the current force of the false hypothesis of the International System of Units, which erroneously attributes the abelian multiplicative group structure to magnitudes, as we have already noted previously. With the volume V we also have the symbolic dimensional equation $[V] = L*L*L = L^3$, so that the dimension of the magnitude affine to the volume is the geometric cube of the length quantity, as before the dyadic cube, not that of R. Similarly, the density D, as an expression of the dyadic quotient between the unit of mass M and that of volume V of a body, will have as its dimensional form the symbolized geometric quotient $[D] = M /\!/ V = M * L^{-3}$, using negative exponents with the usual meaning of divisors or denominators of the same positive power, which means that the density magnitude will be derived from the mass and the length with the indicated dimensional expression with respect to dyadic algebra. In general, every dimensional equation will have the form of the monomial $[X] = L^\alpha * M^\beta * T^\gamma$, where α, β and γ are whole or fractional numbers, therefore, positive or negative, and the derived magnitude X is said to be α dimensions relative to L, β dimensions relative to M and γ dimensions relative to T.

On the other hand, it does not seem doubtful whether the **principle of physical homogeneity** should be admitted or the axiomatic assumption that **the physical formulas symbolize laws that must be admitted are independent of the units in which they are expressed**, which requires that in an equation that includes specific entities the dimensional shapes of the first and second members must coincide, because they refer to the same magnitude, although the respective units may differ, but without ceasing to be homogeneous with respect to the fundamental magnitudes that compose them; in this way the dimensional equations are revealed as a check of the homogeneity of the units that intervene in the members of a formula, so that a homogeneity defect will always reveal a calculation error in its deduction. And this is indicated by stating in synthesis that every physical equation must be dimensionally consistent, which means, in harmony with [4.1], that only two dyadic entities can be equated if they are expressed in homogeneous units, that is, in units of the same magnitude although, it is insisted, they may be different from each other, as would be, for example, the units

155

km∥h and *m∥s* in relation to the derived magnitude called velocity.

In general, a universal law or definition equation can be imagined that is expressed in the following scalar form:

$$a\,U = \left(a_1\,U_1\right)^{\delta_1} * \left(a_2\,U_2\right)^{\delta_2} * \ \ldots\ * \left(a_n\,U_n\right)^{\delta_n} \quad [25.1]$$

Formula [25.1] is nothing more than an algebraic expression assembled with scalar dyadic entities, which in general can come from each of the components of a vector equation. If the units are changed, the same quantity $a\ U$ of the magnitude M corresponding to the unit U can be indicated with the dyad $b\ V$, where V is another unit of the same magnitude M, so by hypothesis U and V will be units homogeneous. And by the principle of physical homogeneity, the concrete $b\ V$ can be expressed with the same form of [25.1], that is, it will have:

$$b\,V = \left(b_1\,V_1\right)^{\delta_1} * \left(b_2\,V_2\right)^{\delta_2} * \ \ldots\ * \left(b_n\,V_n\right)^{\delta_n} \quad [25.2]$$

By hypothesis, the dyads $a\ U$ and $b\ V$ represent the same amount of M, so the second members of [25.1] and [25.2] can be equalized, and then operate with the dyadic algebra, and the real ones when appropriate, from in accordance with [16.2] and [17.1] or, in short, with the single rule that includes all the specific operations, described in section **XXII**; from which it results:

$$\left(\frac{a_1}{b_1}\right)^{\delta_1} \times \left(\frac{a_2}{b_2}\right)^{\delta_2} \times \ldots \times \left(\frac{a_n}{b_n}\right)^{\delta_n} = \left(\frac{V_1}{U_1}\right)^{\delta_1} * \left(\frac{V_2}{U_2}\right)^{\delta_2} * \ldots * \left(\frac{V_n}{U_n}\right)^{\delta_n} \quad [25.3]$$

In equation [25.3] the quotients between units $V_i\,\|U_i$ are real numbers, by virtue of the axiom of continuity and the definition of division of homogeneous concretes, established in section **XI**; so formula [25.3] actually relates real numbers.

Let's analyze the meaning of the exponents δ_i. For this, with M being the derived magnitude associated with the dyad in the first member of [25.1], let M_i be the fundamental magnitudes that correspond to the factors $a_i\ U_i$ of the terms of the second member.

The dimensional equation of M can be written symbolically as the relationship between quantities of the fundamental magnitudes, resulting in the form:

$$[M] = M_1^{\delta_1} * M_2^{\delta_2} * \ldots * M_n^{\delta_n} \qquad [25.4]$$

In conclusion, we can establish that the δ_i terms indicate the dimensions of the magnitude M in the base of fundamental magnitudes $\{M_i\}$, resulting in that the di themselves configure the equation of change of units [25.3], in which it is appreciated that the exponents of the ratios of the measures a_i / b_i are the same as the inverse ratios of the units $V_i /\!/ U_i$ when establishing a unit change from $\{U_i\}$ to $\{V_i\}$.

By dimensional base we have to understand any set of fundamental magnitudes $\{M_i\}$ such that any other derived magnitude can be composed by means of dyadic algebra with those of the base and that these are independent of each other, that is, that none of them can be composed into In no way through the others, because otherwise, according to the mathematical meaning that is attributed to the concept of the basis of a given structure, one could not speak of a basis properly.

These elementary mathematics considerations allow us to imagine two dimensional bases of magnitudes $\{M_i\}$ and $\{M'_j\}$, with i taking the values from 1 to m, and j from 1 to n. In general, there is no reason to require that m and n must be equal. We observe that any magnitude X can be composed with the elements of the two bases and we wonder what relationship will exist between the dimensions of the magnitude X in both systems of basic magnitudes. To solve this question, suppose that X in the base $\{M_i\}$ has the dimensions δ_i, taking i the values from 1 to m, with the dimensional expression:

$$[X] = M_1^{\delta_1} * M_2^{\delta_2} * \ldots * M_m^{\delta_m} \qquad [25.5]$$

In turn, the magnitude X at the base $\{M'_j\}$ will have dimensions δ'_j, with j from 1 to n, and its dimensional equation will be:

$$[X] = M'^{\delta'_1}_1 * M'^{\delta'_2}_2 * \ \ldots \ * M'^{\delta'_n}_n \qquad [25.6]$$

Since, by hypothesis, $\{M'_j\}$ is a base of the magnitude system established in the physical theory considered, the magnitudes of the first base $\{M_i\}$ can be expressed by means of their corresponding dimensional equations:

$$[M_1] = M'^{a_{11}}_1 * M'^{a_{12}}_2 * \ \ldots \ * M'^{a_{1n}}_n$$

$$[M_2] = M'^{a_{21}}_1 * M'^{a_{22}}_2 * \ \ldots \ * M'^{a_{2n}}_n$$

$$\text{...} \qquad [25.7]$$

$$[M_m] = M'^{a_{m1}}_1 * M'^{a_{m2}}_2 * \ \ldots \ * M'^{a_{mn}}_n$$

The coefficients α_{ij} of [25.7], with i of 1 to m and j from 1 to n, indicate the dimensions of the magnitude M_i of the old base with respect to the magnitudes of the new base $\{M'_j\}$, that is, that α_{ij} is, by definition, the dimension of the magnitude M_i with respect to the magnitude M'_j. So, substituting the M_i of [25.7] in [25.5], dispensing with the brackets, given that their meaning is not transcendent for these purposes, we have:

$$[X] = \left(M'^{a_{11}}_1 * M'^{a_{12}}_2 * \ \ldots \ * M'^{a_{1n}}_n \right)^{\delta_1} *$$

$$* \left(M'^{a_{21}}_1 * M'^{a_{22}}_2 * \ \ldots \ * M'^{a_{2n}}_n \right)^{\delta_2} *$$

$$\text{...}$$

$$* \left(M'^{a_{m1}}_1 * M'^{a_{m2}}_2 * \ \ldots \ * M'^{a_{mn}}_n \right)^{\delta_m}$$

Taking into account that the power of a power is another power whose exponent is the product of the exponents and identifying exponents with the expression [25.6] for each M'_j, the new dimensions δ'_j of the magnitude X in the base result $\{M'_j\}$ as a function of the first δ_i with respect to the base $\{M_i\}$ and the dimensions α_{ij}, which are the dimensions of the magnitudes of the first base with respect to those of the second:

$$\delta'_1 = \delta_1 \times \alpha_{11} + \delta_2 \times \alpha_{21} + \ldots + \delta_m \times \alpha_{m1}$$
$$\delta'_2 = \delta_1 \times \alpha_{12} + \delta_2 \times \alpha_{22} + \ldots + \delta_m \times \alpha_{m2}$$

$$\ldots\ldots\ldots\ldots\ldots\ldots\ldots\ldots\ldots\ldots\ldots\ldots\ldots$$

$$\delta'_n = \delta_1 \times \alpha_{1n} + \delta_2 \times \alpha_{2n} + \ldots + \delta_m \times \alpha_{mn}$$

[25.8]

The group of equations [25.8] are relations between real numbers, so the addition and multiplication that appear in them are the composition laws of the body R. They are similar and are deduced with an analogous illation to those that result for changes in base on vector spaces[12]. They can be written in abbreviated form with the form of summation:

$$\delta'_j = \sum_{i=1}^{m} \delta_i \times \alpha_{ij} \quad ; \quad j = 1, 2, \ldots, n \qquad [25.9]$$

To symbolize these sums in algebra, it is usual to use the **concentrated or indexed notation**, which is nothing more than the convention of eliminating sums, considering that the repeated index in the factors of a monomial means or replaces the summation with respect to it, keeping the other subscripts constant.

With this symbology that simplifies writing, the new dimensions with respect to the old ones will be expressed with the following synthetic formula:

$$\delta'_j = \delta_i \times \alpha_{ij}$$

$$i = 1, 2, \ldots, m; j = 1, 2, \ldots, n$$

It is also possible to symbolize the base change equations using **matrix notation**, resulting in the following expression:

---

[12] The change of base in V^3, generalizable to dimension n, is exposed analytically in «Lesson 4» of the volume Mathematize 2.

$$\begin{bmatrix} \delta_1 & \delta_2 & \dots & \delta_n \end{bmatrix} = \begin{bmatrix} \delta_1 & \delta_2 & \dots & \delta_m \end{bmatrix} \begin{bmatrix} \alpha_{11} & \alpha_{12} & \dots & \alpha_{1n} \\ \alpha_{21} & \alpha_{22} & \dots & \alpha_{2n} \\ \dots & \dots & \dots & \dots \\ \alpha_{m1} & \alpha_{m2} & \dots & \alpha_{mn} \end{bmatrix}$$

This section serves as proof of the logical connection with the usual dimensional analysis, once the gap in the algebra of magnitudes has been bridged, giving meaning and foundation to equation [24.1], which usually marks the beginning of Physics texts as an abstract principle and intuitive, and that here we have inferred based on the structures of dyadic entities with their composition laws, and thus we verify that the dimensional aspects of magnitudes and physical equations are described by the dyadic algebra model, based on the geometric algebra.

At the same time, we verify with unquestionable clarity that any dimensional equation with the form $[X] = L^{\alpha} \times M^{\beta} \times T^{\gamma}$ is nothing more than mere symbology empty of true physical-mathematical meaning, as long as the previous and precise definition of the **generating external laws of composition**, as established in sections XII et seq. of this dyadic algebra.

Section XXVI

THE PHYSICAL CONSTANTS

In section XXIV we have analyzed the meaning to be given to physical equations, which must be understood as invariant identities between two dyadic entities, which in the famous case of **Newton's second law**, refers to vector measurements, explicitly written $\overline{F}\ U_F = m\ U_m \odot \overline{a}\ U_a$ (product of section XX) or with symbolic economy $\overline{F}\ U_F = m\ U_m \times \overline{a}\ U_a$ where U_F is the unit of force, U_m the unit of mass and U_a the unit of acceleration. Therefore, for this law, the dyadic meaning must be admitted that the scalar measurement $m\ U_m$ multiplied by the vector measurement $\overline{a}\ U_a$ must be equal to the vector dyad $\overline{F}\ U_F$, where m is a positive scalar of R, called inertial mass, which presents a value specific for each material body.

By virtue of [20.1], which defines the product of a scalar dyad by another vector, and then with [9.3] and [9.4], we can transform Newton's second law in this explicit way:

$$\overline{F}\ U_F = m\ U_m \odot \overline{a}\ U_a = (m\bullet\overline{a})\ (U_m * U_a)] =$$
$$= m\circ[\overline{a}\ (U_m * U_a)] \qquad [26.1]$$

As by the nature of this law m is a scalar, we can apply [11.2] to form the conscious of collinear and homogeneous vector concretes, with which we will have:

$$\frac{\overline{F}\ U_F}{\overline{a}\ (U_m * U_a)} = m \qquad [26.2]$$

So Newton's second law is equivalent to understanding that the primary of the mass, which is a positive real number, is equal to the dyadic quotient between $\overline{F}\ U_F$ y $\overline{a}\ (U_m * U_a)$, and that this number is invariable for each body or point material. So nothing prevents us from considering that m is a **characteristic constant**, without prejudice to the magnitude condition that is attributed to the inertial mass[13]. The scalar dyad that defines the mass of the body will have the form m Um and the unit of mass will not be independent of the units of force and acceleration, because the elements $\overline{F}\ U_F$ y $\overline{a}$ $(U_m * U_a)$ must be homogeneous, due to the definition of equality [4.2] between vector dyads. Furthermore, nothing prevents setting the units so that the first and second members of the physical equations are not only homogeneous, but uniform, and this is usual for convenience. However, for greater generality we will develop the assumption that these dyads are homogeneous and non-uniform, therefore, the continuity axiom [4.3] will determine that there exists $k \in R$ such that $U_m * U_a = k \circ U_F$. Under these conditions, the doubling of the equality of dyads that integrates the physical equation, described in [24.3], applied to equation [26.1], which can be written:

$$\overline{F}\ U_F = m \circ [\overline{a}\ (U_m * U_a)] = (m \bullet \overline{a})\ (U_m * U_a)$$

Produces the primary and secondary relationships, appearing a vector equation and an equality between units:

$$\overline{F} = k \times m \bullet \overline{a}$$

$$k \circ U_F = U_m * U_a \Rightarrow U_m = (k \circ U_F) /\!/ U_a \qquad [26.3]$$

Remember that the quotient $(k \circ U_F) /\!/ U_a$ is not the ordinary one, but the one defined in section XVI. In conclusion, the measurement of the characteristic or specific constant of each body called mass will be given by a scalar dyad whose primary or measure will be determined by the positive real number given by

---

[13] It can be argued that the mass of inertia is considered or not a characteristic constant, since we have it for a fundamental physical magnitude; however, since we admit that it is a constant quantity for each material point, it does not seem unreasonable to consider it that way. In any case, it is only a mere convention. On the other hand, regardless of what is accepted in this regard, the interest of studying Newton's second law, one of the essential laws of Physics, is undeniable, so we have decided to include it in this section, if only to observe the ilative consequences of the algebra of magnitudes in the analysis of such a primordial law.

[26.2] and whose secondary or unit will have the form [26.3], with which it will be that the mass will be expressed by the following homogeneous dyads:

$$m \, U_m = m \circ \frac{k \circ U_F}{U_a}$$

Note that the relationship [26.3] between the units of force U_F, mass U_m, and acceleration U_a means that they cannot be independent. On the other hand, if the units were defined so that that of the first member was uniform with that of the second of the physical equation [26.1], we would simply have the particular and more usual case $k = 1$.

Another famous example of constant characteristic is found in the well-known **Hooke's law**. Formulated in common language, this law establishes that stress, defined as force per unit area, is for each body proportional to deformation, understood as variation in length per unit length. Although it is vector in nature, it can be applied in scalar terms, as we justified in section XXIV, and we will do so to vary the reasoning with respect to the previous example. To do this, consider a wire of length $L \, U_L$ and section v, subjected to a force of component $F \, U_F$ in the direction of L, so it will experience a length variation $E \, U_E$, also in the direction of L, and where $E \, U_E$ the proportionality factor, called **Young's modulus**, and thus Hooke's law will have the form of the following dyadic algebra physical equation:

$$\frac{F \, U_F}{S \, U_S} = E \, U_E * \frac{\Delta L \, U_L}{L \, U_L}$$

The dyads $\Delta L \, U_L$ and $L \, U_L$ are homogeneous and uniform by hypothesis, because both represent lengths in the same unit, so that, according to section XI, their dyadic quotient must be a real number, so we can dispense with the U_L unit in numerator and denominator, and Hooke's law can be written in the form:

$$\frac{F \, U_F}{S \, U_S} = E \, U_E \circ \frac{\Delta L}{L}$$

Let $\sigma = F/S$ be the measure of stress or force per unit area in the associated compound unit $U_F/\!/U_S$; and let $\varepsilon = \Delta L/L$ be the measure of the deformation or variation of length per unit of length, which lacks its own unit. With this notation and by the definition of multiplication by a scalar [9.1] and [9.2], we will have dyadic equality:

$$\sigma \, \frac{U_F}{U_S} = \left(E \times \varepsilon \right) \, U_E \qquad\qquad [26.4]$$

In general, given the definition of equality [4.1] between scalar dyads, in which the identified elements do not have to be uniform, and because of the continuity axiom [4.3], there will be $k \in \mathrm{R}$ such that:

$$U_E = k \circ \frac{U_F}{U_S}$$

Under these conditions, the doubling theorem [24.2] for the equality of dyads that make up the physical equation [26.4] can be written, relating primary and secondary:

$$\sigma = k \times E \times \varepsilon$$

$$k \circ \frac{U_F}{U_S} = U_E \qquad\qquad [26.5]$$

On the other hand, according to the definition of division between homogeneous scalar dyads [11.1], or indistinctly with [16.2], and defined the product by a scalar through [9.1] and [9.2], Hooke's law [26.4] will with the form:

$$\frac{\sigma \, \dfrac{U_F}{U_S}}{\varepsilon \, U_E} = E$$

This means that the ratio between the scalar dyads $\sigma \, (U_F/\!/U_S)$ and $\varepsilon \, U_E$ will be an invariable real number E for each type of wire. Young's modulus E thus becomes a characteristic constant that specifically relates for each material object the invariant relationship between the applied stress and the deformation produced. With the generic hypothesis that the physical equations are homogeneous dyadic equalities, in this case with the scalar form [4.1], the Young's modulus will be represented by the scalar dyad:

$$E \ U_E = (k \times E) \ \frac{U_F}{U_S}$$

Also here we have that the relationship [26.5] between the units U_E, U_F and U_S determines the dependency between them, so they cannot all be arbitrarily established. If the units were defined to be uniform, we would have $k=1$.

Apart from the characteristic or specific constants, such as the mass of inertia and Young's modulus, we find others that are independent of the nature of the bodies, remaining invariable in any case, and we will call them **universal constants**. A prominent example is the **mechanical equivalent of heat**. Let $T \ U_T$ be the mechanical work measured with the unit U_T, let us symbolize $Q \ U_Q$ the amount of heat equivalent to the previous work in the unit U_Q and write $M \ U_M$ for the constant of proportionality between these other two quantities with the unit U_M. The universal law that determines the equivalence between heat and work can be written in dyadic algebra with the form of the expression:

$$T \ U_T = (M \ U_M) * (Q \ U_Q)$$

By its nature, it is a scalar physical equation, so we will have to compose the dyads that intervene with the scalar operations that we have defined. The definition of multiplication [12.1] leads us to the expression:

$$T \ U_T = (M \times Q) \ (U_M * U_Q) \tag{26.6}$$

In general, the dyads of the first and second member should be homogeneous, without the need for them to be uniform, given the definition of equality [4.1] for specific scalars. The axiom of continuity [4.3] determines that there will be $k \in R$ such that $U_M * U_Q = k \circ U_T$. The doubling theorem [24.2], applied to the scalar physical equation [26.6], allows us to determine the two relations that must be satisfied between its two primary and two secondary ones:

$$T = k \times M \times Q$$

$$k \circ U_T = U_M * U_Q \qquad\qquad [26.7]$$

On the other hand, according to the definition of division between homogeneous scalar dyads [11.1], or indistinctly with [16.2], and with the product by a scalar of [9.1] and [9.2], the universal law [26.6] will remain with the shape:

$$\frac{T\, U_T}{Q \left(U_M * U_Q \right)} = M$$

This means that the ratio between the scalar dyads $T\, U_T$ and $Q\, Q\, (U_M \times U_Q)$ will be a real number M invariable and independent of the nature of the bodies, so it is said to be a universal constant called the mechanical equivalent of heat. With the generic hypothesis that the physical equations equal homogeneous concretes, this constant will be represented by the scalar dyad:

$$M\, U_M = \left(k \times M \right)\, \frac{U_T}{U_Q}$$

Also here we observe that the relationship [26.7] between the units U_T, U_M and U_Q determines the dependency between them, so the three cannot be arbitrarily established. If the units were defined so that they were uniform, which would be the most comfortable, we would have $k = 1$.

Another example of the same type is found in the **constant of universal gravitation**. Although it is considered only one, the law of gravitation is actually two: first, which describes that two material points attract each other with a force, with the direction of the line that joins them, directly proportional to the product of their masses gravitational and inversely proportional to the square of the distance that separates them; and second, that for each material point the ratio between the gravitational mass and the mass of inertia of Newton's second law is constant, and that both are manifestations of the same magnitude called mass.

Although strictly speaking the law of gravitation has a vector formulation, as it can be reduced to another scalar with a single component in the direction of the line that joins the material points, we will follow this path. To distinguish the two masses of each material point, the inertial mass will be symbolized by m and the gravitational μ. Let U_F be the unit of force, U_m the unit of mass, both gravitational and inertial, and U_L the unit of length to measure the distance d between the material points. Under these conditions, the universal law of gravitation can be written in concrete algebra using an expression like the following:

$$F U_F = \frac{\left(\mu_1 U_m\right) * \left(\mu_2 U_m\right)}{\left(d U_L\right)^2}$$

The subscripts 1 and 2 are used to distinguish each of the two material points of gravitational masses μ_1 and μ_2, as well as the inertial masses m_1 and m_2. In turn, being H the ratio of proportionality between the gravitational and inertial masses, they will be written for both material points:

$$\mu_1 U_m = H \circ (m_1 U_m)$$

$$\mu_2 U_m = H \circ (m_2 U_m)$$

Since H is a quotient between homogeneous dyadic numbers, by reason of [11.1], it must be a simple real number, so H lacks unity and is such that $H \in R$. Substituting these two equations in the previous one and, operating with [9.1] and [9.2], the formulation of the law of gravitation with the inertial masses results:

$$F\,U_F = \left(G\,U_G\right)* \frac{\left(m_1\,U_m\right)*\left(m_2\,U_m\right)}{\left(d\,U_L\right)^2}$$

Being $G = H^2$ the factor known as the **constant of universal gravitation**, and U_G the corresponding unit. To deduce it, applying the definition [12.1] of multiplication of scalar dyads, we easily arrive at the dyadic algebra equation:

$$F\,U_F = \left(G \times \frac{m_1 \times m_2}{d^2}\right)\left(\frac{U_G * U_m^2}{U_L^2}\right)$$

As in the previous examples, where k is the scalar relationship between the units of both members, which would be the unit $k=1$ if there was uniformity, in the generic case that the concretes of the equation are not uniform, the doubling of [24.2] brings us the two relationships between primary and secondary:

$$F = k \times G \times \frac{m_1 \times m_2}{d^2}$$

$$k \circ U_F = \frac{U_G * U_m^2}{U_L^2}$$

From which it turns out that the concrete that represents the constant of universal gravitation is:

$$G\,U_G = \left(k \times G\right) \circ \frac{U_F * U_L^2}{U_m^2} \qquad [26.8]$$

It might be thought that it would be contradictory that, being H a constant without units, it turns out that $G = H^2$ does have dimensions. However, there is nothing disconcerting in this, since what happens is that while H is indicated only by a real number, given the definition with $\mu\,U_m = H \circ (m\,U_m)$, which for each material point relates its gravitational mass with that of inertia; instead G is determined by $G = H^2$ and also by [26.8], which means that its nature must be represented by a specific scalar entity, not only by a real number, so that the equality $G = H^2$ only applies refers to its primary element and [26.8] determines its secondary.

In the International System of Units, the value of G has been calculated and established based on the meter m as a unit of length, not to be confused with the same symbol used for mass, the kilogram kg as the unit of mass and the newton N as a unit of force:

$$G = 6{,}67 \times 10^{-11} \circ \frac{N * m^2}{kg^2} = 6{,}67 \times 10^{-11} \circ \frac{m^3}{s^2 * kg}$$

To end this section, we must note that in all the constant analyzes carried out, the constant k has appeared, related to the homogeneity and the uniformity axiom of the dyadic entities identified by the physical equations considered. Hence, this kind of invariants must be considered, which we will call **homogeneity constants**, and they are very important, because they intervene in the splitting of the physical equations of concrete algebra into their corresponding two algebraic and dimensional formulations, which are derived respectively for their primaries and secondary, in accordance with the provisions of [24.2] and [24.3].

In this regard and in light of the preceding analyzes, it is clear that the traditional omission of dyadic or magnitude algebra has forgotten the constants of homogeneity, limiting itself to the case where $k = 1$, and thus the doubling of the physical equations in its algebraic and dimensional components are reduced to a specific assumption. And this could contain a critical flaw that could distort the appreciation of the true nature of physical equations. Perhaps the origin of this traditional defect could be found in the importation by Physics of the mathematical method, which in its metric structures operates with abstract and, therefore, uniform units. The doubling theorem evidences and fully justifies that every mathematical equation remains invariant in this process, because, being $k = 1$, it happens that the primary and the corresponding dyadic formulation remain identical.

In the preceding analysis we have abstracted from the indirect rule of section XXII, so that to base the logical steps of all the reasoning we have directly used the laws of composition and properties of the algebra of magnitudes, as corresponds to the way of running more precise and typical of strict logic, reserving for this rule the function of simple verification of the conclusions or to examine its own validity a posteriori in the cases described.

We once again verify that the study of physical constants, like dimensional analysis and that all equations and laws, would not make any sense without having previously defined the **laws of external composition generating** sections XXII et seq.

On the other hand, in the preceding analysis we have tacitly assumed the classical **isometric hypothesis**, which assumes that the quantity of magnitude implicit in every unit is always invariable. We leave the opposite option for the second part of this work, where the **«dysmetric» axiom** is considered and the peculiarities of the spaces in which it is fulfilled are studied, with very relevant consequences for the survival of the physical constants.

Section XXVII

PHILOSOPHICAL CONSEQUENCES

We have already referred in section XXIV to the meaning of the equations of Physics, whether it is about universal laws or definition formulas, because one of the most important consequences of the algebra of magnitudes is to observe that physical equations do not relate entities Ordinary algebraic, either real numbers or vectors, but establish relationships between dyadic entities, defined in section III. This fact would seem to overturn the way of operating described in all the texts, because in them, although it is striking and even alarming, this unfailing circumstance is not taken into account. However, dyadic algebra turns out to be in such a way that it is reduced to the single rule of section XXII, so that the composition laws defined for these entities, which are inherent to Physics, allow the common algebraic part to be grouped on the one hand and, on the other, the dimensional part, being authorized to operate with the fiction or rather imposture that the symbols of the units of the related magnitudes are vulgar algebraic elements, although this is not really the case at all. And this isomorphic property saves with a lot of luck the negligent and pancist praxis that has prevailed since ancient times from having fallen into a crass error with disastrous consequences, for having forgotten to establish as a principle the necessary algebra of magnitudes that defines the laws of composition with numbers specific to science, which are dyadic entities. Such practice has also been legitimized by the International System of Units and its erroneous hypothesis in order to consider that the quantities of magnitudes present an abelian multiplicative group structure, a gross defect that no one who has followed this work carefully would excuse or allow it to remain in force knowing its falsehood. Anyone who is not a faciliton and has methodically followed the motivations set out

171

here, will have no doubt that whoever claims that unit operations are obvious would not have understood that in science everything must be based on experimentation or definition, nothing has to remain unverified or defined. How, then, could the lack of an algebra of magnitudes be justified? It being the case that Physics deals with parity entities with two ordered elements: first, the abstract algebraic entity; and second, the unitary element of some magnitude, which is by no means a numerical entity in the classical sense.

At the same time, no attentive and responsible reader will understand that others qualify the composite forms of physical quantities as mysterious and even mystical, because they are nothing but the result of generalizing the algebra of geometric segments in the abstract. And in this we do not observe any arcane dressing unattainable for the human understanding, rather the opposite, because, after having defined the algebra of magnitudes, taking advantage of that of the geometry of lengths, there is light and it is verified with inevitable suspicion how much dangerous arbitrariness must hide the composite forms of physical quantities.

Therefore, in view of the First Algebra of Magnitudes of this monograph, we must get used to interpreting physical equations with the meaning of relations between dyadic entities, which are the true elements that science uses, distancing ourselves in this aspect from the pure Mathematics of abstract algebraic structures; although, yes, these are those that operate with the primary element or measure of each magnitude, and this always in accordance with the composition laws defined for the dyads specifically.

To get into the philosophy of the algebra of magnitudes, let us first analyze the case of the work of a force, as a more elementary manifestation of what is understood by energy. By definition, **the work of a force is conceived as the scalar product of the vector concretes that represent the magnitude called force by the concrete that refers to the magnitude of the displacement of its point of**

application, according to the definition of the scalar product of the section XIX. In dimensional terms, such a scalar product can be stated in the abstract as the force magnitude multiplied by the length magnitude. Since every quantity of the force magnitude results from the multiplication of a quantity of the mass magnitude by a quantity of the acceleration magnitude, mechanical work is a derived magnitude, which in the International System will have per unit that given by the following dyadic expression, which composes unit magnitudes:

$$\frac{1\,kg*1\,m^2}{1\,s^2}$$

The result of the previous dyadic operation, formed by a product, two powers and a quotient, all dyadic in nature, of the three fundamental units: kilogram, meter and second, is called, by definition, «joule» and is symbolized by the letter J. The compound magnitude called mechanical work, defined as indicated, does not coincide with the common notion of this concept, associated with physical effort.

For example, to hold a heavy object hanging from a pulley it is necessary to counteract its weight by holding firmly one end of the rope; whoever resists the action of the weight in this way will feel that they are making a great effort; but from a mechanical point of view, if the body is not moving, no work will have been done.

Therefore, for Physics, work is derived from the movement of the point of application of all force and leads to the ambiguous notion of **energy**, defined as the **ability to produce work**, or perhaps better said, the **ability to transform itself into work**.

We observe in nature a multitude of phenomena that reveal an infinity of what we could understand as energy deposits, such as the wind, capable of moving a wind turbine and producing electricity; or bodies in motion, which when colliding with others move them and produce work; or the gasoline of a vehicle that, consumed by an engine, will propel it from one place to another, producing work; and many other similar cases easy to imagine. All these energy stores have in common that the stored energy does not manifest itself until, by means of some suitable artifice, it is released or transformed into work.

Thus, work can be considered as one of the many manifestations of energy that exist for Physics. Well, even if the essence of what we call energy is completely ignored with such joy, we understand that we can measure it through some of its manifestations, and specifically through work, so **we have to admit that energy is a measurable magnitude in the same units as work**.

Let's take some examples. Let's think about a weight clock, we observe that, when they are in the highest position, the clock's machinery will work by the action of the descent due to the weight to the lowest position of its career, overcoming the resistance of the various internal mechanisms; at the lowest position the watch will stop and at an intermediate point it will only run for a fraction of the time from the highest position; and this would mean that the clock weights contain different energy simply because of the position in height they occupy at each moment. From which it is inferred that bodies can be depositories of energy simply by reason of the situation in a gravitational field. This magnitude is usually indicated by the product of three dyads $(M \ kg)*(g \ m/\!\!/s^2)*(z \ m)$, where M is a measure of mass in kilograms, g is the measure of acceleration of gravity in meters per second squared dyadic and z the vertical elevation measure in meters; it is a derived magnitude with dimensions of the work magnitude; however, if z were constant, the mass would not move and there would be no work. Now, if it were allowed to fall freely from a given position, the mass would set in motion and do work, as in the case of the weight clock.

This means that the magnitude $(M \ kg)*(g \ m/\!\!/s^2)*(z \ m)$ must be recognized as having the capacity to generate work or to

174

transform into it, so it is admitted that it is a manifestation of energy called potential energy[14], name which suggestively alludes to the quality of possibility, in the sense that it does not exist in action, but will certainly manifest itself if sufficient conditions are present.

On the other hand, rational mechanics considers that a mass M kg moving at speed v $m/\!\!/s$ carries an amount of energy by reason of its movement that has been established in the well-known dyadic formula ($\frac{1}{2}m{\times}v^2$ $kg*m^2/\!\!/s^2$), considering it another manifestation of energy, called kinetic energy, since its dimensional expression is the same as that of mechanical work. This form of energy arises from the observation that a mass in motion is such that, if it is opposed by a force, for example, by arranging a weight hanging from a pulley and attached to the moving mass, the force will slow down its movement and make raise the weight, performing work that is equivalent to the loss of kinetic energy of the moving mass. Therefore, it seems that a moving mass behaves like a deposit of energy that can be transformed into work by means of suitable devices[15].

Mechanical theory deduces the well-known conservation of energy theorem, concluding that the sum of kinetic and potential energy remains constant, and this justifies the general principle of conservation of energy, with the well-known statement that **in nature energy is neither created nor destroyed, but is transformed**[16]. A somewhat different but equivalent wording of this principle is

---

[14] The mathematical model of the function of forces that supports the concept of potential energy is exposed in the author's syllabus, Mathematize 3, pp. 219 and following.

[15] Rational mechanics provides the mathematical justification for this physical fact, through the well-known theorem of living forces or kinetic energy, found in Mathematize 3, pp. 224, 242 and following.

[16] The mechanical justification of the principle of conservation of energy is exposed in Mathematize 3, pp. 225 and following.

the **impossibility of the perpetual mobile**, that is, there is no mechanism that produces energy without altering itself and without taking an equivalent amount from the outside. This means that no device will be capable of producing, without external input, not even the energy necessary to overcome the inevitable internal friction between its various elements. Repeated proofs of this principle are the innumerable failures of so many chimerical inventors of all times and places who have unsuccessfully pursued the continuous movement, the ideal seductive solution to the energy needs of humanity.

Notwithstanding the above, experience shows countless situations in which it seems that the conservation of energy is not fulfilled. If we drop a stone from a certain height, we observe that when it reaches the ground it stops and remains at rest, having lost its potential energy and without showing the least kinetic energy. Or a cyclist descending a hill at a constant speed, applying the brakes, will lose potential energy and, however, will not increase his speed and, therefore, his kinetic energy. In such cases it would appear that energy deviates from the law that marks its conservation. However, in these and in all the cases that can be imagined, it is observed that there has not been the slightest loss of energy, but that it has been transformed into another kind of it that we call heat: the stone and the ground will have been heated in In the first example, as in the case of the cyclist, the brakes and the rim will heat up.

At present we define **heat** as the **energy that passes from one body to another and that causes its dilation and changes of state**. However, the recognition of heat as another way of manifesting energy was not easy for Physics. Until 1780 the magnitudes of heat and temperature were considered similar, which was a serious obstacle to understanding thermal phenomena. Instead, today we distinguish them clearly, understanding by **temperature** the **magnitude that expresses the degree or level of heat of the bodies**. Until the end of the 18th century, the prevailing theory to explain the nature of heat assumed that it was an imponderable fluid called caloric. The current thermodynamics has abandoned the caloric and measures the amount of heat that passes from one body to another by establishing a criterion of equality, defining the addition and choosing a unit. Thus, it is said that two quantities of heat are equal when, absorbed by the same body under the same conditions of pressure and temperature, it turns out that the changes in it are identical. The criterion of addition is conceived with the hypothesis that the amount of heat necessary to produce a certain transformation in a given body is proportional to its mass. Finally, as a unit of heat, the **calorie**, or amount of heat necessary for a gram of water at the normal pressure of an atmosphere to raise its temperature from 14,5 °C to 15,5 °C, has been established.

The experiment of James Prescott Joule (1818-1889) connected thermodynamics with mechanics and showed that heat should be recognized as a main manifestation of energy, because experience strongly shows that for the transformation of mechanical work into heat, or vice versa , the conservation principle is always verified, so it has been possible to observe the mechanical equivalent of heat and establish that the amount of heat in the unit called calorie is equal to 4,186 joules. This connection is perhaps the direct link that elevates the derived magnitude called mechanical work to the category of energetic magnitude, because before finding it, work was rather a merely abstract definition of this mathematical manifestation of energy.

The concept of heat is not enough, however, to formulate a complete statement of the **principle of conservation of energy**, because we appreciate cases that require something more, such as, for example, it is not explained that a coal locomotive starts up, increasing its mechanical energy, without receiving heat from the outside. And this must occur because the coal that you house and burn in your home must be a depository of energy in some way, an energy class that is called **internal energy**. The same phenomenon is observed, for example, if a certain amount of water is heated, supplying it with a little energy, the increase in volume produced by the thermal increase will be very small, so the work done by the expansion against the The forces of atmospheric pressure will be negligible, and if a lower-temperature body is subsequently immersed in the water, it will heat up and the water will lose the energy stored in the initial heating, suggesting that it would have been somehow stored in its interior. And, although this internal energy cannot be known in absolute terms, its variations can be observed. And so, on this basis, in 1847 Helmholtz enunciated the principle of conservation in its most general form, admitting that **the amount of heat transferred to a body is used to increase its internal energy and to produce external work**, a statement known as the **first law of thermodynamics**.

This brief disquisition will serve to appreciate two things: first, that we are referring to energy and its manifestations without having precise knowledge or perhaps without having the least idea of the essence of that magnitude; and second, that the observation of physical magnitudes is not only a matter of analyzing their mathematical or dimensional form, but also requires experimentation and theorizing.

Let's see an illustrative example: let's consider a pair of forces, remember that they are two equal and opposite coplanar forces separated by a certain distance; let M be its moment, the dimensions of the moment are the same as those of a job, because it is the product of a force and a length; the question is, can it be concluded that the moment of the couple is a manifestation of the energy magnitude, given that its dimensional expression is that of a work? The elementary work dT of a pair of forces is given by the product of its moment M and the rotated differential angle $d\theta$, expressed in radians[17], according to the primary differential formula between measures $dT = M \times d\theta$. Recall that the radian is a way to measure angles that is defined as the length of the circumference arc equal to a radius, so any measurement of this magnitude is the quotient between two lengths, the length of the

---

[17] The detailed analysis of this case can be found in the volume of applications of the syllabus, Mathematize 3, p. 220.

arc between the length of the radius, Thus, according to section XI, an abstract real number will result for the measurement of the angles with radians, without any unit; and this motivates that the work of the moment has the same dimensional shape as the moment itself. So what is the difference between the moment magnitudes of a pair and the work of a pair?; we will have to seek the answer in the analysis of the physical fact, and thus, we must understand that diversity must be similar to the distinction between a force and its work, because it seems appreciable that a force cannot be transformed into any form of energy, since it is by the movement of its point of application that a job is developed. So, similarly, it is not the moment of the couple that produces a job, but its rotation; and thus, we will have to conclude that the moment of the pair cannot be characterized as a form or manifestation of energy; while the work of the pair should be. This reflection alerts us to the prudence with which derived magnitudes must be judged, since only with their dimensional equation it is not possible to establish what their nature is, something else is required, and that plus must be sought in the direct physical observation of phenomena, combined with precise reflection on what is observed.

On the other hand, we must warn in this section about certain insidious notations, which induce confusion and which are born from the traditional forgetfulness of dyadic algebra. In the texts and in the International System of Units we can find isolated symbologies such as s^{-1}, m^{-1} or similar, which would seem to suggest that they refer to the inverse units of the second or the meter or any other pattern. However, we discovered in section XIV that the unit and inverse elements do not exist for the multiplication of dyads or scalar measurements, so that the notation U^{-1}, U being any unit, cannot mean the inverse unit of U, but which is another way, analogous to numerical powers, of writing a divisor or denominator of a dyadic fractional notation. Let's see an example: the International System indicates with the isolated notation s^{-1} the compound unit with a divisor or denominator equal to one second s and a dividend or numerator

179

without dimension, such as the number of cycles of a wave or the number of radians or the number of revolutions; therefore, although the dividend or numerator is not explicit, it must be understood as present depending on the context of the equation in which it appears; so, for example, if you refer to a wave frequency, s^{-1} will indicate cycles per second; or if it is related to an angular velocity, it will mean radians per second or revolutions per second.

Recapitulating: in the First Algebra of Magnitudes we have tried to justify the traditional way of operating with physical measurements, in accordance with the unique rule of section XXII, deduced in this monograph based on coherent laws of composition between dyadic mathematical entities, considered the representatives ideal quantities of magnitudes. To arrive at this rule we have had to admit certain axioms and make hypotheses or assumptions such as that the quantity of every physical magnitude is represented by an abstract geometric segment and that thus the geometric algebra of segments is applicable to any magnitude. This shows that the aforementioned rule, subliminally infused into the intelligences, because it is not even mentioned by texts in any field, and thus not by teachers, who tacitly assume it without the least explanation or motivation, going directly to operate with the symbols of the units as if they were ordinary algebraic elements, infecting the intellects with this unconscious vice and degrading the teaching quality, it is by no means evident that it is a correct way of composing magnitudes, not even after having founded the laws of composition that they justify, that rather they contribute to put it in quarantine; because, after having revealed the illusions that it hides and, having born such a habit of a crass outrageous forgetfulness, as it is that the operations with measurements have remained until now epistemically undefined, depriving the operational symbols of the physical equations of the exact meaning that corresponds to them, it is inevitable to feel distrust about the dubious suitability of the algebraic simulations that have been accepted to justify the traditional operation and out of a mere sense of responsibility to produce the minimum intellectual disorder to the careless current of thought prevailing in this matter.

On the contrary, if these antecedents were questioned, the doors would be opened wide to new research on the appropriate ways of composing magnitudes, an area that to this day remains unexplored by the sclerosis of tradition. And this, because it could well happen, and perhaps it is the most probable thing, that the composite magnitudes do not respond at all well to the algebra of Euclidean geometry, which would be causing an insidious stagnation in the development of Physics, which could be saved by laws composition of magnitudes that more accurately reflect natural reality. And this will undoubtedly be the way to go towards new dyadic algebras, which follow the course of this first one and allow us to carve out the forgotten pillar of science, perhaps leading us to new horizons in the search for laws and definitions that better represent that the current phenomena of nature.

For our part, we believe we have contributed to this with the development of dyadic algebra and especially with the precise formulation of the **generating external composition laws** of sections XII et seq., which give meaning to all the laws and equations of physics, Therefore, once presented, they become essential to us and it seems incredible to us that they have not been established before and that we have accepted without altering this symbolic pseudo-algebra that means nothing.

And with this reflection, which alerts us to this primordial pending and forgotten question of traditional Physics, this First Algebra of Magnitudes is concluded, which will undoubtedly admit improvements, extensions and substantial changes in future editions; therefore, in the case of a virgin and pioneering subject, we apologize for the inevitable imperfections or omissions that we may have incurred, typical of any original work, which in its first edition aims above all to prevent the tenacious and toxic forgetting of a support primordial of science, as is without question the algebra of magnitudes. After all, the possible faults that our explanations suffer will always be milder and less harmful than the silent prejudices of the current carelessness. We hope that the possible shortcomings inherent in all innovation will be compensated by the honest attempt to unveil and carve this foundation of science, and with this we trust that we have humbly contributed to recycling the physical knowledge of faithful readers, adding a fundamental foundation to its intellectual baggage and allowing them to gain insight into all their prior knowledge, because for them the best intention has been put without sparing efforts.

Not surprisingly, a first innovative and fascinating contribution of this First Algebra of Magnitudes are the **«dysmetric» spaces**, which arise naturally from the simple observation of the dyadic elements, in which it is not prohibited at all to consider that the secondary, or unitary element of the pair, contains different quantities of the associated magnitude depending on its position in space and time, even without materially varying the body or phenomenon taken as a physical unit, and due to whatever causes. This observation, which we could call the **«dysmetric» axiom forecast**, is a new, unappealable mathematical tool, which must be capable of explaining and describing an infinity of natural phenomena, as outlined in the second part of this work.

Section XXVIII

DYADIC OR PHYSICAL ALGEBRA COMPENDIUM
NEED AND BASIC DEVELOPMENT

This section summarizes the matter developed in the First Algebra of Magnitudes with a course aimed at university professors and students. One more didactic effort to convince that it is necessary to correct the deplorable situation of Physics in its operational foundations. The research is inspired by the philosophical legacy of the fathers of modern Physics, whose testimony warns of the harmful **underlying presuppositions in the application of common algebraic operations to physical magnitudes**, summarized in the current false hypothesis of the International System of Units, which attributes to the quantities of magnitudes the abelian multiplicative group structure, with the practical consequence that it is assumed arbitrarily and with a good deal of absurdity that the operations with magnitudes correspond to those of rational numbers or ordinary fractions.

Crass and harmful error of the highest scientific institutions, which contaminates everything, for which they should hurry to remedy this insidious contamination of Physics, unbecoming of modern times. Here the nature of this congenital malformation is revealed, which insidiously intoxicates teaching, severely curtailing educational quality and scientific excellence with confusing principles, depriving everyone of their right to receive complete information, free from latent or tacit assumptions.

This work aims to shed light on how to overcome this unworthy deficiency. For this, an epistemic physical algebra based on Euclidean geometry is developed as a remedy, which can be extended to other more complex ones. The compendium collects the fundamental principles of this incredibly absent matter, depriving Physics of a fundamental pillar. Something similar to

183

that Mathematics would have been built without arithmetic. The compendium is organized into the following articles:

Article 1

INTRODUCTION, THEORETICAL FRAMEWORK
AND BACKGROUND

From the beginning in the elementary study of Physics, it is customary to use operations with entities that indicate concrete quantities of magnitudes and, by suggestion of arithmetic operations with abstract numbers such as real ones, it is naturally believed that concrete operations should follow the same slide rules and it is normal not to question tradition.

So what unscrupulously is learned and taught to do from a young age unconsciously, seeming so natural, in reality, it is not only not obvious, but it is totally incorrect, since it dispenses with something capital: epistemic definitions of the laws of composition between entities that represent measurements or quantities of magnitudes and their units. So it is not strange that the effects of this omission worried and continue to disturb the sages of Physics of all times; what is more, what is striking is that it must be the prominent ones who philosophize about it and that no one else discusses the tradition, because the root of the problem is very elementary. This is what R. M. Cooke and J. Hilgevoord, The Algebra of Physical Magnitudes, refer to, summarizing the debates of the classics as follows:

> Philosophers have long been interested in the question of the physical presuppositions underlying the application of algebraic operations to physical magnitudes, and this interest has quickened as a result of the existence of hidden variables underlying quantum mechanics. (p. 363)

These presuppositions allude to the gap raised by the distinguished Spanish physicist, Professor Julio Palacios, reflected in the prologue of his Dimensional Analysis (second edition, 1964, Espasa Calpe). This is how he describes the current traditional unknown:

> A widely held opinion, which goes back to Clerk Maxwell, and in which many physicists of my generation have participated, is that these symbols —refers to the right parentheses that enclose

185

the names of the different quantities— and, therefore, the formulas Dimensional refers to units, and is written like this, for example:

$$1\,erg = \frac{1\,g \times 1\,cm^2}{1\,s^2}$$

without realizing that we would be in a bind if an inquisitive student asked us how to multiply a square centimeter by a gram and divide the product by a second squared. (p. 12)

This incongruous omission pending to resolve in this matter of the operations with quantities of physical magnitudes is disturbing for scientific logic, and for this reason it has caused the proliferation of diverse and contradictory opinions regarding its nature and formulation, discussions that would simply be put to an end defining the necessary composition laws. A group of authors such as R. C. Tolman ascribe to the symbols of dimensional expressions a certain impenetrable or mystical character and consider that «The true essence of magnitudes, from the physical point of view, is represented by their dimensional formula» (Physics Review, p. 25, 1917). This hypothesis does not seem to be true, because it would suppose that such disparate magnitudes as the moment of a force and its work, which can both be expressed in «newton*meter», were essentially manifestations of the same magnitude, energy, which seems clearly an unacceptable delirium. Great authors such as Planck indicate that «It is as meaningless to speak of the "real" dimension of a magnitude as it is of the "real" name of an object», which would mean that physical magnitudes should be hidden from the understanding. Planck seems to indicate to us that we must not forget that physical magnitudes are mental entities and that, like any other name that indicates an extramental object, they are the result of the arbitrariness of thought. The positivist faction of the Vienna Circle, headed by Bridgman, states that «Dimensions do not have absolute value at all, but must be defined precisely from the process used to measure the respective magnitude» (Dimensional Analysis, Yale, University Press). Bridgman seems to suggest again that in the realm of magnitudes there must be a good deal of arbitrariness; which made Planck so uncomfortable that he criticized positivism in the famous conference entitled Religion und Naturwissenschaft:

> The views of the positivists cannot be fought from a purely logical point of view. And yet a careful examination of them reveals that they are inadequate and sterile, because they dispense with a circumstance that is of decisive importance for scientific progress. As much as positivism boasts of being free from prejudice, it has to start from a fundamental premise if it is not to degenerate into an unintelligible solipsism. This premise is that every physical measurement can be reproduced in such a way that the result is independent of the observer's personality, the place and time in which the measurement is made, and any other circumstance. All of this simply reveals that the decisive factor for the measurement result lies outside the observer and that, consequently, the measurements pose problems involving causal connections in an objective reality independent of the observer..

The philosophical pandemonium that results in this matter pushes conformism with the usual way of operating with quantities of magnitudes without even wondering if it is compatible with scientific logic and without becoming aware of the incongruity that the lack of epistemic definition of its laws of composition, so that for the majority this capital vice, which is real, does not exist, seriously affecting the teaching quality and the complete training of the students who, at least, have the right to a curriculum that explains the gap and to decide their position intellectual about it. And, for their part, physicists and scientists have the responsibility to base their work on solid and coherent bases. What kind of physicist is he who does not know what he is doing when operating with magnitudes and is satisfied with it?

Article 2

METHODOLOGY AND RESULTS

In the same way that there are algebras for numbers and abstract vectors, universally accepted, an algebra of dyadic entities, representatives of quantities of magnitudes, should be established, because only in this way would the prevailing confusion and ignorance and the **current educational dogma** be ended, being better clarified the meanings of the different composite magnitudes, as humbly outlined in this compendium.

There is a well-known and intuitive magnitude, which is **geometric length**, and this should be the starting point to base a generic algebra of physical magnitudes. To do this, it is necessary to first establish the laws of composition of the segments, insofar as they are the most elementary figures, and define a **geometric or graphical algebra** to compose them before moving on to their corresponding **analytical algebra**.

Geometric addition does not offer any difficulty, it is enough to conceive it as the graphic juxtaposition of segments, which analytically requires measuring them with the same unit of length. On the other hand, geometric multiplication can be conceived in such a way as to produce new magnitudes from length, area with two factors, volume with three, or hypervolumes with more than three; and the multiplied segments can be expressed in any length units, which do not have to coincide, as the addition requires. The length is therefore said to be a fundamental magnitude and the others are called derivatives or compounds through multiplication, and it should be noted that this operation, in principle graphical, differs substantially from the notion of the arithmetic product, given by the addition of a multiplying as many times as the multiplier indicates.

When segments are multiplied in this way, it is not possible to attribute the arithmetic multiplier function to any of them, which shows that this operation must be clearly differentiated from ordinary multiplication.

189

Once the algebra of segments is established, nothing prevents associating any quantity of any magnitude with the quantity of length of a segment. To do this, it would be enough to identify the empirical unit of the magnitude considered with an arbitrary or abstract unit of length. In this way, a one-to-one correspondence could be established between the set of all quantities of the given magnitude and that of all abstract lengths, that is, without real scale. The **affinity postulate** consists in admitting the previous operation and handling the quantities of magnitudes as if they were abstract geometric segments, which is to suppose that, although the quantities are different by nature, their quantities are affine to those of the length quantity. This postulate, in combination with the composition of areas and volumes in Euclidean space, allows defining the dyadic multiplication between any scalar magnitudes.

Starting as the foundation of the algebra of geometric segments, the algebra of lengths is developed and, by reasonable generalization, the precise definition of the laws of composition for any magnitude is reached with relative ease. This reveals the hidden frameworks of the derived units and the meanings that can be attributed to them can be judged more accurately. So the notion of dimension of all magnitude has to be considered after and not before having conceived an algebra of magnitudes, whose analytical mathematical expression is the concrete entities or dyadic elements.

Hence, the method followed in this exposition should be presented according to the following sequence: first, to establish the basic concepts of physical magnitudes in general; then, assign them a mathematical entity and create the concrete or dyadic entities; then define an algebra for such special entities, precise representatives of the quantities of measurable natural magnitudes; then investigate the meaning of definition equations, universal laws and other physical entities; and, finally, explain the principles of dimensional analysis. Well, this is how it is done in the preceding nuclear part of this text.

However, in a didactic compendium such as the one presented here, it is necessary to limit the length and select the substantial, so that the possibilities of the resulting complex algebra are glimpsed together without diminishing the clarity of the exposition. Hence, here only the fundamental aspects can be exposed in the most concise way possible: the definition of the concrete or dyadic entities of Physics, the elementary criterion of equality, the dyadic sets and the basic operations of addition, subtraction, and some multiplication and division of magnitudes.

The results of what **should be taught to science students**, in the degree that corresponds to their level of studies, are included in the following articles, replacing the **traditional toxic dogma** by an **algebraic logic for Physics, reasoned and grounded with coherence as science demands.**

Article 3

DEFINITION OF CONCRETE OR
PHYSICAL DYAD AND EQUALITY

It is convenient to call **measurement** the quantity, extension or portion of a magnitud expressed in the form $q\ U$, as a symbol of the times q, a real number, that a unit quantity U is present in a phenomenon, calling q measured with the unit U of the magnitude included in the observed event. And similarly if the measure were a vector $\overline{q}$ of R^3.

The magnitudes whose measurements are such that $q \in \mathrm{R}$ or that $|\overline{q}| \in \mathrm{R}$ and that can take any value are called **continuous**, on the other hand, those in which the measurements can only be whole numbers, with $q \in \mathrm{Z}$ or $|\overline{q}| \in \mathrm{Z}$, they are called **discrete**. It is observed that the operations with discrete magnitudes are included in the continuous ones, since their measurements will be represented by whole numbers, a subset of the real numbers, so that the continuous ones present greater generality than the discrete ones; and the continuous ones will be explained in the abstract in many cases by means of the length, which fictitiously represents them all, because any of them can be assimilated to the real line, resulting in any case the same reasoning scheme.

In this way, every pair formed by a real number or a vector, followed by a unit that reflects a certain quantity of some magnitude is, by definition, a **concrete entity or physical dyad**.

If the primary q is a real number, the concrete will be called **scalar** and it will mean the quantity of a magnitude equal to q times the quantity of the same contained in the unit, which can only be described in the abstract by means of some symbol empirically associated with some phenomenon; so that the notation $(q\ U)$ or (q, U) or $q\ U$, without superfluous parentheses, to represent the measurement of a magnitude by the unit U with the real number q has parity in nature, hence the name of physical dyad.

Similarly, if the primary is characterized by a vector $\overline{q}$, the dyad will be said to be **vector**.

The scalar parity entities associated with each unit U form sets that can be symbolized $\{R, U\}$, and are capable of being composed together by internal composition laws, establishing applications of the Cartesian product $\{R, U\} \times \{R, U\}$ in $\{R, U\}$; and they can also be composed with the elements of other sets, such as R, through laws of external composition, with applications of $R \times \{R, U\}$ in $\{R, U\}$, so the task of establishing for them an adequate algebra, which must be tried to be as isomorphic as possible with the structure of the field of real numbers, because these are the universal model. We will frame these types of operations in the so-called **additive scalars**, indicating them with the sign «$\oplus$».

On the other hand, when there are diverse dyadic sets associated to different units U_1 and U_2, such as $\{R, U_1\}$ and $\{R, U_2\}$, external composition laws can be defined, which we will call **multiplicative scalars** and symbolize with the asterisk «$*$», through applications of $\{R, U_1\} \times \{R, U_2\}$ in $\{R, U_1 * U_2\}$. **They are laws that generate new magnitudes.**

In turn, vector dyads form sets with each U unit that can be symbolized $\{R^3, U\}$ and are capable of being composed of each other by internal composition laws, with applications of the Cartesian product $\{R^3, U\} \times \{R^3, U\}$ in $\{R^3, U\}$; and they can also be composed with the elements of other sets, such as R, by means of external composition laws, with applications of $R \times \{R^3, U\}$ in $\{R^3, U\}$, ensuring that it is isomorphic with the structure of the vector space R^3 on R. We will frame these types of operations in the **additive vector** calls, marking them with the sign «$\oplus$».

In the same way as with scalar dyadic sets, for the vectors associated with different units U_1 and U_2, such as $\{R^3, U_1\}$ and $\{R^3, U_2\}$, we can define external composition laws, called **multiplicative vectors**, which we will symbolize with the asterisk «$*$», by applying $\{R^3, U_1\} \times \{R^3, U_2\}$ in $\{R^3, U_1 * U_2\}$. Like scalars, these external laws are also **generators of new magnitudes**.

194

Since dyadic entities are made up of pairs of elements linked together and inseparable, a **mathematical primary** of ordinary algebra and a unitary **physical secondary**, their specific algebra must obey operational criteria with **dyads**, so proper composition laws must be established that allow the construction of a sui generis structure similar to the forms of classical dyadic algebra, a precursor of tensor algebra. On the other hand, the dyadic nature of the concrete entity would justify its being indicated with terms such as concrete dyad, physical dyad, dimensional dyad or other similar nomenclature.

Likewise, a **criterion of dyadic equality** is required, which will have to identify two concrete elements when the quantity of the magnitude to which they refer is the same, which in analytical terms will mean that, if they correspond to the same unit of measurement, the primaries must match. Thus, given two scalar dyads (q_1, U_1) and (q_2, U_2), it will be said that they are equal if and only if they represent the same quantity of the magnitude to which the units U_1 and U_2 belong, indicating said equality with the form $(q_1, U_1) = (q_2, U_2)$. Similarly, if the dyads were vector, the equality $(\overline{q}_1, U_1) = (\overline{q}_2, U_2)$ would be written. Note that equality can be expressed with any of the forms admitted for the pair notation: $(q_1, U_1) = (q_2, U_2)$ or $(q_1\ U_1) = (q_2\ U_2)$ or $q_1\ U_1 = q_2\ U_2$, and so similar for vector dyads: $(\overline{q}_1, U_1) = (\overline{q}_2, U_2)$ or $(\overline{q}_1\ U_1) = (\overline{q}_2\ U_2)$ or $\overline{q}_1\ U_1 = \overline{q}_2\ U_2$.

At the end of article 7 of this same section, this criterion of equality is generalized, once the multiplication of a scalar by a dyad and the dyadic division are defined, which are necessary to give it complete and precise meaning.

Article 4

THE DYADIC ADDITION

There is a physical magnitude that can be inspiring of how it should be operated with all the others, it is the longitude. And it is precisely Euclidean geometry that has resolved the way of composing lengths, through the **geometric algebra of segments**, which establishes the addition and subtraction of lengths by means of the adequate graphic juxtaposition of the segments to be composed, analytically by means of the prior requirement that the measurements of the components are referred to the same unit or **axiom of uniformity**, simply so that they can be counted as equal elements. Thus, given two quantities of length expressed, for example, in centimeters, $(q_1\ cm)$ and $(q_2\ cm)$, the juxtaposition of these lengths will have a length equal to $[(q_1+q_2)\ cm]$. Identifying this operation with the addition of segments, the dyadic sum of lengths could be symbolized with the proper sign «$\oplus$», and the analytical definition of segment addition would be:

$$(q_1\ cm) \oplus (q_2\ cm) = [(q_1+q_2)\ cm]$$

This definition can be generalized to any magnitude, **idealizing every physical quantity with the fiction that it is a quantity of length**, what we have called the **affinity postulate**. And so it is easy to arrive at the generic dyadic addition as a **law of internal composition** or application of the Cartesian product $\{R,U\} \times \{R,U\}$ in $\{R,U\}$, for scalar magnitudes, or application of $\{R^3,U\} \times \{R^3,U\}$ in $\{R^3,U\}$, for vectors. Thus the two respective analytical definitions result, which, expressed with superfluous parentheses, for greater significance and distinction of their elements, are:

$$(q_1\ U) \oplus (q_2\ U) = [(q_1+q_2)\ U]$$
$$(\overline{q}_1\ U) \oplus (\overline{q}_2\ U) = [(\overline{q}_1+\overline{q}_2)\ U]$$

Note that, although in both the same «$\oplus$» sign appears for the dyadic addition, in the first equation it represents the scalar dyadic sum and in the second the vector one, in the same way

197

that the « + » sign of the first is the addition of R and that of the second is the vector sum of R^3.

All of which is nothing more than a reflection of the existence of an isomorphism between the structures of the dyadic sets $\{R, U\}$ and $\{R^3, U\}$.

Article 5

DEDUCTION OF DYADIC SUBTRACTION

The **dyadic subtraction** can be deduced from the addition and based on the generic subtraction criterion, which establishes the difference between a minuend and a subtrahend as that remainder or difference such that added to the subtrahend gives the minuend. To do this, let the scalar dyadic sum $(d\ U) \oplus (s\ U) = (m\ U)$. The parentheses are dispensable, but they are kept to mark the dyads well, and the symbology of the dyadic addition has simply been adapted to indicate with the letters m a minuend, s a subtrahend and d for the difference that corresponds to them. The usual subtraction criterion, as an operation that, given an addition, allows óne of the addends to be obtained as a function of the sum and the other by adding, is just another way of writing the initial sum, which allows establishing the uniform dyadic difference , distinguished with the sign «⊖», by the equation:

$$(m\ U) \ominus (s\ U) = (d\ U)$$

The definition of dyadic addition applied to the initial sum allows it to be written in the form $[(d+s)\ U] = (m\ U)$. The simple criterion of equality of dyads consists in considering them equal when their primaries and secondaries coincide. The equality of primaries leads to the relation $(d+s) = m$. For its part, the definition of subtraction in R leads to $d = m - s$. So, substituting d in $(m\ U) \ominus (s\ U) = (d\ U)$, we finally have:

$$(m\ U) \ominus (s\ U) = [(m-s)\ U]$$

And this is the analytic definition of dyadic subtraction, and it means that the dyadic difference between two uniform scalar dyads, called minuend and subtrahend, is a dyad called difference whose primary is the subtraction in R of the primaries and with the same secondary as them. It is a law of internal composition of $\{R, U\} \times \{R, U\}$ in $\{R, U\}$.

199

The subtraction of uniform vector dyads is isomorphic with the scalar, given the abelian and additive group structure of R^3, which presents the same formal properties for the sum of vectors that occur with the real numbers.

DYADIC MULTIPLICATION BY A SCALAR

The dyadic addition allows to conceive the case in which all the addends of a sum are equal, that is to say that, referring to the same magnitude or being **homogeneous**, in addition, they are referred to the same unit, that is, they are **uniform**. So, being p a real number and, given a scalar physical dyad $(q\ U)$, with q belonging to R, the dyadic addition $(q\ U) \oplus (q\ U) \oplus \ldots \oplus (q\ U)$ can be formed with p adding us. Well, the result of this sum, which means p times the quantity represented by the physical pair $(q\ U)$, can be briefly symbolized as a multiplication indicated by $p \circ (q\ U)$, where the symbol «$\circ$» indicates the operation of adding equal dyads a certain number of times. The similarity of this operation with the arithmetic multiplication «$\times$» is evident, because the factor p acts as a multiplier; however, their difference is notable, because this refers to operations with real numbers, while the present one operates with real numbers and physical dyads, then, the sets that relate these forms of multiplication do not coincide, hence, strictly speaking, they must be assigned different operational symbols, so as not to favor the algebraic confusion of disparate laws of composition.

It is clear that the same sum of identical dyadic elements can be represented with the form $(q\ U) \circ p$, without more than attributing p the multiplier function on both sides, which is equivalent to axiomatizing the **commutative property** of this kind of multiplication, and with it its analytical definition can be written like this:

$$p \circ (q\ U) = (q\ U) \circ p = (q\ U) \oplus (q\ U) \oplus \ldots \oplus (q\ U) \text{ with } p \text{ addends}$$

The definition of dyadic addition allows you to write the second member of the definition equation above like this:

$$(q\ U) \oplus (q\ U) \oplus \ldots \oplus (q\ U) = [(q + q + \ldots + q)\ U] \text{ with } p \text{ addends}$$

The definition of multiplication «$\times$» in R allows us to write the abbreviated sum $q + q + \ldots + q$ with p addends, with the form $p \times q$ or $q \times p$, because the multiplication of R is commutative.

Therefore, in conclusion, the **multiplication of scalar dyads with real elements of R** is established by the equation:

$$p \circ (q\ U) = (q\ U) \circ p = [(p \times q)\ U] = [(q \times p)\ U]$$

If the dyad were of a vector nature $(\overline{q}\ U)$, being $\overline{q}$ an element of R^3, the development is analogous. To do this, using the same sign «∘» for this composition law, even knowing that it is a different operation, the sum to describe in this case would be $(\overline{q}\ U) \oplus (\overline{q}\ U) \oplus \ldots \oplus (\overline{q}\ U)$ with p addends; but here the symbol «⊕» indicates the vector dyadic addition, which allows us to write the following:

$$(\overline{q}\ U) \oplus (\overline{q}\ U) \oplus \ldots \oplus (\overline{q}\ U) = [(\overline{q} + \overline{q} + \ldots + \overline{q})\ U] \text{ with } p \text{ addends}$$

Note that the «+» that appears in this equation is not the addition of R, but the addition of R^3, even though the same sign is used. And, as R^3 has a vector space structure over R, designating the product of a scalar by a vector with the sign «•», we arrive at the formulation described in the final definition equation:

$$p \circ (\overline{q}\ U) = (\overline{q}\ U) \circ p = [(p \bullet \overline{q})\ U] = [(\overline{q} \bullet p)\ U]$$

The **multiplication of real numbers by scalar dyads** is nothing but a law of external composition or application of the Cartesian product $R \times \{R, U\}$ in $\{R, U\}$ on the left and the symmetric on the right of $\{R, U\} \times R$ in $\{R, U\}$.

In turn, the **multiplication of real numbers by vector dyads** defines an external composition law or application of the Cartesian product $R \times \{R^3, U\}$ in $\{R^3, U\}$ on the left and the symmetric on the right of $\{R^3, U\} \times R$ in $\{R^3, U\}$.

It is observed that the forms deduced for this kind of multiplication justify that it can be operated symbolically as if all the symbols were elements of R, without actually being it, commuting and grouping them with the usual rules, which is due to the isomorphism that is established between the different algebraic structures that participate in reasoning. However, it would be wrong to understand such a way of operating as

202

something immediate, because it is not, it requires justification, otherwise it would be to contaminate logical development with inadmissible presuppositions.

Note that in the preceding arguments it has been tacitly assumed that the multiplying number p belonging to R is also an integer, since we always speak of p addends. However, the logical deductions can be repeated considering any rational number p/q, with p and q integers, simply by forming the sums $q/h+q/h+ \ldots +q/h$ with p as addends, that is, also with p as multiplier. And thus the reason for the definition is completed with any number of R, leaving aside the incommensurables, for which limit concepts would have to be introduced, which would also include these, covering the entire spectrum of real numbers.

As we already did at the end of section IX and we repeat here due to its importance and inclusion in this alternative formulation, we must ask ourselves what happens to a dyad $(q\ U)$ when its unit is multiplied by a number p. If the dyad $(q\ U)$ indicates the quantity of magnitude equal to q times the quantity indicated by U, the new dyad $(q\,p\circ U)$ must represent the quantity of p times the quantity $(q\ U)$, which allows us to complete the definition of the multiplicative operation in this section with the following equivalent analytic forms:

$$(q\,p\circ U)=p\circ(q\ U)=(p\times q\ U)=(p\times q)\circ U$$
$$(p\ U)=p\circ(1\ U)=p\circ U=(1\,p\circ U)$$

Another very obvious property that can be useful in some deductions is obtaining the multiplicative form of every dyadic expression. In this regard, given any dyad $(q\ U)$, the following reasoning can be put together:

$$(q\ U)=q\circ(1\ U)=q\circ U$$

Therefore, the quantity of magnitude that any dyad symbolizes is the product of its primary by its secondary.

Article 7

DEDUCTION OF THE UNIFORM DYADIC DIVISION

The dyadic multiplication by a scalar, which relates concrete entities referred to the same unit, taking advantage of the generic criterion of dividing, allows us to deduce the analytical form of the kind of division in which dividend and divisor are uniform, scalar or vector dyads. In the first case, given two dyads $(q_1 \ U)$ and $(q_2 \ U)$, by the properties of R, there will always be a real number p such that $q_1 = p \times q_2$. This allows us to affirm the dyadic equality $[(p \times q_2) \ U] = (q_1 \ U)$. So, considering the definition of multiplication of a scalar by a concrete, the first member can be written in the form $[(p \times q_2) \ U] = p \circ (q_2 \ U)$. Combining both equations, we have $p \circ (q_2 \ U) = (q_1 \ U)$. And this expression, with respect to the multiplicative form «$\circ$», can be interpreted as a division between the dividend $(q_1 \ U)$ and the divisor $(q_2 \ U)$, resulting in the quotient $p = q_1 / q_2$.

Distinguishing this operation with a **double bar**, inclined or horizontal, we arrive at the result that the **dyadic division of uniform scalar elements**, that is, referred to the same unit, must give as a result a real number, which will be precisely the quotient in R of the primaries between dividend and divisor. Analytically we will have:

$$\frac{\left(q_1 \, U\right)}{\left(q_2 \, U\right)} = \frac{q_1}{q_2} = p$$

The observation of this equation proves that the form of division established here operates like that of R, allowing us to simplify the equal terms U that appear both in the numerator and in the denominator. However, such a substantial property is not due to the dogmatic application of the algebra of real numbers, but to the dyadic definitions that are being formulated. What happens is that the resulting isomorphism allows us to operate with the various symbols as if they were all elements of R, without being it, as the units of magnitudes are not.

For vector dyads, it should be noted that the algebra of R^3 is such that multiplication by a scalar relates **collinear vectors**, so that uniform dyadic division will only be possible when the dividend and divisor primaries are in turn collinear. So, let us now be the vector dyads $(\overline{q}_1\, U)$ and $(\overline{q}_2\, U)$, such that vectors $\overline{q}_1$ y $\overline{q}_2$ are collinear. Given the algebra of the vector space R^3, the existence of a scalar p of R such that $\overline{q}_1 = p \bullet \overline{q}_2$ is certain. Operating with vector algebra and with the definition of a dyadic product by a scalar, the following reasoning can be written with full justification: in R^3, the initial equality $\overline{q}_1 = p \bullet \overline{q}_2$ allows establishing the identity of the dyads $(\overline{q}_1\, U) = [(p \bullet \overline{q}_2)\, U]$; the product «∘» converts the second member into $[(p \bullet \overline{q}_2)\, U] = p \circ (\overline{q}_2 U)$, from which it follows that $(\overline{q}_1\, U) = p \circ (\overline{q}_2 U)$, and in this equation, in relation to the multiplicative operation «∘», we can apply the generic criterion of division and consider the factor $(\overline{q}_1\, U)$ as a dividend, the $(\overline{q}_2\, U)$ as a divisor and p as a quotient, with which it is concluded that the quotient between two vector dyads with collinear and uniform primaries, that is, referred to the same unit U, is a real number p such that $\overline{q}_1 = p \bullet \overline{q}_2$. Expressing this analytically, we have the definition equation of this uniform **vector dyadic quotient**:

$$\frac{\left(\overline{q}_1\, U\right)}{\left(\overline{q}_2\, U\right)} = p \quad of \quad R \quad \text{and such that} \quad \overline{q}_1 = p \bullet \overline{q}_2$$

Here we also observe the permissiveness of simplification of the identical symbols U that appear both in the numerator and in the denominator, but with the difference that the quotient of the second member is not that of R, but that of R^3, and **only for collinear vectors**. With this, the presuppositions that are admitted without any rigor are made manifest when the symbolic tradition is applied without further ado, based on the principle of economy of operational signs, which handles all the elements that relate quantities of physical magnitudes with the unfounded tacit assumption that they are behave as elements of R.

On the contrary, with the algebra of magnitudes, the different transformations allowed for the dyadic equations are justified one by one, establishing them with the **quality due to scientific, logical and didactic rigor.**

In the event that the dividend and divisor units are not uniform but homogeneous, there will be a dividend $(q_1\ U_1)$ and a divisor $(q_1\ U_2)$. The axiom of continuity guarantees that there exists a real number k of R such that $U_1 = k \circ U_2$. Now it can be formed and operated dyadic with the following quotient:

$$\frac{(q_1\,U_1)}{(q_2\,U_2)} = \frac{\left[q_1\left(k \circ U_2\right)\right]}{(q_2\,U_2)} = \frac{\left[(q_1 \times k)U_2\right]}{(q_2\,U_2)}$$

Once the numerator and denominator have been reduced to uniform units, that is, equal, we can apply the property of the uniform quotient and eliminate the unit U_2 from the numerator and denominator, resulting in:

$$\frac{\left[(q_1 \times k)U_2\right]}{(q_2\,U_2)} = \frac{q_1 \times k}{q_2} = p \times k$$

We have, then, that the dyadic quotient of two homogeneous elements $(q_1\ U_1)$ and $(q_2\ U_2)$, is a real number, without dimension, which is given by the ordinary quotient of the second member of the previous expression.

For homogeneous and non-uniform vector dyads $(\overline{q}_1\ U_1)$ and $(\overline{q}_2\ U_2)$, with $U_1 = k \circ U_2$, following the same reasoning as above, with vectors $\overline{q}_1$ and $\overline{q}_2$ being collinear, we will verify $\overline{q}_1 = p \bullet \overline{q}_2$, with $p \in$ R , and the quotient of collinear vectors $\overline{q}_1 / \overline{q}_2 = p$ can be formed. Under these conditions, the following reasoning can be made:

$$\frac{\left(\overline{q}_1\,U_1\right)}{\left(\overline{q}_2\,U_2\right)} = \frac{\left[\overline{q}_1\left(k \circ U_2\right)\right]}{\left(\overline{q}_2\,U_2\right)} = \frac{\left(\overline{q}_1\,U_2\right)}{\left(\overline{q}_2\,U_2\right)} \times k = \frac{\overline{q}_1}{\overline{q}_2} \times k = p \times k$$

In conclusion, the dyadic quotient of two homogeneous, non-uniform and collinear vector measurements is given by the factor $k \times p$, where k is the real number that represents the dyadic ratio between the homogeneous units U_1 and U_2 of the secondaries, and p the real number which indicates the ratio between the collinear vectors $\overline{q}_1$ and $\overline{q}_2$ of the dyadic primaries.

Thus we are now in a position to complete the **dyadic equality criterion** in section IV. It is enough to take into account that the axiom of continuity guarantees the existence of the real number k such that $U_1 = k \circ U_2$. So, if two dyads $(q_1\ U_1)$ and $(q_2\ U_2)$ are equal, which is denoted $(q_1\ U_1) = (q_2\ U_2)$, their dyadic ratio must obviously be the unit of real numbers. Therefore, we can conclude the following:

$$\frac{\left(q_1\ U_1 \right)}{\left(q_2\ U_2 \right)} = p \times k = 1 \ \Rightarrow \ p = \frac{1}{k}$$

We have verified that exactly the same relationship between p and k holds for vector dyads with collinear measures. Consequently, for both scalar and vector magnitudes, it can be stated that, **if two dyads are equal, the algebraic ratio p of their primaries is the inverse of the dyadic ratio k of their secondaries.**

These properties of the definition of equality of quantities of magnitudes, which we remember only make sense for homogeneous dyads, that is, representative of the same magnitude, although the units to which they refer are different, are essential for the construction and interpretation of the laws and equations of Physics.

At this point, we will move on to the precise definition of the **binary relationships** between quantities of a given magnitude and the establishment of the **mathematical definition of magnitude quantity**.

The reader is assumed to have basic knowledge of this fundamental part of modern algebra. To do this, let us take any quantity m, which we remember is not a number, but at most we can assimilate it to a segment of given abstract length, by virtue of the affinity postulate, and form the set $M = \{m\}$, which represents the complete repertoire of all possible quantities of the magnitude considered. Let us first establish the equivalence relation of homogeneous dyads $(q_1 \ U_1)$ and $(q_2 \ U_2)$ and then the relations of total order «less than or equal to» and «less than».

We have said in an informal way that two dyads are equal if they represent the same quantity of magnitude and we have come to formulate the relationship that the primaries and the secondary ones must maintain based on the division of real numbers and the homogeneous dyadic division, so that equality requires that the following proportion be met:

$$\left(q_1 \, U_1\right) = \left(q_2 \, U_2\right) \ \Rightarrow \ \frac{q_1}{q_2} = \frac{U_2}{U_1}$$

As we have indicated before, this means that dyadic equality requires that the ratio of the real numbers q_1/q_2 be a real number equal to the ratio inverse dyadic of the units $U_2/\!/U_1$. Well, this property will allow us to define equality as a relationship of equivalence between dyads.

Let us take the set $D = \{(q \ U)\}$ of all dyads $(q \ U)$ of any magnitude, with $q \in R$ y $U \in M$. D can also be conceived as the Cartesian products $R \times \{U\}$ or $R \times M$. And in an analogous way for vector dyads. In any case, in D we will ideally find all the possible dyads that can be formed to represent quantities of the given magnitude.

We will say that two dyads $(q_1 \ U_1)$ and $(q_2 \ U_2)$ are equivalent, if their elements verify the equation $q_1/q_2 = U_2/\!/U_1$ and we will write $(q_1 \ U_1) \sim (q_2 \ U_2)$ or, if preferred, $(q_1 \ U_1) = (q_2 \ U_2)$. Note that this equality sign does not mean numerical equality, as occurs with common algebraic expressions, but rather equality of quantities of magnitudes, which we will not tire of repeating are not numbers, but at most affine segments, by virtue of the

209

affinity postulate. The relation thus defined, symbolized «~», is a subset of the Cartesian product D×D and establishes in the set D an equivalence relation and, therefore, a classification of the dyads $\{(q\ U)\}$ of D in all their classes and the corresponding partition of this set. It is very trivial to demonstrate that, in effect, the relation «~» verifies the reflexive, symmetric and transitive properties, so it is an equivalence relation. Like any equivalence relation, the partition into classes of the set D has the meaning that each quantity of magnitude m of M can be indicated by any element of the corresponding class, which can be indicated $m = \mathcal{C}(q\ U)$, that is, the quantity of magnitude m is that which corresponds to any dyad of the class $\mathcal{C}(q\ U)$, for example, its representative $(q\ U)$. In other words, any dyad of the class $\mathcal{C}(q\ U)$ indicates the same quantity of magnitude m. In analytical terms the definition can be written:

$$m = \mathcal{C}(q\ U) = \{\text{all dyads } (x\ X) \text{ such that } q/x = X/\!/U\}$$

Forming the set whose elements are all the equivalence classes $\{\mathcal{C}(q\ U)\}$ in D with «~», we will have arrived at the definition of M as the set of classes of equivalent dyads $M = \{\mathcal{C}(q\ U)\}$. Thus it turns out that M is the partition that corresponds to the quotient set of D as a function of the equivalence relation «~», which in algebra is written:

$$\frac{\mathrm{D}}{\sim} = \mathrm{M}$$

In summary, a quantity of magnitude m is defined by a certain equivalence class $\mathcal{C}(q\ U)$ of the set D of all dyads. In turn, the set of all classes $\{\mathcal{C}(q\ U)\}$, which is a partition of D, established by the equivalence relation «~», is by definition the set $M = \{m\} = \{\mathcal{C}(q\ U)\}$ of all quantities m and dyadic equivalence classes of the given magnitude.

We must note that the definition of dyadic equality is synonymous with the equivalence relation «~» defined in D. The equality of dyads does not require that their primaries and secondaries coincide, but rather assumes membership in the same

210

equivalence class. Obviously, as a particular case, the reflexive property guarantees that two dyads with the same primaries and secondaries are equal, because they belong to the same class.

Let's see how we can analytically characterize the equivalence classes of the set D. To do this, let's observe that the axiom of continuity guarantees that any unit U' can be expressed in terms of another unique given U through the product $y \circ U = U'$, with y being a number real. Therefore, any dyad $(x\ U')$ can be expressed with the form $(x\ y \circ U)$ and the set D of all dyads will be formed with a single unit U through numerical pairs (x,y) with $D = \{(x\ y \circ U)\}$. It is easy to check that the dyads $(x\ y \circ U)$ such that $x \times y = h$, with h being any real number, are equivalent. In effect, let's take the dyads $(x_1\ y_1 \circ U)$ and $(x_2\ y_2 \circ U)$ and carry out the following reasoning:

$$\frac{x_1}{x_2} = \frac{\dfrac{h}{y_1}}{\dfrac{h}{y_2}} = \frac{y_2}{y_1} = \frac{y_2 \circ U}{y_1 \circ U}$$

Therefore, the ratio of the primaries is the inverse of the ratio of the secondary ones, so the dyads $(x_1\ y_1 \circ U)$ and $(x_2\ y_2 \circ U)$ satisfy the equivalence condition and belong to the same class. As it turns out that $x \times y = h$ must be verified by initial hypothesis, we have that x and y are related by the function $y = h/x$. If we translate this function to a Cartesian system, for each value of $h \in R$, the function $y = h/x$ will be represented graphically by a hyperbola or inverse proportionality function. Then, the graphical form of the partition into equivalence classes established in $D = \{x\ U'\} = \{(x\ y \circ U)\}$ is a set of infinite hyperbolas associated each of them with its corresponding value $h \in R$. The following figure represents this infinite family of hyperbolas for $h > 0$ and $h < 0$. Each curve indicates a set of dyads equivalent to each other, constituting the partition of D whose classes are the elements of the quotient set defined above:

Graphical expression of the partition of the set of all dyads D into their equivalence classes

$$D=\{x\,U'\}=\{(x\,y\circ U)\}$$

$$\frac{D}{\sim}=M$$

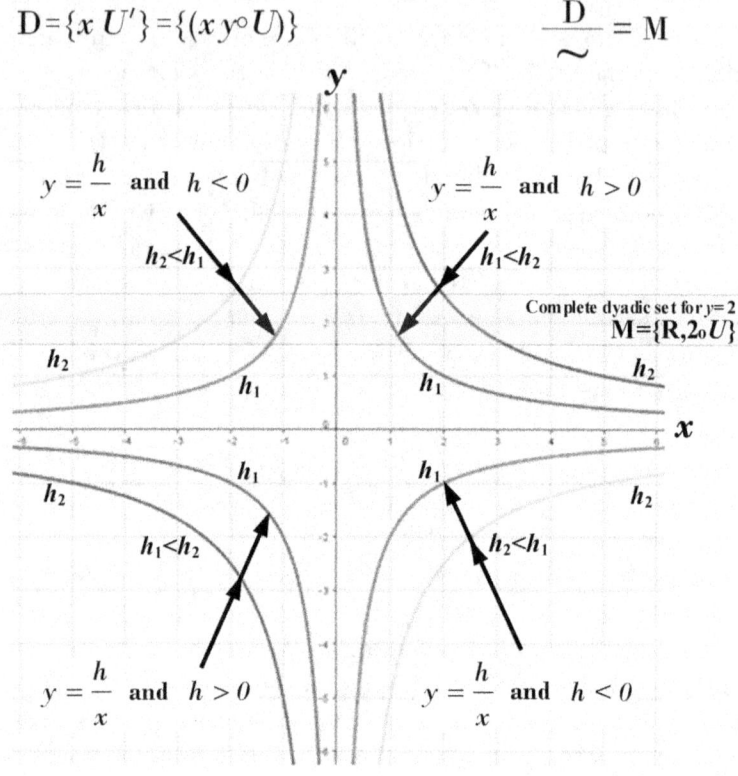

The previous figure requires the following precisions: first, we generally consider quantities of positive and negative algebraic magnitudes; second, each hyperbola links the equivalent dyads to each other and represents a certain quantity of magnitude or element of the set M depending on the numerical pairs (x,y) of each hyperbola; Third, each line parallel to the abscissa represents a complete dyadic set $\{R,y\circ U\}$, which includes without repetition all the possible quantities M of the given magnitude, so both sets are identified, resulting in $M=\{R,y\circ U\}$ for any standard unit U and $\{R^3,y\circ U\}$ for vector quantities.

Let us now define the relation of order «less than or equal to», indicated «≤», for the elements of M. The criterion that must satisfy this relation, for any two dyads $(q_1 \ U_1)$ and $(q_2 \ U_2)$ will be, by definition:

$$(q_1 \ U_1) \leq (q_2 \ U_2) \quad \Leftrightarrow \quad \frac{q_1}{q_2} \leq \frac{U_2}{U_1}$$

If this condition is met for the dyads $(q_1 \ U_1)$ and $(q_2 \ U_2)$, it is trivial that any other of the same classes must verify it, which can be indicated analytically with the form:

$$\mathcal{C}(q_1 \ U_1) \leq \mathcal{C}(q_2 \ U_2)$$

Therefore the magnitude quantities given by $m_1 = \mathcal{C}(q_1 \ U_1)$ and $m_2 = \mathcal{C}(q_2 \ U_2)$ will verify $m_1 \leq m_2$ if $q_1/q_2 \leq U_2 /\!/ U_1$ and since q_1/q_2 and $U_2 /\!/ U_1$ are real numbers, **the set M of quantities of every magnitude is ordered as R**. The relation «≤» for quantities of magnitudes is, therefore, a relation of total order.

In effect, the reflective and transitive properties in M are immediate. The antisymmetry is easily verified, then, given two quantities m_1 and m_2 of M, if $m_1 \leq m_2$ and $m_2 \leq m_1$, then, $m_1 = m_2$. These three properties characterize the relation «less than or equal to» as an order relation. Furthermore, since given any two quantities m_1 and m_2 of M, it is $m_1 \leq m_2$ or $m_2 \leq m_1$, but not both, all the elements of M are comparable to each other and thus the relationship «less than or equal to» is of total order. Therefore, for any magnitude, the set M of all possible quantities m is completely ordered by the relation «≤», defined by the condition $m_1 \leq m_2$ if and only if $q_1/q_2 \leq U_2 /\!/ U_1$.

It is worth noting that dyadic sets admit multiple representations. Here we have started from the set of all possible dyads of a given magnitude, and we have called it D. This set includes equal quantities for all dyads of the same equivalence class. However, we can also construct complete dyadic sets from any unit U of a certain magnitude. To do this, it is enough to allow the primaries of the dyads to take all the values of R. Thus, the generic dyad $(q \, U)$ will allow us to represent any quantity m by simply varying q throughout R. Obviously these sets do not repeat quantities m and are included in D, but they do not have to be equal to D. In this work we represent them with the notation $\{R, U\} \subset D$. And the same can be said about vector dyads, simply replacing R with R^3.

The analysis of the relation «less than», symbolized «<», is completely analogous to the relation «≤» or «less than or equal to» and, since these relations reduce to those of R, the relation «less than» in M is also as in R of strict order, meaning that it is irreflexive, antisymmetric, and transitive; or what is the same, neither reflective nor symmetrical, a property called asymmetry, and transitive. So, given any quantities m, m_1, m_2 and m_3 of M, it cannot be $m < m$, an irreflexive property; if $m_1 < m_2$, then $m_1 \neq m_2$, asymmetric property, which is equivalent to if $m_1 < m_2$, then, it cannot be $m_2 < m_1$; on the other hand, if $m_1 < m_2$ and $m_2 < m_3$, then, $m_1 < m_3$, transitivity; finally, if $m_1 \neq m_2$, it is $m_1 < m_2$ or $m_2 < m_1$, but not both, so all the elements of M are comparable and therefore the «less than» relationship is of total order. In conclusion, for any magnitude, the set $M = \{m\}$ of all possible quantities m is, like R, totally ordered by the relation «<», defined by the condition $m_1 < m_2$ if and only if $q_1/q_2 < U_2 /\!/ U_1$.

On the other hand, we know that for the dyadic notation $(x_1 \, y_1 \circ U)$ and $(x_2 \, y_2 \circ U)$ the equivalence condition is $x \times y = h$. That is, all dyads such that $x_1 \times y_1 = h_1$ belong to the same class and represent the same amount of magnitude m_1. And similarly for $x_2 \times y_2 = h_2$ and m_2. So $(x_1 \, y_1 \circ U) \leq (x_2 \, y_2 \circ U)$ implies that it is $x_1/x_2 \leq y_2/y_1$. If $h > 0$, we have $x_1 \times y_1 \leq x_2 \times y_2$ and $h_1 \leq h_2$ with $m_1 \leq m_2$. If $h < 0$, it is $x_1 \times y_1 \geq x_2 \times y_2$ and $h_1 \geq h_2$ with $m_1 \geq m_2$. Since $x \times y = h$ is the measure of m in the unit U, the order of h defines the order of m. And the same is true for the «less than» relationship. Then, the order in $M = \{m\}$ is given by the order of h and its associated hyperbolas, which define the equivalence classes of m for each h, and thus the always non-numerical quantity of magnitude m can be quantified by its associated real number $h \in R$.

214

Interpreting the hyperbolas $x \times y = h$ as contour lines with respect to the plane (x,y) for a given elevation h, results in a hyperbolic surface that represents all possible quantities of magnitude in relation to any magnitude.

Thus, each level curve $x \times y = h$ will indicate the equivalence class of the quantity $(x \, y \circ U)$, with U being any standard unit of the given magnitude.

The x and y axes for $h=0$ represent the null magnitude quantity class.

In this way, the path of all quantities of any magnitude is indicated by a hyperbolic surface such as the one indicated in the following figure:

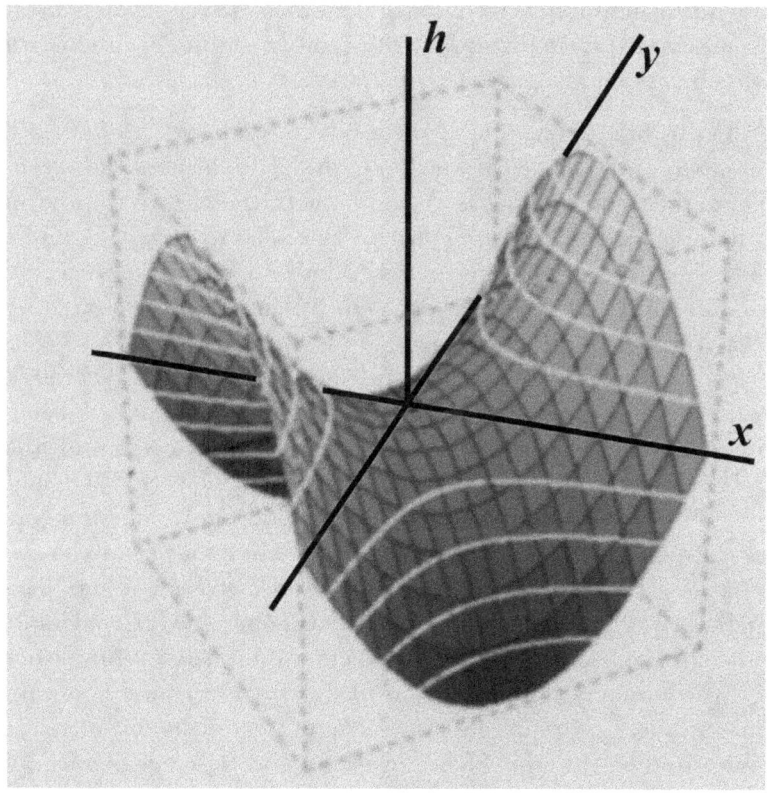

We observe in the image that in the first and third quadrants the magnitude measurement h is positive, it is above the plane (x,y) and increases when the contour lines move away from the center of coordinates; while in the second and fourth quadrants h is below the (x,y) plane and decreases the further the curves are from the center, since these are negative quantities that increase in absolute value.

In sum, given a standard unit U, the real number h indicates the equivalence class of measurements of quantities of length, or any other magnitude, such that every pair of dyads $(x_1 \, y_1 \circ U)$ and $(x_2 \quad y_2 \circ U)$ are equivalent if the equivalence condition $x_1 \times y_1 = x_2 \times y_2 = h$ is verified. Therefore, every quantity of magnitude is determined by the pair of real numbers (x,y) and the hyperbolic function $f(x,y) = h$ such that $x \times y = h$. In this way we complete the evolution of numbers from natural numbers to these pairs of real numbers that formalize quantities of any magnitude, also known as concrete numbers.

Let us briefly recall that the elements of the set N of natural numbers are conceptualized as the equivalence classes of countable sets. This set is expanded with the negatives to form the integers Z, which are conceived in pairs (a,b), with a and b being natural, defining $a - b = z$ if $a > b$ and $b - a = -z$ if $a < b$, with $z > 0$; and defining the integer as each equivalence class generated by the relation $(a_1,b_1) \sim (a_2,b_2)$ if and only if $a_1 - b_1 = a_2 - b_2$ or $b_1 - a_1 = b_2 - a_2$. In turn, the integers lead to the rationals Q through pairs (c,d) or in fractional notation c/d, with c and d being integers, conceiving them as the equivalence classes of the relationship $(c_1,d_1) \sim (c_2,d_2)$ if and only if $a_1 \times d_2 = d_1 \times c_2$, with d_1 and d_2 not null; which leads us to define the set Q of rational numbers as the quotient set of $Z \times Z^*$ with respect to the defined equivalence relation. Expanding the set Q with the irrational numbers indicated by the convergent sequences that have no rational limit, we arrive at the set R of real numbers. And, finally, taking pairs of real numbers (x,y) and an arbitrary standard unit U of any magnitude, we represent the concrete numbers of physical quantities by the equivalence relation between dyads defined by

$(x_1 \ y_1 \circ U) \sim (x_2 \ y_2 \circ U)$ if $x_1 \times y_1 = x_2 \times y_2 = h$ is verified, as established above.

So, finally, we can conclude that this last equivalence relation symbolized as all with the same sign «~», such that, given the pairs of real numbers (x_1, y_1) and (x_2, y_2) of R×R, they are related if and only if $x_1 \times y_1 = x_2 \times y_2$, constitutes in the set product R×R of the real numbers with itself a partition into equivalence classes or quotient set, which can be assumed equivalent to the set M of all possible quantities of magnitude and this is identified with the dyadic set $\{x \ y \circ U\}$ in terms of any arbitrary standard unit U, x and y being any real numbers. So the following formal definition can be established:

$$\frac{\textbf{R} \times \textbf{R}}{\sim} \equiv \textbf{M} = \{x \ y \circ U\}$$

From all this it follows that we have defined the set of all quantities of magnitude M in three different ways, but all of them equivalent: first, from the quotient set $\dfrac{D}{\sim} = M$ of all dyads D; second, through the dyadic set $\{x \ y \circ U\}$, which uses an arbitrary standard unit U and all pairs (x, y) of real numbers; and third, using only the real numbers, as just explained above, with the quotient set of the Cartesian product R×R.

It does not seem doubtful that this result is very significant for Physics and reveals how important it is for it to find a true algebra of magnitudes, which has historically been denied. However, we have proven here that with solid algebraic techniques something as invisible, elusive and intangible as the quantities of physical magnitudes can be observed and reduced to pairs of totally precise and concrete real numbers, linked by a specific equivalence relation defined by multiplication common numeric.

Article 8

GEOMETRIC MULTIPLICATION OF LENGTHS

In the same way that dyadic addition based on the geometric sum of segments has become generalized, the dyadic multiplication of magnitudes, pending definition, must be inspired by the **geometric product of lengths**. To do this, the first thing to do is to observe how this operation works, which will be denoted by the mathematical asterisk «*», to differentiate it from the arithmetic product and break the erroneous illusion caused by the traditional identity of symbols with which different composition laws are represented. Thus, given two segments S_1 and S_2, geometric multiplication consists, by definition, in forming with them a rectangle whose dimensions are the segments themselves. The product of two segments is thus not another length, but a different magnitude called area or surface, whose analytical form can be expressed with a specific symbol such as the product $S_1 * S_2$. In this way, the factors of geometric multiplication are two lengths, those that correspond to the segments S_1 and S_2, while the product $S_1 * S_2$ is a determined quantity of the new magnitude derived from the length and called surface or area.

It is observed in the previous definition that geometric multiplication does not require that the factors be expressed in uniform units, as occurs with addition, because whatever the segments S_1 and S_2 are, the corresponding rectangle can always be formed with them. In turn, since it is a geometric operation with figures, which is not expressed numerically, none of the factors can act as a multiplier, which highlights the «arithmetization» that is arbitrarily assumed, revealing the incoherence unacceptable of current formulations with magnitudes.

This operation has a remarkable property that will be the basis of the generalization adopted for the multiplication of any magnitudes. It is the following **geometric fact**: let the lengths of the segments S_1 and S_2 be given by the dyads $(L_1 \ U_{L1})$ and $(L_2 \ U_{L2})$, where $L_1 \in R$ and indicates the measure of S_1 in the unit of length U_{L1} and $L_2 \in R$ and indicates the measurement of S_2 in the unit of length U_{L2}. It is recalled that **the parentheses are superfluous, but are expressed to mark the dyadic factors**. Under these conditions, it is materially verified that the surface of the rectangle symbolized by the product $S_1 * S_2$ is such that it can be measured in units of area equal to the area of the unit rectangle formed by the units of length when multiplying them geometrically, that is, the indicated surface by the geometric product $U_{L1} * U_{L2}$, and such measurement is equal to the arithmetic product of the dyadic primaries $L_1 \times L_2$. Expressed this property analytically, we have the capital fundamental **equation of geometric multiplication**:

$$(L_1 \ U_{L1}) * (L_2 \ U_{L2}) = [(L_1 \times L_2) \ (U_{L1} * U_{L2})]$$

It is necessary to observe the difference and the relationship that this epistemic equation establishes between the geometric product of lengths, symbolized by a mathematical asterisk « * », and the ordinary arithmetic product, indicated by the typical cross « × », as well as the correspondence between the factor lengths and the resulting area by multiplying them, that is, by composing them with this operation. This relationship is what justifies that this composition law is considered a multiplicative operation, but this does not mean much less that it is ordinary multiplication. On the contrary, it is a law of composition clearly different from the arithmetic product of R. In figure 10 the examples that graphically clarify the crucial property described above are visualized and developed.

In the case that there are three segments to be multiplied, S_1, S_2 and S_3, the logical development is totally analogous, with the difference that its geometric product generates, by definition, instead of a surface, a symbolically designated straight parallelepipedic volume $S_1 * S_2 * S_3$ such that its dimensions are precisely the three multiplied segments. Also here the **fundamental equation of geometric multiplication** with three factors is verified:

$$(L_1 \ U_{L1}) * (L_2 \ U_{L2})) * (L_3 \ U_{L3}) = [(L_1 \times L_2 \times L_3) \ (U_{L1} * U_{L2} * U_{L3})]$$

Geometric experiment of the areas

Given two lengths expressed in the same unit U_L, if an **abstract rectangle without scale** is formed with its numerical parts, it is observed that, dividing it into ideal squares with sides equal to one, the number of these is equal to the product of the measures of the lengths given relative to the unit. This observation of geometry allows defining the product of two lengths a a U_L and b U_L or two concrete numbers with the same unit, interpreting it as an area that is symbolized:

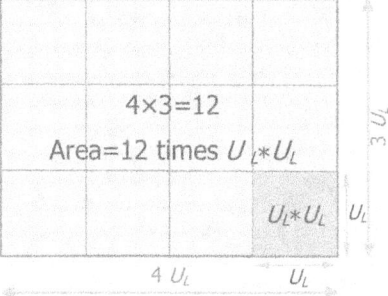

$$a\, U_L * b\, U_L = [(a \times b)\, (U_L * U_L)] = [(a \times b)\, U_L^2]$$

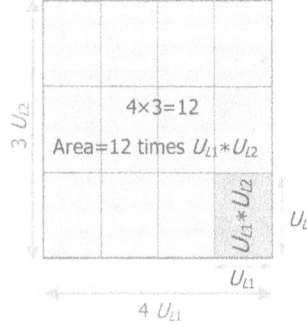

On the left, the case in which the lengths or dyadic are not expressed in the same unit as U_{L1} and b U_{L2}, in the abstract rectangle built with them, it is observed that their product can be associated with the quantity called area, which is measured by means of rectangles equal to the unit of area symbolized $U_{L1} * U_{L2}$, justifying the same product definition:

$$a\, U_{L1} * b\, U_{L2} = [(a \times b)\, (U_{L1} * U_{L2})]$$

$$(3/5)\ U_{L1} * (2/3)\ U_{L2} =$$
$$= [(6/15)\ (U_{L1} * U_{L2})]$$

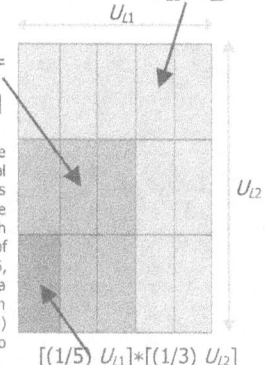

On the right the product of two lengths with fractional measure $[(3/5)\ U_{L1}] \cdot [(2/3)\ U_{L2}]$. Dividing one of the dimensions into five equal segments and the other into three, results in a set of equal rectangles whose sides measure 1/5 of U_{L1} and 1/3 of U_{L2}, the number of these equal elements that make up the unit is equal to $5 \times 3 = 15$, which coincides with the product of the denominators, and the number of equal elements that fit in the assumed fractional measure is $3 \times 2 = 6$, which coincides with the product of the numerators; the fractional area will be 3×2 elements of the 5×3 total rectangles, which is the fraction $(2 \times 3)/(3 \times 5)$, which is equal to the product of fractions $(3/5) \times (2/3) = 6/15$, so here the form of the definition of dyadic multiplication also holds.

Figure 10

In figure 11 the definition of this composition law and its fundamental geometric property are clarified with an example, which are the basis of the generalization that will define the dyadic product of any physical magnitudes.

221

Experimental significance of the product three-length geometric

$4 \times 2 \times 3 = 24$

Volume = 24 times ($U_{L1} * U_{L2} * U_{L3}$)

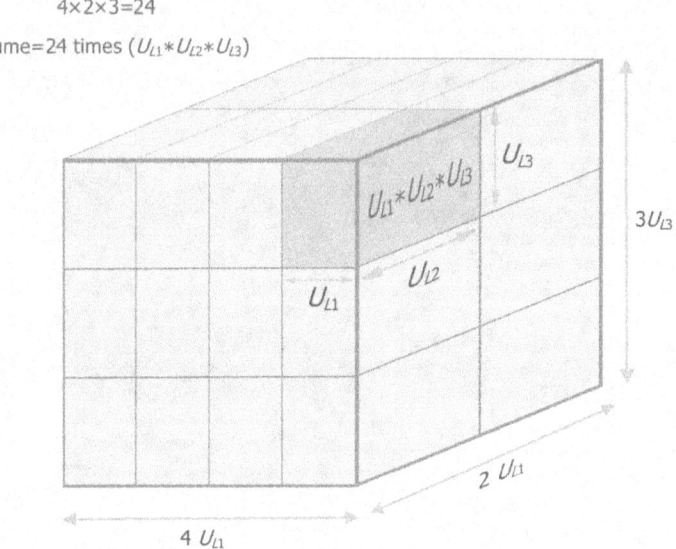

Given three lengths 4 U_{L1}, 2 U_{L2} and 3 U_{L3}, an **abstract straight parallelepiped without scale** can be formed with them and ideally decomposed by delimiting the corresponding symbolic length on each edge. Thus, they result in a series of parallelepipeds with the same ideal unit measurements, so they are congruent and equal. The new magnitude that results from composing three lengths is called volume, and the fact that the number of elementary parallelepipeds is equal to 24 makes it possible to refer to the quantity of volume indicating that one of these elements measures 24 times, which nothing prevents symbolizing with the Notation similar to the algebraic $U_{L1} * U_{L2} * U_{L3}$, writing this result [24 ($U_{L1} * U_{L2} * U_{L3}$)]. With this, the operation of composing three lengths consisting of forming a straight parallelepiped with them can be called multiplication of the concrete numbers or initial dyads given by three lengths, and this operation is symbolized (4 U_{L1})·(2 U_{L2})·(3 U_{L3}) = [(4×2×3) ($U_{L1} * U_{L2} * U_{L3}$)], resulting in that the numerical part is equal to 4×2×3=24. So it can be defined that multiplying lengths is to obtain another quantity of the magnitude called volume whose measure is the arithmetic product of the numerical parts of the factors and whose unit of volume is expressed as the geometric product of the units of the factors. Since the unit elements are composed in the same way regardless of the order in which the factor units are composed, the commutative and associative properties of geometric multiplication must be axiomatized..

Figure 11

Given the current erroneous «arithmetization» of the composition of lengths and other physical quantities, which is tolerated by the false hypothesis of the International System of Units, which attributes the abelian multiplicative group structure to magnitudes, it is necessary to insist once again that the multiplication of lengths does not correspond to the ordinary product in R, but rather it is a graphical operation, it is a law of composition with geometric segments, as straight figures of one dimension, in such a way that with two segments it produces a surface and with three a volume. It is, therefore, a **law of generating external composition.**

We are, then, before a misconception in the physical foundations, which is saved with full coherence through the algebra that is established here.

Article 9

DYADIC MULTIPLICATION OF
SCALAR MAGNITUDES

The geometric multiplication of segments allows and justifies the generation of the product between all the scalar magnitudes. It is enough to simulate that the quantity of a certain magnitude is indicated by a segment or a quantity referred to an **abstract unit of length** that resembles the unit considered as a reference quantity of that magnitude, which is equivalent to the observation that, given a unit of length U_L and another U_M of any magnitude M, between the dyadic sets $\{R, U_L\}$ and $\{R, U_M\}$ it is possible to establish one-to-one correspondence without more than imaginatively identifying the U_M unit with the U_L, since formally any element $(q\ U_M)$ of the second set may be associated biunivocally with the element $(q\ U_L)$ of the first. We will say in this case that the magnitude M is affin to the length, and the attribution of this quality to the physical magnitudes we will call it the **affinity postulate**.

Since we have defined physical magnitudes as those properties of nature that can be measured, and we have seen that dyadic sets arise from measurement, this definition is equivalent to having as physical quantities those natural properties affine to length.

In this way, given two scalar physical dyads $(q_1\ U_1)$ and $(q_2\ U_2)$, the dyadic multiplication is defined, by generalization of the geometric one, as the abstract rectangle of surface indicated by $(q_1\ U_1) * (q_2\ U_2)$ such that the measure of its area expressed in the **abstract unit** $U_1 * U_2$ is fixed by the following epistemic definition equation:

$$(q_1\ U_1) * (q_2\ U_2) = [(q_1 \times q_2)\ (U_1 * U_2)]$$

This equation determines a composition law that applies the Cartesian product $\{R, U_1\} \times \{R, U_2\}$ on $\{R, U_1 * U_2\}$. It is an **external generating law**, which creates a new magnitude whose unit is $U_1 * U_2$.

If the dyadic factors to multiply were three, the dyadic multiplication would form an **abstract parallelepipedic volume**, whose measure would be given by the definition equation:

$$(q_1 \; U_1) * (q_2 \; U_2) * (q_3 \; U_3) = [(q_1 \times q_2 \times q_3) \; (U_1 * U_2 * U_3)]$$

In this case the external composition law applies the Cartesian product set $\{R, U_1\} \times \{R, U_2\} \times \{R, U_3\}$ in $\{R, U_1 * U_2 * U_3\}$.

And, in general, for any number of factors n, the dyadic multiplication of quantities of scalar physical magnitudes will be defined by a **hypervolume**, whose measure would be determined by the following analytical equation:

$$(q_1 \; U_1) * (q_2 \; U_2) * \; \ldots \; * (q_n \; U_n) =$$

$$= [(q_1 \times q_2 \times \; \ldots \; \times q_n) \; (U_1 * U_2 * \; \ldots \; * U_n)]$$

The external composition law thus defined represents an application or function of the Cartesian product set indicated by $\{R, U_1\} \times \{R, U_2\} \times \; \ldots \; \times \{R, U_n\}$ in $\{R, U_1 * U_2 * \; \ldots \; * U_n\}$.

The external nature of these multiplicative composition laws, as explained in section XIV, irrefutably denies the existence of unit or inverse elements of dyadic entities, so the negative exponents that result in algebraic expressions should not be interpreted as inverses of other measurements, but as denominators of dyadic fractions. Thus, the noxious and scandalous gap in the International System of Units in this matter is overcome, which admits notations such as m^{-1}, kg^{-1} or s^{-1}, without defining at all what kind of entities these symbologies refer to, taking for granted childishly that they obey vulgar algebra, an erroneous and garrafal presupposition that vitiates all physical content from the root. Crass mistake produced by not asking what is the meaning of the inverse of a meter, a kilogram or a second, as would correspond to any physicist responsible for his science.

Article 10

DEDUCTION OF THE DYADIC SCALAR DIVISION

It is possible to deduce the analytical definition of the dyadic quotient without more than attending to the generic concept of division. To do this, just imagine an abstract rectangle whose surface is identified with a dyadic dividend $(a\ U_1)$, one of its dimensions with the divisor $(b\ U_2)$, with $b \neq 0$, and the other with the dyadic quotient $[c\ (U_1 /\!/ U_2)]$. The unit associated with c must be identified with the dyadic ratio of units $[c\ (U_1 /\!/ U_2)]$, because the unit rectangle must have the unit U_1 by area and by dimensions U_2 and $U_1 /\!/ U_2$. In the same way, the three indicated dyads cannot be independent, but must satisfy the division condition, that is, the quotient multiplied by the divisor must equal the dividend; or, in other words, the dyadic product of the two dimensions of the abstract rectangle must be equal to its surface; and it will be written analytically like this:

$$(a\,U_1) = (b\,U_2) * \left(c\,\frac{U_1}{U_2} \right)$$

The parentheses are superfluous, but as usual they are kept to mark well the dyads involved in the formula, which can be interpreted according to the generic division criterion, which leads to consider the factor $[c\ (U_1 /\!/ U_2)]$ as the quotient between the total surface of the abstract rectangle $(a\ U_1)$ and the other of its two dimensions $(b\ U_2)$. And this analytically can be described as follows:

$$\left(c\,\frac{U_1}{U_2} \right) = \frac{(a\,U_1)}{(b\,U_2)}$$

The geometry of the abstract rectangle is such that $a = b \times c$, given the fundamental property of geometric multiplication, so that $c = a/b$ with the algebra of R. So, substituting $c = a/b$ in the first member of the last equality, we will finally have this other dyadic expression:

227

$$\left(c \frac{U_1}{U_2} \right) = \frac{(a\,U_1)}{(b\,U_2)} = \left(\frac{a}{b} \frac{U_1}{U_2} \right)$$

And between the second and third terms of this equation the **analytical definition of the dyadic division** between the scalar dyadic concretes $(a\ U_1)$ and $(b\ U_2)$, deduced by the preceding reasoning, is already observed. So it can be concluded that the quotient of these two dyads is equal to a dyadic element whose primary is the quotient of the primaries of the factors and whose secondary is the dyadic division of the units of the dividend and the divisor. Expressed analytically:

$$\frac{(a\,U_1)}{(b\,U_2)} = \left(\frac{a}{b} \frac{U_1}{U_2} \right)$$

In this way, we verify, as for the rest of the operations previously analyzed, that the symbols of the units behave ideally like the other elements of R, but this consequence is not due to the traditional symbolic logic, and it is insisted, it would be an error It is crass and inadmissible to consider it this way, because it has been irrefutably justified that this formal behavior is due to the concept of dyadic multiplication through abstract rectangles, it is not the result of the properties of the operations in R. It is simply an isomorphism, not an identity in absolute.

On the other hand, it is noted that with the **double bar** the division of scalar concrete analyzed in this article has been symbolized, an operation different from the quotient of homogeneous dyads, which it has been agreed to represent with the same sign. And it is that the diversity of algebraic laws is such that, although symbolic exhaustiveness is sought for pedagogical clarity, it is inevitable and even sometimes convenient to resort to a certain degree to the principle of symbolic economy, but this without it being permissible to confuse the different operations indicated with the same signs.

Article 11

MULTIPLICATION BETWEEN SCALAR
AND VECTOR DYADS

Newton's second law relates quantities of the vector magnitude called force to the product of the scalar magnitude called mass and the vector magnitude known as acceleration. However, a multiplication like this has never been defined, but an isomorphic behavior with the algebra of R has been tacitly assumed, which is an unacceptable misadventure that forces it to be established epistemically, as has been done with the others that the they precede in this work, because it appears constantly in the physical equations and, without such definitions, the meanings of the scientific laws are denatured and gloomy, apart from being expropriated of all logical foundation and scientific consistency.

So the multiplication of scalar dyads by other vectors has to compose scalar measurements $(a\ U_1)$ of $\{R, U_1\}$ with vector elements $(\overline{b}\ U_2)$ of $\{R^3, U_2\}$. You can distinguish this composition law using any sign, for example, «$\odot$». To establish the appropriate definition of this product form, the external law of the vector space R^3 over R and dyadic multiplication can be counted on, so there is no better formulation than these two equations:

$$(a\ U_1) \odot (\overline{b}\ U_2) = [(a \bullet \overline{b})\ (U_1 * U_2)]$$

$$(\overline{b}\ U_2) \odot (a\ U_1) = [(\overline{b} \bullet a)\ (U_2 * U_1)]$$

The sign «$\odot$» of the first members of these expressions symbolizes the composition law that is being defined in this article, the product of a scalar physical pair by another vector; the sign «$\bullet$» placed in the factors $(a \bullet \overline{b})$ y $(\overline{b} \bullet a)$ of the second members indicates the external law of R^3 on R or product of a scalar by a vector; and the multiplications of the terms $(U_1 * U_2)$ and $(U_2 * U_1)$ mark the dyadic product of two scalar dyadic elements, defined in article 9.

However, as we are observing with insistent repetition, the stubborn vice of using the same «$\times$» sign for all these multiplicative laws still prevails, in application of the easy tacit principle of symbolic economy, whose fatal illusory and equivocal effects are noted throughout of this work and in more detail in article 15.

Although it may be idle to the attentive reader, it is noted here that the product defined in this article allows, like any other, to establish division as a derivative operation. It is enough to consider the second member of the previous definition equations as a dividend and any of the factors of the first member as a divisor, resulting in the other factor of this being the quotient. Thus we find that the quotient between the vector dyad of the second member and the scalar dyad of the first member gives a vector quotient; as well as the quotient between the same vector dyad of the second member and the other vector of the first one results in a scalar quotient. Although yes, by the very definition of this product, the vectors of the first and second member must be collinear.

This division could be symbolized by any arbitrary sign; but, in order not to increase the operational symbology to infinity, it is enough to distinguish it as one more dyadic quotient, for example, with the double bar «//». However, the algebraist must know how to distinguish the various operations not by their symbols but by the elements they relate. Detailed symbology is more of a didactic element than necessary and can be cumbersome, hence the need for a certain symbolic economy, provided that some operations are not confused with others.

Article 12

DEFINITION OF SCALAR PRODUCTS
AND VECTOR OF VECTOR DYADS

In Physics, vector magnitudes use mathematical vectors and the products between vectors called scalar product and vector product. It is usual to indicate the scalar with a mathematical point «·» and to distinguish with the angle «∧» the vector product of vectors. In turn, for the dyadic homonyms of quantities of vector magnitudes, the circle with a point «⊙» will be reserved for the scalar dyadic product and the circle with an asterisk «⊛» for the vector dyadic product.

There is no better way to define both the scalar product and the vector product of vector magnitudes than in terms of their counterparts of mathematical vectors. Starting with the scalar product of two vector concretes $(\overline{a}\ U_1)$ and $(\overline{b}\ U_2)$ of the dyadic or concrete sets $\{R^3, U_1\}$ and $\{R^3, U_2\}$, the first elements $\overline{a}$ and $\overline{b}$ the pairs are distinguished here as vectors of R^3. And with this, the **dot product of two vector dyads** must be defined with the epistemic equation:

$$(\overline{a}\ U_1) \odot (\overline{b}\ U_2) = [(\overline{a} \cdot \overline{b})\,(U_1 * U_2)]$$

That is, by definition, the scalar product of two vector dyads measures a scalar magnitude with the scalar pair of the set $\{R, U_1 * U_2\}$ such that its primary is the scalar product of the vector primaries of the factors and the unit is the product dyadic or concrete of the factor units.

Regarding the **vector product of two vector dyads**, the following definition equation will be had in an analogous way:

$$(\overline{a}\ U_1) \circledast (\overline{b}\ U_2) = [(\overline{a} \wedge \overline{b})\,(U_1 * U_2)]$$

In this case, by definition, the vector product of two vector concretes measures a vector magnitude with the dyad of the set $\{R^3, U_1 * U_2\}$ such that its primary is another vector equal to the vector product of the vectors that make up the primaries of the given concrete or dyads, and whose unit is the dyadic product of the factor units.

The two definition formulas above will facilitate the correct interpretation of the physical equations in which these composition laws intervene, such as the magnitudes of work and moment of a force, the work for the scalar product and the moment for the vector product.

On the other hand, these composition laws will not make sense for the case of scalar dyads.

Artículo 13

ALGEBRAIC STRUCTURE OF
THE DYADIC SETS

The scalar and vector dyadic sets associated with a given unit U, respectively symbolized $\{R, U\}$ and $\{R^3, U\}$, endowed with the corresponding additive internal law, scalar or vector, and indicated in both cases with the sign «$\oplus$» , form the **algebraic structures R, $U, \oplus$} and {$R^3, U, \oplus$} with the abelian group properties**, because it is relatively easy to prove that they are defined everywhere with uniqueness and the commutative and associative properties are verified in both cases , existence of a neutral element and existence of a symmetrical element, as we do in the first part of this work.

The same does not happen with the dyadic multiplication «$*$», because in the case of a law of external composition, even if it is commutative and associative; however, unitary and inverse elements cannot exist in the same set. So **the structures $\{R, U, *\}$ and $\{R^3, U, *\}$ do not satisfy the group conditions.** In turn, with the two laws of composition indicated, **the structures $\{R, U, \oplus, *\}$ and $\{R^3, U, \oplus, *\}$ do not satisfy the properties of the field.**

Associating the abelian groups $\{R, U, \oplus\}$ and $\{R^3, U, \oplus\}$ with the field R of the real numbers and considering the respective external laws, both of which have been identified with the sign «$\circ$» in article 6, For both scalar and vector magnitudes, it is easily verified that these external composition laws are defined everywhere and they verify the following properties: they are associative with respect to the multiplication of the field R; they are modular, which means that the unit element of the body R leaves every element of R and R^3 invariant; they are distributive with respect to the additive laws of R and R^3; and they are distributive with respect to the additive law in the field R.

Consequently, **the abelian groups $\{R, U, \oplus\}$ and $\{R^3, U, \oplus\}$ present respective vector space structures on the field R of the real numbers.** This is why the resulting isomorphisms allow the dyadic elements to be operated as if their various components were

233

elements of R, even though they are not really so and the stubborn tradition presumes it subliminally or arbitrarily. In this way, the question of why it is possible to operate with the magnitudes as is usually done is resolved and this insidious gap is overcome, which suggested diverse and rather esoteric explanations, lacking foundation and alien to the scientific method.

In short, it all comes down to solving the latent incongruous omission by establishing the necessary composition laws and in harmony with the usual algebraic structures. This leads to the isomorphism between the structure of the geometric segments $\{S, \oplus, \circ, *\}$ with the field of real numbers R, which with its two internal laws, additive and multiplicative, is also a vector space with its own body R as domain of operators and the same multiplication.

For its part, the set of geometric segments $\{S\}$ with its internal additive law «$\oplus$», its external multiplicative law by a scalar «$\circ$», together with the external generating multiplicative operation «$*$», constitutes a homologous structure of R.

In this way, the bijective map f that makes each segment $S \in \{S\}$ correspond to the real number $x \in R$ that indicates its measure with a certain unit such that $x = f(S)$ represents the isomorphism between $\{S\}$ and R. With the laws of composition defined in $\{S\}$ the map f is such that, given any segments S, S_1 and S_2 of $\{S\}$, it makes them correspond their measures x, x_1 and x_2 of R and the following properties are obtained for all α of R:

$$f(S_1 \oplus S_2) = x_1 + x_2 = f(S_1) + f(S_2)$$

$$f(\alpha \circ S) = \alpha \times x = \alpha \times f(S)$$

$$f(S_1 * S_2) = x_1 \times x_2 = f(S_1) \times f(S_2)$$

Thus the arithmetic of R and the non-arithmetic algebra of $\{S\}$ are connected, becoming isomorphic structures.

234

Note that the inverse or reciprocal application f^{-1} of R in $\{S\}$ is also an isomorphism, characterized by the relation $f^{-1}(x) = S$ for all $x \in R$ and all $S \in \{S\}$.

Article 14

DYADIC ANALYSIS OF THE *THALES' THEOREM*

Thales' Theorem[18], in its classical formulation, is based on the theory of ratios and proportions, with its specific notions of equality and addition of segments. In sum, **the traditional theory assumes replacing the segments by their measure in a certain unit of length**, which can be any or even abstract, but always the same for all the segments considered. That is, the segments or lengths are assumed to be uniform. With this artifice the ratios and proportions between segments, which are geometric, are reduced to abstract numerical ratios and proportions.

However, it is clear that a segment is not just a number, but a quantity of length. It is a dyad with the form $(a\ U_L)$, where the primary a is R and indicates the measure of the segment in the unit of length U_L. Well, given two segments expressed with uniform dyads $(a\ U_L)$ and $(b\ U_L)$, classical Mathematics defines the ratio between them, identifying it with the arithmetic of its primaries a/b. With this, geometric algebra is replaced by symbolic algebra, described by the measures of the segments in a certain unit of uniform length.

In the same way, the geometric proportionality of segments is replaced by the proportionality of real numbers, so that two given segments will be said to be proportional to two others when the numbers that express their measures in the same unit of length form a proportion in the set R That is, the segments $(a\ U_L)$ and $(b\ U_L)$ are said to be proportional to the $(p\ U_L)$ and $(q\ U_L)$, if and only if the arithmetic proportion $a/b=p/q$ is verified. And this is where this traditional method is childish, because it is not at all justified why the unit of length is outside the defined proportion. And this even despite the fact that, as is known, and it will be recalled here, the conclusion is correct, even if it is unfounded and is not epistemic, which leads to overlooking certain intuitive and

---

[18] The classical deduction of Thales' Theorem can be found in «Lesson 26» of Mathematize 1.

latent presuppositions, which become evident when a dyadic analysis of the geometric proportionality of segments is performed.

The first question to elucidate is which dyadic operations affect it. Well, taking into account that it relates only lengths, it should be pointed out to the addition of segments of article 4, the symbol «⊕», to its derivative operation or multiplication by a scalar of article 6, with the symbol «∘», and to the dyadic division associated with this one of the symbol «⫽» of article 7, where it has been proven that the **dyadic quotient of uniform scalar elements**, that is, referring to the same unit, must result in a real number, which will be precisely the quotient in R of the primaries between dividend and divisor. So that the proportionality of segments will be deduced in $\mathscr{D}$ by the following analytical argumentation:

$$\frac{\left(a\,U_{L}\right)}{\left(b\,U_{L}\right)} = \frac{\left(p\,U_{L}\right)}{\left(q\,U_{L}\right)} = \frac{a}{b} = \frac{p}{q}$$

So the geometric or dyadic proportionality is, in effect, reduced to the arithmetic proportionality of the measures of the segments in the same unit of length. But with the nuance, nothing banal from an epistemological point of view, that, just as the mathematical tradition presupposes or postulates it, the dyadic algebra deduces it unequivocally, without giving the option to subjective intuition or arbitrariness.

So the geometric or dyadic proportionality is, in effect, reduced to the arithmetic proportionality of the measures of the segments in the same unit of length. But with the nuance, nothing banal from an epistemological point of view, that, just as the mathematical tradition presupposes or postulates it, the dyadic algebra deduces it unequivocally, without giving the option to subjective intuition or arbitrariness.

A question that arises immediately is whether in $\mathscr{D}$ the proportions also verify as in R the property that the product of the means is equal to the product of the extremes. To verify this, it must be remembered that the product of segments is the geometric one of article 8, generalized to any magnitude in article 9, so that, given the proportion of segments in the last equation, the geometric products $(a\,U_L)*(q\,U_L)$ and $(b\,U_L)*(p\,U_L)$ to check if they are the same. Well, operating according to the dyadic laws, we have the equations:

$$(a\,U_L)*(q\,U_L)=(a{\times}q)\,(U_L*U_L)$$

$$(b\,U_L)*(p\,U_L)=(b{\times}p)\,(U_L*U_L)$$

Both products are uniform, because they refer to the compound unit $U_L*U_L=U_L^2$, so to be equal, according to the dyadic criterion of equality, they must have the same primaries, and, in effect, they do, because in R the property under study $a{\times}q=b{\times}p$ is satisfied, which allows to conclude that the proportions of segments also satisfy the condition that the product of the means is equal to that of the extremes:

$$\text{If } \frac{(a\,U_L)}{(b\,U_L)}=\frac{(p\,U_L)}{(q\,U_L)}, \text{ then, } (a\,U_L)*(q\,U_L)=(b\,U_L)*(p\,U_L)$$

It can be easily verified that the converse statement also holds, so that if four segments satisfy the equality $(a\,U_L)*(q\,U_L)=(b\,U_L)*(p\,U_L)$, they must be in the corresponding dyadic proportion.

Segment proportionality reveals an important fact, and it is that it establishes a subtle relationship between the uniform dyadic division of article 7 and the dyadic multiplication between scalar magnitudes of articles 8 and 9.

On the other hand, it should not be forgotten that geometric proportionality is the basis of metric geometry and, therefore, also of trigonometry. Today this fundamental fact has been forgotten, with the somewhat tricky simplification of substituting the segments that are made up in the different operations by their measurements in the same unit of length. And this, although operationally correct by chance, is intellectually and pedagogically toxic, because it loses sight of the true significance of geometric algebra which, although isomorphic in many respects with that of **R**, is a different algebra due to its specific nature. So, it is convenient not to forget things as simple as that the trigonometric ratios are not in themselves arithmetic ratios, but geometric or dyadic divisions of article 7, which analytically, in the right triangle of figure 12 would be expressed as follows:

The trigonometric ratios
Paradigm of dyadic algebra

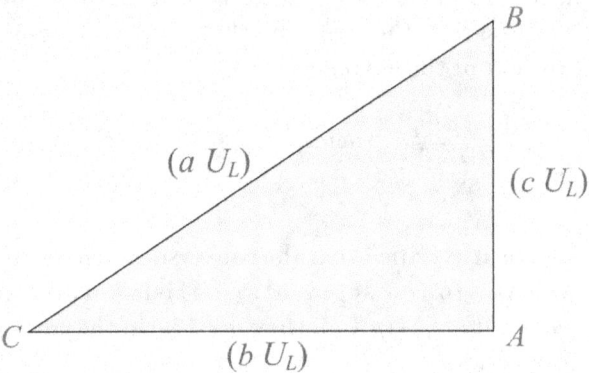

Trigonometric ratios such as the sine, the cosine or the tangent are not arithmetic divisions in themselves, because they do not relate pure numbers, but segments or quantities of lengths, so they strictly belong to geometric algebra and indicate dyadic divisions between magnitudes, in accordance with the operation analyzed in section 7, which serves as fundamental support for the proportionality of segments.

Figure 12

$$sen\ C = \frac{AB}{BC} = \frac{(c\,U_L)}{(a\,U_L)} = \frac{c}{a}$$

$$cos\ C = \frac{AC}{BC} = \frac{(b\,U_L)}{(a\,U_L)} = \frac{b}{a}$$

$$tg\ C = \frac{AB}{AC} = \frac{(c\,U_L)}{(b\,U_L)} = \frac{c}{b}$$

With capital letters A, B and C the angles and vertices of the triangle have been designated, with the pairs of letters AB, AC and BC the segments that form the sides of the triangle, which in turn are expressed by the dyads $AB=(c\ U_L)$, $AC=(b\ U_L)$ and $BC=(a\ U_L)$.

Article 15

THE PYTHAGOREAN THEOREM:
THE FIRST DYADIC FORM OF MATHEMATICS

Egyptian builders and surveyors used a very simple and ingenious instrument, a knotted string that marked equal lengths or segments. With it they formed a triangle with sides 3, 4 and 5 of those segments, which turned out to be right, and thus they were able to draw perpendicular alignments. The Greek mathematician Pythagoras, born in Samos in 580 BC, investigated this property, known to the Egyptians for that singular triangle with sides proportional to 3, 4 and 5, and generalized it to all triangles such that one of its angles be straight, formulating his famous Pythagorean Theorem[19].

Pythagoras used geometric algebra to compose segments and thus concluded that **the area of the square built on the hypotenuse of a right triangle is equal to the sum of the areas of the squares built on the legs.** Such a way of operating with segments, consisting of constructing squares with them, is the same way that the geometric multiplication of lengths is defined in article 8, which is generalized to the dyadic multiplication of any scalar magnitudes in article 9. Therefore The Pythagorean Theorem seems to be the first dyadic formulation of Mathematics, as explained in figure 13.

However, although the classics differentiated in this way the geometric operations with segments from the arithmetic ones, at present the property analyzed in article 8, on the geometric experiment with areas and volumes, has been used to substitute geometric algebra for the arithmetic, operating only with numbers that represent the measures of the segments, ignoring the segments themselves. With this, the Pythagorean Theorem is usually formulated referring to R operations with this statement: **«In every right triangle the hypotenuse squared is equal to the sum**

---

[19] Various deductions of the Pythagorean Theorem can be found in «Lesson 29» of Mathematize 1.

of the squares of the legs, analytically $a^2=b^2+c^2$». And this arithmetic simplification, which omits the geometric origin of such an important property, loses much of the meaning it encompasses.

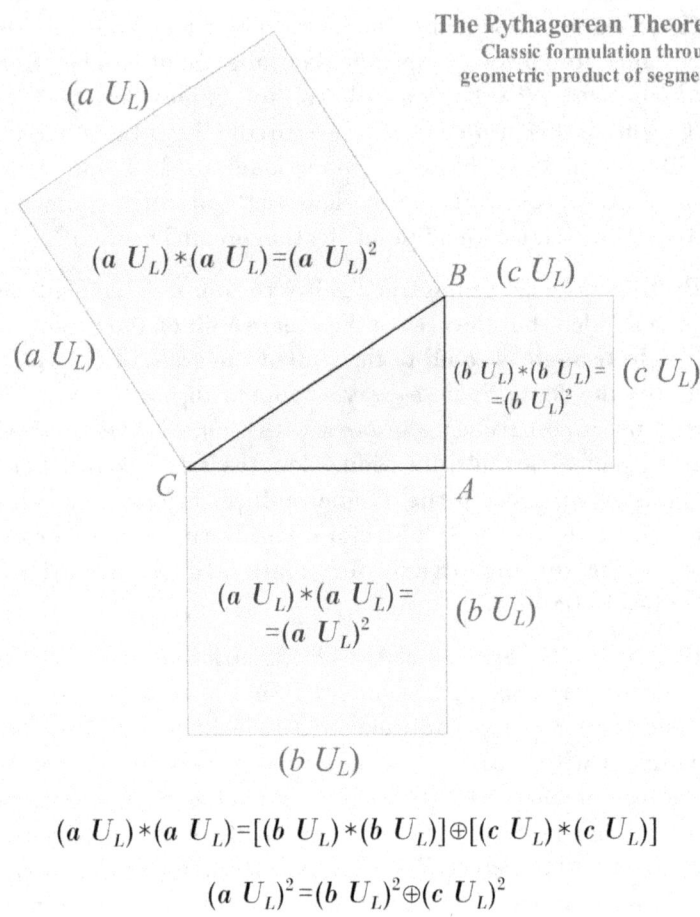

The Pythagorean Theorem
Classic formulation through
geometric product of segments

$(a\ U_L)$

$(a\ U_L)*(a\ U_L)=(a\ U_L)^2$

$B \quad (c\ U_L)$

$(a\ U_L)$

$(b\ U_L)*(b\ U_L)=$
$=(b\ U_L)^2$ $\quad (c\ U_L)$

C $\qquad A$

$(a\ U_L)*(a\ U_L)=$
$=(a\ U_L)^2$ $\quad (b\ U_L)$

$(b\ U_L)$

$$(a\ U_L)*(a\ U_L)=[(b\ U_L)*(b\ U_L)]\oplus[(c\ U_L)*(c\ U_L)]$$

$$(a\ U_L)^2=(b\ U_L)^2\oplus(c\ U_L)^2$$

In every right triangle, the area of the square built on the hypotenuse is equal to the sum of the areas of the squares built on the legs.

Figure 13

The importance that the Greeks conferred on geometric algebra is evident in Euclid's Book II of the Elements. Although it is one of the shortest, with only 14 propositions, none of which is present in modern textbooks, for the classics its statements had great significance. Such disagreement between the ancient and modern criteria is due to the imposing hypnotic force of symbolic logic, which has led to the foundations or meanings of symbols being ruined, more than arbitrarily, spuriously giving them their own substantivity and undermining the mathematical quality.

The first proposition of Book II states: **«If we have two straight lines and we cut one of them into any number of segments, then the rectangle contained by the two straight lines is equal to the rectangles contained by the straight line that was not cut and each one of the previous segments»**. The meaning of this proposition can be better understood with the example in figure 14:

Let AB and AD be two segments; take AD and form the arbitrary sum with, for example, the three segments S_1, S_2 and S_3, so that we have $AD = S_1 \oplus S_2 \oplus S_3$, where «$\oplus$» indicates the geometric addition of lengths; the rectangle $ABCD$ will be the geometric product of the segments AB and AD, which is symbolized $AB*AD$ or $S*AD$, if $S = AB$ is identified, and where the symbol «$*$» indicates the geometric multiplication of segments; the rectangle $S*AD$ is the sum of the interior rectangles described analytically by $S*S_1 \oplus S*S_2 \oplus S*S_3$; so that we have the conclusion:

$$S*(S_1 \oplus S_2 \oplus S_3) = (S*S_1) \oplus (S*S_2) \oplus (S*S_3)$$

And this is nothing but the distributive **property of multiplication with respect to dyadic addition** for segments or areas, operations proper to geometric algebra.

Another didactic case of the effect of loss of geometric meaning caused by symbolic algebra is the **square of a binomial**. Modern algebra simply writes it abstractly in the form $(a+b)^2 = a^2 + b^2 + 2 \times a \times b$. Well, in ancient geometric algebra this property is described in the fourth proposition of the referred

245

Classical geometric algebra
Book II of Euclid's Elements
First proposition or distributive property of multiplication
regarding the addition of segments

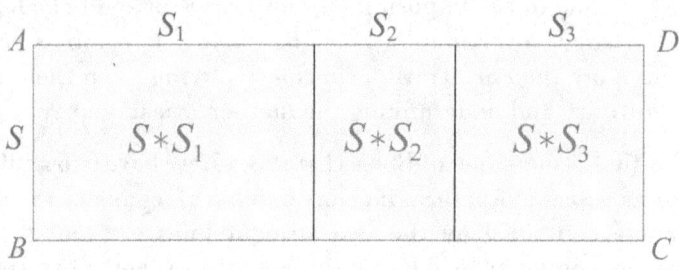

$$S*(S_1 \oplus S_2 \oplus S_3) = (S*S_1) \oplus (S*S_2) \oplus (S*S_3)$$

The geometric multiplication of two segments, which consists of forming rectangles with them, satisfies the distributive property with respect to the dyadic addition of lengths or areas.

Figure 14

Book II of Euclid: **«If a straight line is cut in an arbitrary way, then, the square built on the total is equal to the squares on the two segments and twice the rectangle contained by both segments».** The meaning of this proposition is clarified with figure 15:

Let segment AB and decompose into the sum of two arbitrary segments S_1 y S_2; form the square $ABCD$, which is decomposed into the two squares and two rectangles indicated in the figure. The area of the square ABCD represents the geometric product $((S_1 \oplus S_2)*(S_1 \oplus S_2) = (S_1 \oplus S_2)^2$ and is equal to the sum of the area indicated by $S_1*S_1 = S_1^2$, of the area $S_2*S_2 = S_2^2$ and twice the area

246

of the rectangle identified with $S_1 * S_2$; this sum can be symbolized by the following equation:

$$(S_1 \oplus S_2)^2 = S_1^2 \oplus S_2^2 \oplus 2 \circ (S_1 * S_2)$$

Note that the operation «$\circ$» corresponds to the multiplication of a scalar by a dyad, defined in article 6. And thus we have the geometric or dyadic shape of the square of a binomial without losing an iota of its geometric meaning.

There is no doubt that the visual evidence of geometric reasoning sheds a lot of light on the different observed properties, which are hidden in modern abstract expressions, which is why the high school students of our time are deprived of the ability to

Geometric meaning of square of a binomial

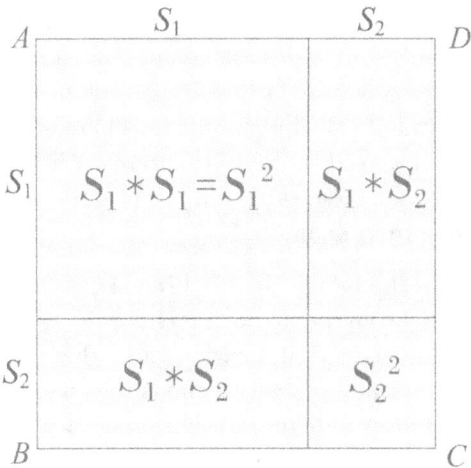

The area of the square $ABCD$ is the sum of the four interior areas into which it is decomposed. In this way the geometric or dyadic expression of the square of an abstract binomial $(a+b)^2=a^2+b^2+2\times a \times b$ results:

$$(S_1 \oplus S_2)^2 = S_1^2 \oplus S_2^2 \oplus 2 \circ (S_1 * S_2)$$

Figure 15

247

understand the rules algebraic that apply with a mechanism almost unworthy of intelligent beings.

With these examples it is clear that **in Euclid's time the magnitudes were already represented as segments subject to the axioms and theorems of geometry**. It is true that at that time algebraic structures, as they are conceived today, did not exist, which has led some to affirm that the Greeks lacked algebra, which is not entirely true, because, although incomplete and primitive, algebra Geometric is there, but yes, **formulated in propositions of ordinary language** and, therefore, curtailed the power of abstraction inherent to the symbology of modern Mathematics. However, that same symbolic and logical power provoke a fascination that is very difficult to contain, inadvertently inducing almost invisible presuppositions, as would be the case of the arbitrary reduction of operations with magnitudes only to their numerical part, losing the geometric meaning along the way. of the symbolic abstractions thus formed, which can simplify the manipulation of these graphic elements, but clearly detrimental to the complete knowledge of mathematical and physical phenomena. There is no doubt that a Greek geometrist versed in the 14 theorems of **Euclid's geometric algebra** would be far more skilled in the practice of measurement than a modern expert geometrist.

Hence, the methodology of this First Algebra of Magnitudes tries to combine the virtues of both techniques, the classical, with its eloquent geometric visualization, and the symbolic, which provides structures and logical methods that were very valuable before unknown, in order to weave a modern physical algebra, abstract and ilative, but well grounded in basic geometric algebra, and thereby **substantiate, explain and define operations with magnitudes without omissions, inconsistencies or latent presuppositions**.

Article 16

DYADIC FORM OF PHYSICAL EQUATIONS

The dyadic multiplication of scalar magnitudes of article 9 and its derived operation, the scalar dyadic division of article 10, are those that correspond to physical quantities such as, for example, density. Its analysis could be done like this: let $(M\,kg)$ be the dyad that expresses the mass of a body indicated in kilograms, let $(V\,m^3)$ be the volume of the body in cubic meters. It is not forbidden to define a compound magnitude, which is called a density and indicated, for example, with the dyad $(d\,U_d)$, where U_d represent the unit in which density must be measured to meet the definition of this magnitude. Defining the density as the dyadic quotient between the mass of a body and the volume it occupies, analytically the following reasoning will be had:

$$\left(d\,U_d\right) = \frac{\left(M\,kg\right)}{\left(V\,m^3\right)} = \left(\frac{M}{V}\,\frac{kg}{m^3}\right)$$

The uniform equality criterion of the dyads of the first and last member requires that the respective primaries and secondaries be equal, which leads to two identities, the first of R and the second of dyadic algebra, which to understand each other we have been symbolizing with the letter $\mathscr{D}$, and thus, if the U_d unit of the first member is identified with that of the second $kg/\!/m^3$, it results:

$$\text{If } U_d = \frac{kg}{m^3} \text{ in } \mathscr{D}, \text{ then, } d = \frac{M}{V} \text{ in } R$$

The physical meaning of density, which is a compound magnitude, by definition, from mass and volume, is obtained without more than considering the unit volume of a cubic meter and finding out its mass, resulting in that, being $V=1$, the density will indicate precisely the mass that corresponds to each cubic meter, which is commonly expressed as «mass per unit volume».

Other forms of physical laws are those involving products of scalar and vector magnitudes, such as Newton's second law. This is the multiplication of article 11. The logical scheme in this case should be the following: let it be a mass given by the scalar dyad $(M\ kg)$ and let it be the vector dyad $(\overline{a}\ m/\!/s^2)$ that indicates the acceleration of its movement . Note that the unit of acceleration is a dyadic quotient. Let be the vector dyad $(\overline{F}\ U_F)$, where U_F indicates the unit of uniform force that corresponds to satisfy the equality represented by Newton's second law, an identity that should be written in dyadic algebra $\mathscr{D}$ with the form $(\overline{F}\ U_F)=(M\ kg)\odot(\overline{a}\ m/\!/s^2)$, where the multiplication indicated «$\odot$» corresponds to the composition law of article 11. The definition of this operation allows us to write the initial equation with the form $(\overline{F}\ U_F)=[(M\bullet\overline{a})\ (kg*m/\!/s^2)]$, in which three laws of composition appear: the multiplication «$\bullet$» of a real number by a vector, the scalar dyadic multiplication «$*$» and the scalar dyadic division «$/\!/$». The criterion of equality of uniform dyads, that is to say, referring to the same unit, allows the last formula to be divided into its two components, one from the algebra of R and the other from the dyadic algebra $\mathscr{D}$, resulting:

$$\text{If } U_F = \frac{kg*m}{s^2} \text{ in } \mathscr{D}, \text{ then, } \overline{F} = M \bullet \overline{a} \text{ in } \mathrm{R}^3$$

It is customary to name the unit v with the term newton, abbreviated N, whose physical meaning corresponds to the force that must be applied to the mass of one kilogram to induce an acceleration of one $m/\!/s^2$ in it, which will be read «meter per second squared», meaning scalar dyadic division.

Another capital concept of Physics is the **work of a force**, defined as the dyadic scalar product of the force magnitude by the length magnitude traveled by its point of application. What in $\mathscr{D}$ will be written explicitly, without the traditional symbolic presuppositions, $(T\ U_T)=(\overline{F}\ N)\odot(\overline{e}\ m)$, where $(T\ U_T)$ denotes the scalar dyad that measures the work done by a quantity of force indicated with the dyad vector $(\overline{F}\ N)$ when its point of application is displaced as determined by the vector dyad $(\overline{e}\ m)$, here m symbolizing the standard length unit called meter, not the mass magnitude. The corresponding multiplication is «$\odot$», that is, the scalar dyadic of article 12. The definition of this composition law allows us to write the work $(T \ U_T) = (\overline{F \cdot e}) \ (N * m)$. The multiplication of the mathematical point «·» is the scalar product of vectors in $\mathbb{R}^3$, while the asterisk «*» symbolizes the scalar dyadic product. As in the previous cases, the equality criterion in $\mathscr{D}$ determines the formulation of the compound unit U_T and the resulting primary:

$$\text{If } U_T = N * m \text{ in } \mathscr{D}, \text{ then, } T = \overline{F \cdot e}$$

The compound unit $U_T = N * m$ in Physics is called the joule, and it means the work or energy produced by a force of one newton when its point of application is displaced by an amount of length equal to one meter.

The last significant example of a physical equation to be analyzed is that of the **moment magnitude of a force.** In dyadic algebra $\mathscr{D}$ the moment $(\overline{\mu} \ U_\mu)$ of a force with respect to a point must be defined as the vector dyadic product between the vector concrete that indicates the position vector $(\overline{r} \ m)$ of the point of application of the force, with respect to that other to which refers to the moment, where m is the standard meter, and the vector dyad that represents the force $(\overline{F} \ N)$. In $\mathscr{D}$ analytics it will be written with the explicit form $(\overline{\mu} \ U_\mu) = (\overline{r} \ m) \circledast (\overline{F} \ N)$, where «$\circledast$» is the vector dyadic multiplication of article 12. According to the definition of this composition law, it is permissible to put $(\overline{\mu} \ U_\mu) = [(\overline{r} \wedge \overline{F}) \ (m * N)]$. The multiplication «$\wedge$» is the vector product of vectors in $\mathbb{R}^3$, while the asterisk «*» symbolizes the scalar dyadic product. The equality criterion in $\mathscr{D}$ sets the compound unit U_μ and its dyadic pair:

$$\text{If } U_\mu = m * N \text{ in } \mathscr{D}, \text{ then, } \overline{\mu} = \overline{r} \wedge \overline{F} \text{ en } \mathbb{R}^3$$

Article 17

EFFECTS OF THE PRINCIPLE OF SYMBOLIC ECONOMY

Throughout the preceding compendium it has been observed how the geometric algebra of segments or lengths is generalized in the abstract, giving rise to the generic algebra of concretes or dyads, as mathematical representatives of the quantities of physical magnitudes. It has also been warned about the hypnotic effect that can be produced by availing itself of the symbolic economy, understood as the simplification of signs for the different operations of the same species, such as the additive ones, all denoted with the typical cross «+», the multiplicative ones indicated generically, for example, with the sign «×», subtraction with the dash «−», or divisions with the slash «/». To break this spell and warn for pedagogical purposes about how easy it is to be fascinated by it and believe that what really remains in the dark is understood, an effort has been made of symbolic detail, to make explicit the maximum number of distinguishable laws of composition each other, as well as the relationships that arise between them; although, given their large number, as the symbology is limited and it would not be useful to take such differentiation to the absolute extreme, it is inevitable and even convenient that some share common signs, which is not an obstacle for the phenomenon to be explained with sufficient didactic clarity. Take the expression in $\mathscr{D}$ as an example:

$$[p\circ(\overline{a}\ U_1)]\circledast[q\circ(\overline{b}\ U_2)]=(p\times q)\circ[(\overline{a}\ U_1)\circledast(\overline{b}\ U_2)]$$

The principle of symbolic economy allows to symbolize all the laws of composition of the same multiplicative species with the same character «×», and with this the traditional notation results:

$$(p\times\overline{a}\ U_1)\times(q\times\overline{b}\ U_2)=(p\times q)\times[(\overline{a}\ U_1)\times(\overline{b}\ U_2)]$$

Observing this last expression, unless one has algebraic expertise, it is difficult to escape the illusion caused by the constant sign of the sign «×» and it is easy to believe that the property that describes equality is evident by the laws of R^3. However, this is not the case, because what are related are

253

physical dyads, and the complete meaning of equality is given by the different composition laws that comprise the equation itself and specifically defined between the sets R, $\{R^3, U_1\}$, $\{R^3, U_2\}$ y $\{R^3, U_1 * U_2\}$, so the formula must be interpreted based on them.

For a better overview of the **spell of symbolic reduction, so toxic for the learning of Physics, the scientific accuracy and the precise meaning of the compound magnitudes**, the symbols of the operations that intervene in the dyadic algebra can be detailed, represented with the sign $\mathscr{D}$, unlike the traditional structures of R, C and R^3 or any other. In this way the following synoptic scheme results:

Type of dyadic composition law / Article of section XXVIII		Ordinary number algebra (see note)		Dyadic algebra or physical In $\mathscr{D}$	With the principle of symbolic economy
		In R y C	In R^3		
Scalar magnitudes and vector	Addition (4)	$+$	$+$	$\oplus$	$+$
	Subtraction (5)	$-$	$-$	$\ominus$	$-$
	Multiplication by a number (6)	$\times$	$\bullet$	$\circ$	$\times$
	Homogeneous division (7)	$/ \div$		$/\!/ \doteqdot$	$/ \div$
Scalar magnitudes	Heterogeneous multiplication (9)	$\times$		$*$	$\times$
	Heterogeneous division (10)			$/\!/ \doteqdot$	$/ \div$
Vector magnitudes	Product of mixed magnitudes (11)			$\circledcirc$	$\times$
	Scalar product (12)		$\cdot$	$\odot$	$\cdot$
	Vector product (12)		$\wedge$	$\circledast$	$\times$

(Note) The symbols of the operations in R, C and R^3 obviously refer to the addition, subtraction, multiplication and division of these algebraic structures, not to the dyadic or concrete ones that are defined in the articles of the first column.

Article 18

DISCUSSION AND CONCLUSIONS

Physics teachers take for granted that unit abbreviations operate with the same algebra of abstract numbers, and on this tacit assumption, without justifying it in any way, they teach their classes and completely omit all specific algebra for magnitudes, ignoring , as if they did not exist, the philosophical problems related to the magnitudes and their composition laws, teaching the physical operations in an intuitive, subjective and arbitrary way, sowing in the students, even without knowing it, seeds of ignorance and confusion that vitiate all knowledge acquired with this lagoon pending clarification. Thus the teaching quality is debased, because the key to a thorough understanding is not to advance at all without first having precisely defined all of the foregoing, and even more so, if possible, in the case of something so fundamental to understand and develop natural laws such as magnitudes, your measurements and your operations.

And even more serious is the stubborn negligence of the International System of Units and of all the great scientists who propitiate and tolerate the already repeated false hypothesis that physical quantities behave with the abelian multiplicative group structure, hypothesis that we have put in evidence with the unappealable configuration of multiplicative operations as external composition laws, and therefore lacking in unitary and inverse elements. All this makes up a poisoned panorama that traps the foundations of Physics and prevents the development of coherent and precise models.

Let's look at the following experiment: imagine a mass of one kilogram arranged materially by means of a weight, a length of one meter measured with a ruler, and a quantity of time of one second marked by a clock; it is essential to have prepared teaching answers to explain the multiplication of a kilogram by a second, for example, or even to solve if it is possible to divide a kilogram by a second or a meter by a kilogram, or if such divisions can only

255

be conceived in certain cases or if, on the contrary, they will always be possible.

And this is necessary because, if these omnipresent operations are not rigorously justified, then what foundation and meaning could be attributed to compound maagnitudes and units, scientific laws and physical equations? Is not all knowledge empty? imparted without having saved this gap? Physical algebra responds to these kinds of questions, which gives full meaning to the various compound magnitudes and formulations of natural laws, saving that pernicious omission of the essential **generating external composition laws**. We will not get tired of repeating it, because the need for these algebraic elements is very evident.

Hence, it is **convenient for the quality of education** to solve this **capital pedagogical defect**, whose permanence denatures scientific language and its real meaning, insidiously depriving students of their rights not to be intoxicated with **latent presuppositions** and to receive complete information on any subject of curricular interest, since it is a core content of the sciences, which stimulates creative and free talent, saving the prevailing ignorance and confusion.

It is not an accessory or superfluous subject, physical algebra is essential, it is an necessary principle. Its omission disqualifies and precarious all the scientific knowledge built without that nuclear pillar. Science should not allow such a serious incongruity, as Mathematics would not accept that arithmetic be thrown away, because, just as the laws of numerical composition solidly base all mathematical structures, from the most basic to the most abstract, the algebra of magnitudes is the origin of all physical formulations, without exception. So they lose all their meaning when their composition laws are disregarded.

So why has the physical tradition indulged in this insidious and elemental **epistemological heresy**? It's a mystery. But the unequivocal thing is that, once the incongruity and its resolution have been discovered, it must be saved, first, in the interests of the logical coherence of scientific theories, and second, to chart the course of new investigations oriented to non-Euclidean algebras, who knows in this case? first moment of dyadic innovation to what new domains, models or discoveries it can lead.

Section XXIX

THE BLACK LAGOON OF MATH
ORIGIN OF THE «ARITHMETIZATION» OF PHYSICS

Here we reveal the vice of «arithmetization» of modern mathematics, which has been forgotten by supine ignorance of the geometric algebra handed down by the Greeks and analytically updated in the present work, giving it the form of a dyadic algebraic structure. This alarming and intolerable vice has intoxicated and severed Physics with the same gap for the rest of the measurable and and those affine to length. Obviously, it is not sustainable to deprive Physics of such a fundamental tool as its own algebra, as all awake and honest intelligences will easily appreciate, examining the development of the following articles:

EPITOME OF THE BLACK LAGOON

In this section, in which the reader is supposed to have sufficient knowledge of mathematical foundations, a transcendental gap in modern mathematics is exposed, which debases metric operations from its base. The contamination process is as follows: first, the theory of geometric equality and addition of segments leads to prove the theorem of the mean parallel, and this rationally justifies the proportionality of segments of Thales' Theorem. And this done, the blur is produced, because Mathematics correctly accredits that the proportionality of segments implies the numerical proportionality of their measurements in a determined arbitrary unit of length; but then he concludes without more than that, being such proportions of a numerical nature, as well as in arithmetic the product of the extremes of all proportions must be equal to that of the means, in the same way, the proportionality of segments should meet the same condition , «because if». And this is how one falls into the error of confusing arithmetic multiplication with segment multiplication, which are very different operations, and arises like a charm the undesirable «arithmetization» of the most fundamental magnitude of all: the length. The first is a law of internal composition, since every product of numbers is another number; but not so for the geometric multiplication of segments that, as we will not tire of repeating, is a non-arithmetic operation, it is geometric, and such that it generates new magnitudes, the area with two factors or the volume with three. None of these geometric factors, which are quantities of length, can assume the multiplier function on which arithmetic multiplication is based. Which means that this is a law of external composition totally different from the ordinary product. Following this erroneous scheme, modern mathematics ignores and forgets the indispensable geometric multiplication of segments inherited from the classical Greeks, the basis of the later analytical formulation of the product of lengths and of the entire mathematical metric, which in turn is the basis of multiplicative operations with physical quantities, as we have detailed in detail in what precedes. Here we epistemically expose this substantial error, which debases current Mathematics and spreads insidiously to Physics, recovering the lost values of classical geometry.

Article 20

EQUALITY AND ADDITION OF SEGMENTS

Segment equality is defined in geometry with the condition that the compared segments can be made to coincide by means of a movement, which can be represented by a transformation of the space itself[20].

In turn, the addition of segments is conceived geometrically as the graphical operation that provides the segment sum by

Display of the sum of segments

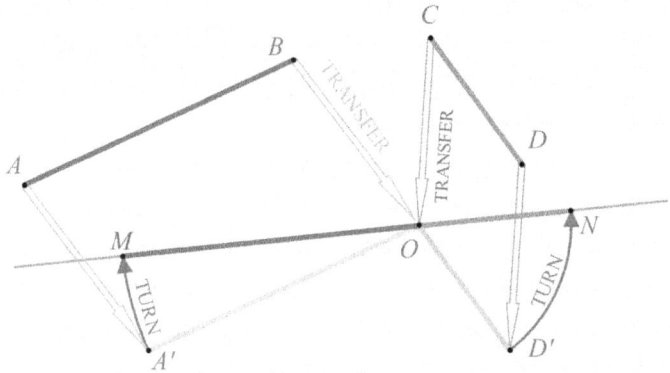

Given two segments *AB* and *CD*, representatives of the classes of the congruent with them, and given any line and a point *O* of it, the sum of the given segments is defined as the segment *MN* that results from the juxtaposition of those on the given line and starting from *O*. Each segment is transported on the line as follows: a translation that carries *B* over *O* and *A* over *A'*, a turn that carries *A'* over *M*, a translation that carries *C* over *O* and *D* over *D'*, and a spin that takes *D'* over *N*. So we say that *MN* is the sum of *AB* and *CD*, we write it *MN=AB⊕CD*. The representative chosen from each class to compose the sum is indifferent, because the segments of the same class are equal. Segments *AB* and *MO* belong to the same class, because they are congruent. Like the *CD* and *ON* segments. Juxtaposition consists, then, in arranging on the line considered two congruent segments with the given addends, one after the other and on a different side of a point *O*, so that the union segment of these two is called the sum.

Figure 16

---

[20]   In Mathematize 1, «Lesson 16, Movements and congruence or geometric equality», this topic can be consulted at length.

juxtaposing the addends[21], as described in figure 16. Symbolizing $\{S\}$ the set of all segments, the analytic form this operation can be symbolized with a specific sign, for example «$\oplus$»; although the usual thing in Mathematics is to apply the principle of symbolic economy and represent all additive operations with the same cross «+». However, the expert mind must clearly distinguish that, depending on the elements that are composed, the corresponding law of composition is not arithmetic, but its own, which must have been expressly defined previously. Under these conditions, keeping the symbolic difference, to avoid confusion, the addition of segments would be analytically defined in the following way: given two segments S_1 and S_2 of the set $\{S\}$, the addition is an internal composition law that applies the Cartesian product $\{S\} \times \{S\}$ in $\{S\}$ and such that the segment sum S of $\{S\}$ is obtained by graphical juxtaposition of the addends, the function indicated by the equation $S_1 \oplus S_2 = S$ being analytically described.

---

[21]   In Mathematize 1, «Lesson 22, Metric Geometry, Sum of segments and angles», the composition law of the geometric addition of segments is developed.

Article 21

THE PROPORTIONALITY OF SEGMENTS
BASES THE MATHEMATICAL METRIC

Once the geometric equality and addition of segments are defined, these being understood as figures formed by certain sets of points, following the dictates of elementary geometry, the mean parallel theorem, described in figure 17, makes it possible to divide a segment into equal parts, as indicated in figure 18, and this operation allows us to conclude Thales's famous Theorem on the proportionality of segments formed on certain lines sectioned by others parallel to each other, in accordance with what is determined by geometry and shown in figure 19^{22}.

Segment proportionality is the daughter of addition, because it arises from a sum in which an addend is repeated. Thus, by reiterating the sum of the same segment S a certain number of times λ, the resulting segment can be indicated analytically with the form $\lambda \circ S$, where the sign «∘» indicates the multiplication of the real number λ by the segment S, making the numerical factor times multiplier and segment multiply. If λ were integer, the interpretation of the product $\lambda \circ S$ does not offer difficulty. If λ were rational, such as $\lambda = a/b$, with a and b integers, the product $(a/b) \circ S$ must indicate the operation of dividing the segment S into b equal segments, taking one of them and adding it a itself sometimes. In this way, the product $\lambda \circ S$ can always be found for any real λ. Note that the operation «∘» is not the arithmetic «×», because it only composes numbers, whose product is another number, and the other one composes numbers and segments to result in a new segment. So arithmetic multiplication is an internal composition law, while the operation $\lambda \circ S$ is an external law.

Let P be the segment resulting from the product $\lambda \circ S$, which will be written with the traditional equal sign with the form

---

22 In Mathematize 1, «Lesson 26, Thales' Theorem», the complete reasoning that concludes the proportionality of segments is developed.

$\lambda \circ S = P$. Nothing prevents observing this expression and considering that P is a dividend, that S is a divisor and that λ is a quotient, with which the product could be written with another equivalent symbolic form such as $P /\!/ S = \lambda$, and thus it would be defined the division of segments «$/\!/$», whose quotient will always be a real number λ. This division has been indicated with the double bar «$/\!/$» to distinguish it with regard to arithmetic «$/$», because this confusion is the cause of the mathematical blur that is being described in this investigation.

Therefore, the ratio of two segments always results in a real number, so that if two ratios are equal, what could be called a proportion is formed with them, but it is not a numerical proportion, but of segments . So the proportionality of Thales, although it is conventionally represented by the arithmetic sign of the dividing bar «$/$», in reality it is necessary to appreciate that it is the geometric operation «$/\!/$», derived from the product of a scalar by a segment. Well, this very elementary confusion is the poison that poisons Mathematics in a tragic way, as will be seen immediately.

The differences between addition and scalar multiplication are also evident by the different algebraic structures that they generate. The addition of segments, which is a mapping of the Cartesian product $\{S\} \times \{S\}$ in $\{S\}$, is a law of internal composition and it can be easily verified that it satisfies the properties necessary to endow the set $\{S\}$ with the abelian group structure. Instead, multiplication by a scalar is an external law, indicated by an application of the Cartesian product $R \times \{S\}$ en $\{S\}$, where R is the field of real numbers. It can be easily verified that this operation is such that it endows the set $\{S\}$ with the algebraic structure of a vector space over R. However, for this, it is necessary to define the concept of an **opposite or negative segment**, and nothing prevents establishing it as that which it is added by juxtaposition in the opposite sense to that considered positive, since in the line there are two directions to carry out an addition; if the first addend is greater than the second, the sum will result in a positive segment; if the first addend is less than the

266

second, the sum will be in a negative segment; and all this by definition of this law of internal composition. This is equivalent to recognizing in the segment the attribute of **sense or linear order**, so that the equality of segments requires not only congruence but also the same order between their points or, in other words, that the segments have the same sense, although they may differ in direction, an attribute that is characteristic of vectors.

Similarly, multiplication by a negative scalar must consider that the sign of the product is opposite to that of the segment that appears as multiplying. This conceptualization is the origin that justifies the Cartesian reference systems that give way to analytical geometry.

Middle parallel of a trapezoid and triangle

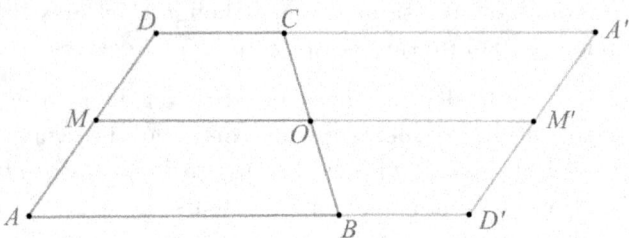

Let be the trapezoid *ABCD*, let *MO* be the segment joining the midpoints of the non-parallel sides *BC* and *AD*; the center *O* symmetric figure of the trapezoid *ABCD* is *A'CBD'*; the central symmetry is such that homologous lines are parallel, so that the segments *DC* and *BD'* are parallel, and the *AB* and *CA'* also, as well as the segments *AD* and *A'D'*; the indicated central symmetries give rise to the parallelogram *AD'A'D* with center of symmetry *O*; The homologous points *A'*, *D'* and *M'* will result and, being by hypothesis *AM=MD*, we have that the central symmetry establishes that their homologues *A'M'=M'D'* are equal; thus it turns out to be *MM'* the mean parallel of the parallelogram, so this mean parallel must be parallel to the sides *DA'* and *AD'*; reaching the conclusion that the segment *MO* that joins the midpoints of the non-parallel sides of the trapezoid *ABCD* must be parallel to *AB* and *DC*; it must also be *MO=OM'*, due to the central symmetry, and resulting in *MM'=AB⊕BD'* or *MM'=DC⊕CA'* indistinctly, since *DA'=MM'=AD'*, and the segments *AB* with *CA'* being homologous and *DC* with *BD'*, *MM'=AB⊕DC* and *MM'=2∘MO*; so, finally we have that, *MO=(AB⊕DC)/2*; and thus we can state that **the segment that joins the midpoints of the non-parallel sides of a trapezoid is the average parallel to its parallel sides, called bases, and measures half their sum.**

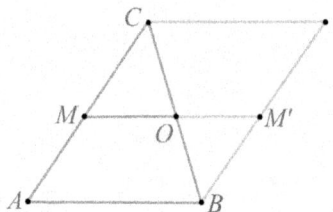

Let us be the triangle *ABC*, let us form the segment *MO* that joins the midpoints of the sides *BC* and *AC*, we are facing a particular case of the previous one, so we could extend to it the deduced property for the trapezoid without further ado. The center *O* symmetry transforms triangle *ABC* into triangle *A'CB*.

The homologous lines of the central symmetry are parallel, therefore, the segments *AB* with *CA'* and *AC* with *A'B* are parallel; it thus turns out that the figure *ABA'C* is a parallelogram, with all its properties; as the segments *A'M'* and *M'B* are equal, because they are symmetric homologous of the equal segments *CM* and *MA*, it turns out that the segment *MM'* joins the midpoints of the sides *AC* and *BA'* of the parallelogram *ABA'C*, then, *MM'* must be parallel to *AB* and *CA'* and equal to them; since *MM'=2∘MO*, it follows that *MO=AB//2*; with which the following statement is true: **the segment that joins the midpoints of two sides of a triangle is parallel to the third side and equal to half of it.**

Figure 17

268

Division of a segment
in equal parts

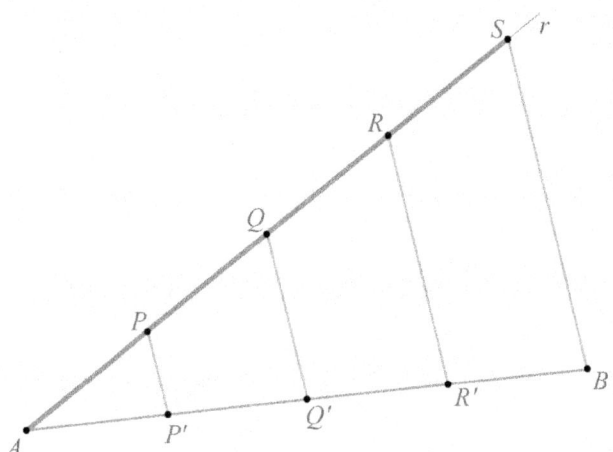

Given a segment *AB*, we try to divide it, for example, into four equal parts. To do this, we will repeatedly use the theorem of the mean parallel in the various triangles that can be conceived. First, we draw any ray *Ar*, which does not coincide with *AB*; second, we take any segment and carry it from *A* and over *r* four times, one after the other, thus we obtain the points *P*, *Q*, *R* and *S*; third, we join the last of these points, which in this case is *S*, with *B*; fourth, we draw by *P*, *Q* and *R* parallel to *SB*, and we obtain the points *P'*, *Q'* and *R'*; It thus turns out that in triangle *AQQ'* the segment *PP'* is mean parallel, with *P* being the midpoint of *AQ*, because that is how we have constructed it, then *P'* is the midpoint of *AQ'*, and it turns out that *AP'=P'Q'*; we observe something similar with the segment *QQ'*, which is the mean parallel of the trapezoid *P'Q'RP*, since *Q* is the midpoint of the side *PR*, because we have established this by carrying equal segments on *r*, then, it also turns out that *P'Q'=Q'R'*; and, finally, in an analogous way it is also obtained that it is also *Q'R'=R'B*, by identical reasoning scheme with the trapezoid *Q'BSQ*; concluding that segment *AB* has thus been divided into four equal segments, *AP'*, *P'Q'*, *Q'R'* and *R'B*. The determination of the segments on the line *AB* by drawing parallels from the line *r* is called the parallel projection of the segments of one line onto the other in the direction of the drawn parallels. This allows us to affirm the following statement: **given two lines, the parallel projection of equal segments of one line determines equal segments on the other.** And from here to Thales' Theorem there is almost nothing left.

Figue 18

Thales Theorem

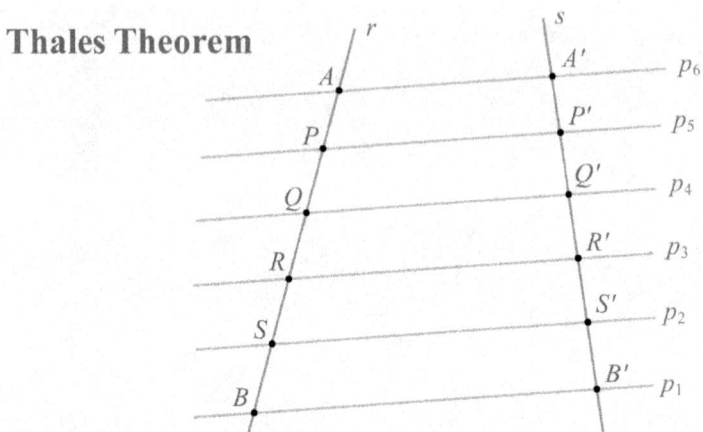

Let the lines r and s be the segment AB of the line r; it is possible to divide it into equal parts, in the figure it has been divided into five; suppose that the points that result from dividing the segment AB in equal parts are the P, Q, R and S; let the points A', B', P', Q', R' and S' be projected in parallel from their homologues A, B, P, Q, R and S, by means of the given system of parallels p_1, p_2, p_3, p_4, p_5 y p_6; under these conditions, segment AB is the sum of segments AP, PQ, QR, RS and SB, by definition of the sum of segments; as all these addend segments are equal by hypothesis, by definition of the product of a segment by a number, we have that sum can be expressed as the product $AB=5\circ AP$, with AP being the segment that we take as representative of all equal addends to express this operation, which would be identical if any other of the equal addends were taken; any other segment of the same line r, such as PS, for example, is the sum of the segments PQ, QR and RS, equal to each other and also equal to AP, then, we can write $PS=3\circ AP$. The quotient of segments is defined with the same appearance as the quotient of numbers, although with its specific meaning, in this case, the quotient between segments $PS//AB$, with a double bar to differentiate it from the arithmetic quotient, is the rational number 3/5, because, by definition of segment division, $PS=3/5\circ AB$. So $3/5\circ AB$ means dividing AB into 5 equal segments and multiplying the resulting segment by 3, or adding it 3 times. Since segment AB divided by 5 is precisely AP, it turns out that $3/5\circ AB=3\circ AP$, and $3\circ AP$ turns out to be PS, then, the expression $PS//AB=3/5$ is correct; that is, the PS and AB segments are in ratio 3/5. Let us now see in what ratio are the segments projected parallel on s by the corresponding parallels. The projected segments are $P'S'$ and $A'B'$; the segments $A'P'$, $P'Q'$, $Q'R'$, $R'S'$ and $S'B$ are all the same, so we can take one of them to represent them all in the sums that we are going to establish, let $A'P'$ this representative; we can write that $P'S'=3\circ A'P'$ y $A'B'=5\circ A'P'$; so the segments projected in parallel from r to s are in the ratio $P'S'//A'B'=3/5$, the same rational number as the ratio $PS//AB$. It turns out, then, that these two ratios form a proportion and we can safely write that $PS//AB=P'S'//A'B'$. As this same result would be had whatever the segments were taken on the line r and their projected in s, we arrive at the following statement, which is the well-known Thales Theorem: **if two lines are cut by other lines parallel to each other, the segments that they determine on one line are proportional to the segments they project onto the other line**. Analytically, multiple proportions will result like these: $AP//AQ=A'P'//A'Q'$, $PR//AQ=P'R'//A'Q'$, $QB//PQ=Q'B'//P'Q'$, etc.

Figure 19

Article 22

THE BLOT OF CLASSIC TEXTS: MATH HAS
BYPASSING SEGMENT MULTIPLICATIONS

Let us see the common plot scheme of geometry texts with the case of the prestigious Course in Metric Geometry, Volume I, by Professor Pedro Puig Adam. In «Lesson 22» (p. 129) on the Pythagorean Theorem he formulates an enigmatic warning: «Preliminary warning about the **product of segments**».

Once the proportion of segments is established, he declares **without demonstrating** that «any proportion between segments can be interpreted as a proportion between their measures», and continues:

> In this way we will interpret the segmental proportions in this lesson and in the following ones, as soon as **we equalize the products of means and extremes in them**. Thus, from now on, wherever the reader sees a product of segments written or enunciated $\overline{AB} \cdot \overline{AC}$ must understand as such the number product of the measures of AB and AC with the same unit.

In this way, the proportionality of segments, which is a singular geometric operation, is confused in the exemplary texts in a disturbing way with the product of segments that generates new magnitudes, which is a **very different composition law, since it relates geometric figures to each other. with different magnitudes: lengths, areas and volumes**.

The algebraic error of the previous approach is obvious: the proportionality of segments only implies that the reasons that make up the relationship are such that they are equivalent to the same real number, according to the homogeneous dyadic division of section XI; but to infer from this that the product of the means is equal to that of the extremes is to fall into an inadmissible assumption, because the product of segments, as elementary geometric figures, cannot be assimilated «just for the sake of it» to the numerical product, but must be expressly defined separately, since the segments are not numbers in themselves, they are geometric figures that include non-numerical quantities of a fundamental magnitude: length.

Puig Adam is not the only one who fell into this trap. None of us have overcome the temptation to «arithmetize» everything without thinking about what we were doing. Even the most illustrious mathematicians like David Hilbert, who sought to give analytical form to the multiplication of segments, was not able to free himself from the spell of arithmetic. His text *Fundamentals of Geometry*, published in 1899, is considered his most important contribution to modern mathematics, incorporating the formal axiomatic method.

In this investigation, Hilbert proposes the product of segments by means of an «arithmetized» multiplication based on the following geometric figure: let us take two secant lines at point O, on one of them we carry the segment $\overline{OA}$ and on the other we carry the segments $\overline{OB}$ and $\overline{OU}$, the latter taken as unit; let us draw through B the parallel to AU, which will intersect the line OA at the point P. Hilber defines the product of the segments $\overline{OA}$ and $\overline{OB}$ as the segment $\overline{OP}$. It is evident that this construction incurs the same error as Puig Adam, since it identifies the segments with their measurements and with this the geometric multiplication of segments remains undefined, which we already know is strictly a law of external generating composition that produces a new geometric magnitude called area, so the product of segments can never give rise to another segment.

This is how Puig Adam and David Hilbert and with them all of us confuse the additive operations of sections V to XI without distinguishing them from the so-called generating multiplicative operations of sections XII to XVII, forgetting the latter, which are essential to compose geometric and physical magnitudes.

In short, given a proportion of segments $S_1 /\!/ S_2 = S_3 /\!/ S_4$, it is not correct to infer that the product of the extremes is equal to the product of the means $S_1 \times S_4 = S_2 \times S_3$, as in the proportions arithmetic, because the products $S_1 \times S_4$ and $S_2 \times S_3$ lack meaning for the segments, if the generating multiplicative laws have not previously been defined to compose magnitudes.

272

Article 23

SAVING THE LAGOON WITH THE OMITTED
GEOMETRIC MULTIPLICATION

It has been previously concluded that the proportions of segments do not allow at all to establish the multiplication of these elementary figures, which represent quantities of lengths, so the assumption that in such proportions the product of the means is equal to that of the extremes, does not it only has no foundation, but must be excluded a priori, without first having defined the multiplicative composition laws that compose segments with mathematical rigor. To do this, without forgetting that what is being handled are geometric figures, the multiplication of two segments can be conceived as the geometric operation consisting of forming with them another rectangular figure whose dimensions are precisely the multiplied segments. By definition of this composition law, the result of such multiplication or product will be precisely that rectangle, which will host a certain quantity of a new magnitude composed or derived from length, called **area or surface**, as graphically described in figure 20.

In turn, if instead of two segments three are multiplied, the result or product can be identified, by definition, with a geometric figure called a straight parallelepiped, whose edges are precisely the multiplied segments, giving rise to a body that accommodates a certain quantity of a new magnitude, composed with the length or derived from it, which is called **volume**, a composition law that is also described in the same figure 20. To represent these operations, so different from the multiplication of numbers, the mathematical asterisk can be used « * », Although the principle of symbolic economy, which uses the same sign « × » to represent all multiplicative laws, tends to indicate them all with this same spelling; but this should not prevent the intellects from distinguishing them and knowing how to differentiate them, because they are independent operations.

273

**Graphic definition of
segment multiplication**

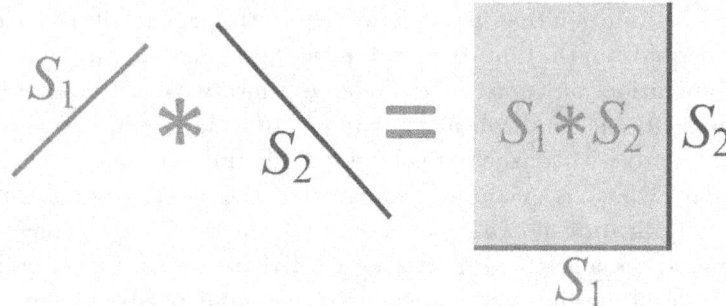

The graphic or geometric product of two segments is not another segment, but, by definition, it is a new geometric figure called a rectangle, giving rise to a compound or derived quantity of length called **surface or area**.

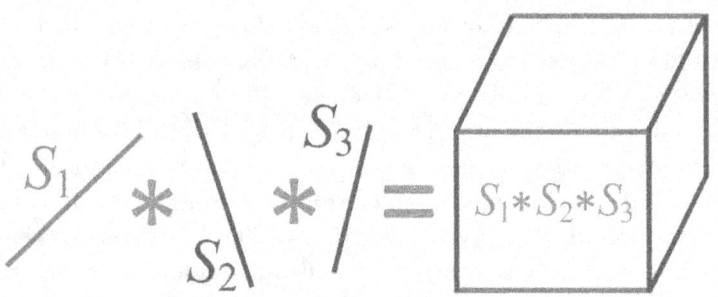

The graphic or geometric product of three segments is not another segment, but, by definition, is a new geometric figure called a right parallelepiped, giving rise to a compound or derived quantity of length called **volume**.

Figure 20

An important observation about these notions of segment multiplication is that, considering that the products do not refer to lengths, but rather indicate areas or volumes, that is, different magnitudes of length, **they are composition laws of an external nature, not internal,** which reveals a notable conceptual difference with numerical operations. This is clear when specifying the applications that define them: with two factors it is an application of $\{S\} \times \{S\}$ in $\{S*S\}$, where simply $\{S*S\}$ indicates the set of all possible rectangles formed with segments, which can also be denoted by the exponential form $\{S^2\}$; in turn, with three factors the geometric product is an application of the Cartesian product $\{S\} \times \{S\} \times \{S\}$ in $\{S*S*S\}$, with this symbol representing all the possible right parallelepipeds formed with segments, in exponential notation $\{S^3\}$.

Article 24

THE TRANSCENDENTAL EXPERIMENT
WITH AREAS AND VOLUMES
(Article 8, figures 10 and 11)

Once the mathematical gap of geometric multiplication has been saved, it is observed that this operation, given its graphic nature, does not by itself provide information regarding the quantities of lengths of the segments or regarding the quantities of surface or volume. Therefore, it is necessary to undertake **measurement** procedures that allow quantifying and operating analytically with such magnitudes.

To establish a criterion for measuring the lengths of the segments, any one of them is used that is taken as a unit, which will be such that it will comprise a quantity of uncountable length, indeterminacy that will be saved by assigning it a symbol that represents it, for example , U_L. With this, the measurement of any segment S can be indicated with the pair or dyad (λ, U_L), where λ is a real number that indicates the times that the unit of length U_L is included in S or, expressed with the symbology that has already been developed, the meaning of (λ, U_L) is to verify the symbolic equation $S = \lambda \circ U_L$, where the operation «$\circ$» indicates the multiplication of a scalar by a segment, defined in article 6. So we can admit the equality of meaning between the dyad (λ, U_L), which represents the amount of length of the segment S, with the multiplication $\lambda \circ U_L$, or what is equivalent, with the sense of equivalence of meanings we can write the equality $(\lambda, U_L) = \lambda \circ U_L$.

Under these conditions, the geometric experiment with the areas is described in figure 10 of article 8 and the one corresponding to the volumes in figure 11. The result of them is evidence that, by analytically describing the length quantities of the segments that are multiplied by dyads referring to certain units of length, it is found that with two factors, the measure of the area that integrates the product of two segments is equal to a dyad in which the first of its elements is the real number obtained by multiplying the numerical elements of the factor dyads, and

277

the second of the product dyad elements is the amount of area, not numerically expressible, of the unit rectangle that results from geometrically multiplying the units of length in which the factors are expressed. In terms of mathematical analytics, it would have been, if the multiplied segments S_1 and S_2 had been expressed in terms of the respective units of length U_{L1} and U_{L2} with the respective dyads (a, U_{L1}) and (b, U_{L2}), where a and b are the numbers real values indicated by the measurements of the segments S_1 and S_2 using the units U_{L1} and U_{L2}, it will be found that the amount of area generated by the geometric product of those segments would be given by the dyad $[(a\times b),(U_{L1}*U_{L2})]$, whose meaning is that said product area is $a\times b$ times the amount of surface area of the unit rectangle generated by the unit segments U_{L1} and U_{L2}, that is, the rectangle symbolized by $U_{L1}*U_{L2}$.

With three multiplied segments, the procedure is totally analogous, although in this case the magnitude generated is a volume instead of a surface, a volume that will be measured with the compound unit with the geometric product of the length units of the factors $U_{L1}*U_{L2}*U_{L3}$. And from all this we can conclude the analytical forms of these two geometric multiplications, which remain in this way for two and three factors:

$$(a, U_{L1})*(b, U_{L2}) = [(a\times b),(U_{L1}*U_{L2})]$$

$$(a, U_{L1})*(b, U_{L2})*(c, U_{L3}) = [(a\times b\times c),(U_{L1}*U_{L2}*U_{L3})]$$

At this point, nothing prevents generalizing the multiplication of segments to the case of n factors, which analytically does not offer the least difficulty, since it is enough to establish the following equality:

$$(a_1, U_{L1})*(a_2, U_{L2})* \ ... \ *(a_n, U_{Ln}) =$$

$$= [(a_1\times a_2\times \ ... \ \times a_n),(U_{L1}*U_{L2}* \ ... \ *U_{ln})]$$

RELATIONSHIP BETWEEN PROPORTIONALITY OF
SEGMENTS AND GEOMETRIC MULTIPLICATION

It has been observed that the proportionality of segments arises from multiplication by a scalar. This operation has been symbolized with the notation $\lambda \circ S = P$, and it has been established that for this reason it is said that the segments S and P are in the ratio $P/\!/S$ of the scalar λ. Thus, when two ratios correspond to the same scalar λ, it will be said that they form a proportion, and $S_1/\!/S_2 = S_3/\!/S_4 = \lambda$ will be written.

Expressing the four segments of a proportion in the same unit of length U_L, being a_1, a_2, a_3 and a_4 the respective measurements of the segments with said unit U_L, we have the identities $S_1 = a_1 \circ U_L$, $S_2 = a_2 \circ U_L$, $S_3 = a_3 \circ U_L$ and $S_4 = a_4 \circ U_L$. In turn, the initial proportionality determines that $a_1 \circ U_L = \lambda \circ (a_2 \circ U_L)$. The meaning of the operation «$\circ$» is that of an abbreviated sum of segments, so we can write $\lambda \circ (a_2 \circ U_L) = (\lambda \times a_2) \circ U_L$, because the measure of the segment $\lambda \circ (a_2 \circ U_L)$ with the unit U_L must agree axiomatically with the arithmetic product $\lambda \times a_2$. All of this leads us to the equality $a_1 \circ U_L = (\lambda \times a_2) \circ U_L$. And, the units of length U_L of both members being equal, it can only be concluded that $a_1 = \lambda \times a_2$, based on the equality of segments. So it can be stated that, if two segments S_1 and S_2 are in the geometric ratio λ, their measures a_1 and a_2 in the same unit of length U_L are in the same arithmetic ratio λ.

In this way, the first happy assumption that makes up the blur of the model texts, represented by Puig Adam and David Hilbert, is evidenced, and it is proven that every geometric ratio of segments measured with the same unit can be said to generate an equal arithmetic ratio between their measurements. And in the same way, if two segment ratios form a geometric proportion, it is certain that their measurements with the same unit of length form an arithmetic proportion.

Next, given a geometric proportion of segments $S_1/\!/S_2 = S_3/\!/S_4 = \lambda$, let's analyze what happens to the geometric product of the extremes $S_1 * S_4$ and that of the means $S_2 * S_3$. Geometric multiplication allows writing $S_1 * S_4 = (a_1 \times a_4) \circ U_L$ and $S_2 * S_3 = (a_2 \times a_3) \circ U_L$. Considering that, by hypothesis, the given segments form a geometric proportion, as has just been proven, their measurements with the same unit will form an arithmetic proportion, and it is verified that $a_1 \times a_4 = a_2 \times a_3$, and from here we have that $S_1 * S_4 = S_2 * S_3$. Then, given the geometric proportion $S_1 /\!/ S_2 = S_3 /\!/ S_4$, it is verified that the geometric product of the ends is equal to that of the means, without forgetting that these two products are not ordinary, but rather represent areas that turn out to be equal.

With this approach, the defects of the assumptions of Puig Adam and David Hilbert and of the classical texts emerge. Their error is evident when considering that the proportion between the measurements of the segments with the same unit of length can be inferred that of the segments, because in no case do they define geometric multiplication as they should and, in the absence of this, such assumption is an inadmissible excess. Hence, they must resort to the rather tricky device of substituting the segments for their measurements. However, with the argument of these articles it is proven that such an agreement, apart from being incorrect, is not necessary, because it is enough to fill in the gap in force until now with the operations that had been unduly omitted, that is, those inherent to geometric multiplication of segments, and with this the range of operations and properties is completed with full logical coherence, the current classic deficiency being resolved.

The very texts of Mathematics, led in this investigation by Puig Adam's *Course in Metric Geometry* and David Hilbert's *Fundamentals of Geometry*, make full proof that Mathematics has bypassed geometric multiplication, ignoring this law of generatrix composition and essential to operate properly with the quantities of a fundamental magnitude: the length.

From this point on, this negligent oversight has blurred all the mathematical developments that rely on metrics: vector spaces, which are based on the concept of the modulus of a vector;

280

Euclidean spaces, which are based on the interior connection or scalar product; the tensor spaces and, in short, as has been indicated, all those innumerable algebraic structures that use the notion of metric or distance between their points.

As Physics makes use of Mathematics almost blindly, hiding itself in that comfort, it has not noticed the malformation that affects its matrix science and, simply conforming to what Mathematics offered it, has left pending justification and development, not only operations with quantities of lengths, which are the most fundamental physical magnitude of all, but also those corresponding to any other magnitude. Due to this neglect, Physics drags from the beginning a fundamental pending subject, which is the epistemic algebra of magnitudes. This is how the mathematical virus has infected Physics and both sciences must heal and create antibodies, which is by no means impossible, because here we demonstrate how to do it. And, as long as the treatment for a disease is known, it does not seem doubtful that it should be prescribed.

With this it will be observed that the algebras of magnitudes will revolutionize Physics, because the current pending subject relegates the compound magnitudes to a merely symbolic plane, giving priority to the search for strictly arithmetic and rather childish numerical proportionalities. However, what Physics needs is that the magnitudes be the object of investigation in themselves, to discover their true nature and properties, and thus be able to adapt the relevant composition laws to reality for each physical experience, structuring them appropriately.

A first notable result of this research is the «dysmetric» observation, which leads Physics to the innovative and unexplored field of «dysmetric» spaces, which are called upon to improve physical models and which constitute a much more powerful tool than the current one tacit isometry to more accurately represent natural phenomena, as outlined in the second part of this work.

Article 26

THE NON-ARITHMETIC COMPOSITE RULE OF THREE
FOUNDATION OF PHYSICAL EQUATIONS

We are going to examine the basic problems known by the name «compound rule of three», which are taught and faulty solved by arithmetic proportions, resulting in misunderstanding errors due to the «arithmetization» error. However, we will observe that, applying the algebra of magnitudes, these cases are solved with great ease and full mathematical rigor, without leaving room for logical gaps, as the reader can verify for himself, solving the problems by the classical method, which is It is known, and comparing it with what is explained in this article. We will also see that these elementary cases reflect with singular clarity the formation of physical equations in general. To do this, we will operate on specific examples and apply the operations that we have called «homogeneous multiplication and division», sections X and XI, or articles 6 and 7 of section XXVIII, as well as «heterogeneous multiplication and division», sections XII and XVI, or articles 9 and 10 of the aforementioned section XXVIII. We will not enter into repetitive demonstrations, but we will refer to the properties already exposed in these sections.

Consider the following «direct compound rule of three» problem: 5 *bottles* of 2 *liters* each filled with a liquid weigh 15 *kilograms*. How much do 2 *bottles* of 3 *liters* each weigh? It is understood that the weight is that of the liquid without the container. The generating magnitudes are the *number of bottles* and the *volume*. The magnitude generated is the *weight*. So the problem equations must have the form *bottles* * *volume* = *weight*. The statement «5 *bottles* of 2 *liters* each filled with a liquid weigh 15 *kilograms*» can be written as a product of dyadic algebra as follows:

$$5 \; bottle \, * \, 2 \; liter = 15 \; kilogram$$

Similarly, the ordinary language statement «How much do 2 *bottles* of 3 *liters* each weigh?», can be written algebraically with the following dyadic equation:

$$2 \text{ bottle} * 3 \text{ liter} = x \text{ kilogram (generated unknown)}$$

Where x is the unknown measure of the weight of the 2 3-*liter* *bottles*. Thus, if these two equations are raised, operating with them, they become these two:

$$5 \times 2 \text{ bottle} * \text{liter} = 15 \text{ kilogram}$$

$$2 \times 3 \text{ bottle} * \text{liter} = x \text{ kilogram}$$

We can dyadically divide these two equations member by member and we get:

$$\frac{5 \times 2 \, botella * litro}{2 \times 3 \, botella * litro} = \frac{15 \, kilogramo}{x \, kilogramo}$$

Thus we have formed two equal dyadic ratios, so they form a non-arithmetic proportion. The two ratios are such that their numerators and denominators consist of dyads with the same secondary, so they are homogeneous, and their quotient must be equal to the arithmetic ratio of their primaries. Thus it turns out that the previous dyadic proportion justifies entering the following arithmetic proportion:

$$\frac{5 \times 2}{2 \times 3} = \frac{15}{x}$$

Solving, we have $x=9$, that is, 2 *bottles* of 3 *liters* each weigh 9 *kilograms*, which is the solution to the problem.

We observe that dyadic algebra allows us to relate units as different as the *bottle*, the *liter* and the *kilogram*. And this phenomenon is the one that occurs permanently in Physics when it relates fundamental units of length, mass or time, such as the *meter*, the *kilogram* and the *second*, or any others.

In such a way that in this example we could conceive with relative value limited to the assumption the dimensional formulas of section XXVIII, such as [WEIGHT]=BOTTLE*VOLUME. We will not escape the incongruity that it may seem for Physics to consider *weight*, associated with *mass*, a fundamental magnitude, as a unit generated by the generating product of *bottles* and *volume*. However, its algebraic validity is unquestionable.

The freedom in the choice of generating magnitudes is evident with a first reverse variant case: 5 *bottles* of 2 *liters* each filled with a liquid weigh 15 *kilograms*. How many liters are the *bottles* if 2 weigh 9 *kilograms*? The generating equations are here: 5 *bottlle* * 2 *liter* = 15 *kilogram* and 2 *botella* * x *litro* = 9 *kilogram*. The arithmetic ratio is $(5×2)/(2×x)=15/9$, so $x=3$. Each bottle will have a volume of 3 *liters*, like the first statement.

The second inverse variant would have the *number of bottles* as unknown. We would ask how many 3 *liter* bottles weigh 9 *kilograms*? The generating equations in this case are: 5 *bottle* * 2 *liter* = 15 *kilogram* and x *bottle* * 3 *liter* = 9 *kilogram*. It turns out $(5×2)/(x×3)=15/9$, so $x=2$. The solution is 2 *bottles*, logically, coinciding with the first statement.

Let's look at another example, in this case of the «inverse compound rule of three»: A passenger transport system is capable of moving 400 *passengers* a distance of 500 *kilometers* in 8 *hours*. How many *passengers* can be transported a distance of 300 *kilometers* in 12 *hours*?

Simplifying the notation by using the abbreviations of units v for *passengers*, *km* for *kilometers* and *h* for *hours*, the two previous statements can be translated into the mathematical language of the algebra of magnitudes by means of the following expressions:

$$400 \, v * 500 \, km = 8 \, h$$

$$x \, v * 300 \, km = 12 \, h \text{ (generating unknown)}$$

Operating with the dyadic product, these two equations are transformed into these others:

$$400×500 \, v*km = 8 \, h$$

$$x×300 \, v*km = 12 \, h$$

285

Both expressions have in their primaries the same compound unit $v*km$ in the first member and the time h in the second. Therefore, dividing them member by member, two equal dyadic ratios result, that is, a non-arithmetic dyadic proportion, which is reduced to the arithmetic proportion of their primaries, that is:

$$\frac{400 \times 500\,v*km}{x \times 300\,v*km} = \frac{8h}{12h} \Rightarrow \frac{400 \times 500}{x \times 300} = \frac{8}{12} \Rightarrow x = 1.000$$

Therefore, the answer to the problem is that the number of transportable *passengers* is 1,000. In this case we also find the same observation about the relative validity of the dimensional equation that operates in the problem with the related magnitudes, whose expression would be:

$$[\text{TIME}] = \text{PASSENGER} * \text{LENGTH}$$

The physical incongruity of an expression like this is manifest, since time is an independent fundamental magnitude and in the previous form it appears as a magnitude composed of two others. Again we come across the necessary delicacy to interpret the relationships between the magnitudes that appear in the mathematical formulations of natural phenomena.

The examples analyzed in this article belong to the «compound rule of three direct inverse proportion», according to ordinary terminology. The classical rules for solving these problems are all abstruse and incomplete, due to the vice of «arithmetization». However, with the algebra of magnitudes, all logical steps are fully justified by mathematical laws and it is not necessary to resort to mysterious old-fashioned rules or intuitive reasoning, solving problems of this type in a very direct and easy way. And this serves very well as a didactic prelude in the handling and interpretation of the most complex physical laws and equations.

Section XXX

«DYSMETRY» OF MAGNITUDES
An impressive physical-mathematical truth

In this brief section, the physical-mathematical discovery that is probably the most important since gravitation is succinctly presented, and the significance of the Copernican turn that is approaching for Physics, science and technology is clearly shown.

The historic advance has global repercussions and begins with the discovery and resolution of the «arithmetization» paradox of Physics. This paradox consists succinctly of the following: today we know that no one can answer questions such as how is a kilogram multiplied by a meter? Which is the multiplier, the kilogram or the meter? Neither of them can be, because the kilogram is a quantity of mass that is impossible to identify with a single number and the meter is, for its part, a quantity of length that is indeterminable arithmetically. How much length is there in a meter? How much mass is there in a kilogram? How much time is there in a second? You can't know. It is impossible to reduce these physical quantities to simple abstract numbers. So, if physical units are not arithmetic numbers, why do we operate with them as if they were? And the answer is that compound expressions with physical units are not defined and are not arithmetic operations, so no one knows their meaning, they are empty and arbitrary notations.

The «arithmetization» paradox of Physics is equivalent to the arbitrary hypothesis of assuming that magnitudes can be operated as if they were an abelian multiplicative group, which is the hypothesis currently in force for the International System of Units and, therefore, for all current theories, which leads to notable inconsistencies, such as the negative exponents of composite units and dimensional equations, because in the Dyadic

Algebra of Magnitudes it is rigorously demonstrated that such inverses cannot exist, since multiplicative operations are external laws of composition, so this supposed abelian multiplicative group structure cannot occur either. So the **aforementioned hypothesis of the International System is false.** And it does not seem reckless to describe it as scandalous and above all a scourge for the development of Physics, as can be seen without more than examining the manual with impartial attention and scientific curiosity.

Therefore, the so-called paradox with a certain excess of humility and consideration towards what is established, is rather an unacceptable primitivism in modern times, especially when there already exists a physical algebra that resolves the gaps and contradictions of the most basic foundations of Physics, which is to operate with its elements, the quantities of physical magnitudes, in a rigorous and coherent manner.

Once the paradox has been identified, a consequence of the **false hypothesis of the International System of Units,** the next necessary step is to form an algebra that operates with those quantities of physical magnitudes, which are not simple mathematical entities, but rather have a specific non-arithmetic nature, for example. which require a unique and broader treatment than that of pure mathematics. In the search for a non-arithmetic algebra we come across a very relevant precursor: the way of composing segments to produce new magnitudes, areas and volumes. These operations combine length quantities to generate area quantities and volume quantities. Therefore, they are an ideal reference to understand the rest of the physical quantities. Not in vain a segment is nothing more than a quantity of length in the physical sense, therefore this geometric algebra also belongs to Physics.

Once the algebra of segments has been established, as an algebra of geometric figures, not of abstract symbols, the next step can be taken, which is its analytical transformation through dyadic elements, formed by a mathematical entity in the primary, number, vector or tensor, and a unit of length in the secondary.

Then, it is enough to observe that the physical quantities can be made to correspond one-to-one with the algebraic structures of the geometric segments, so that, having affinity, the laws of composition of the segments can be generalized, extending them to the other magnitudes.

Additive operations with homogeneous magnitudes do not offer any complications, but even so, they lead us to the perfect rational understanding of non-dimensional magnitudes such as the radian and to other fundamental consequences for operating with the additive laws of Physics.

Resulting in the fact that by composing two segments through geometric multiplication a new magnitude is produced, the surface, and that by composing three, another is produced, the volume, the multiplicative laws of magnitudes, unlike the additive ones, appear to us as **external generative laws**, therefore which cannot exist unitary or inverse elements, which makes it impossible for such external laws to provide abelian group structure to dyadic sets, or sets of quantities of physical magnitudes. It is easy to observe this fact by simply asking yourself, what is the inverse of a meter? That is, what amount of length multiplied by one meter gives one meter? Or also, what quantity of length multiplied by any other gives this same length? There is no such thing, because two quantities of length multiplied produce an area, not a length. With all this, the correct interpretation of the inverses of the physical units and the divisions between quantities of heterogeneous magnitudes is reached, giving complete significance to the negative dimensional exponents and to the composite magnitudes obtained by multiplying or dividing other fundamental ones.

The nonexistence of unitary and inverse elements for multiplicative operations does not prevent dyadic sets, endowed with those additive and multiplicative laws, defined consistently and specifically for the quantities of physical magnitudes, from revealing an isomorphic structure with the body of real numbers.

The development of this magnitude algebra was the initial objective of the research, which is detailed in the manual. At first everything focused on describing the paradox of «arithmetization» and overcoming the contradictions and omissions of the International System of Units. But it happened that the observation of any dyadic element reveals without difficulty that a quantity of magnitude can vary in two ways: because its primary, the mathematical element that represents the measurement, varies, or because the secondary, the physical or non-mathematical element of measurement, varies. the dyad. What does a secondary variation mean? It does not refer to a simple unit change of the same magnitude. It's something much more subtle. To understand it, let's take the unit of length, the meter, as an example. We can imagine a rigid segment measuring one meter. The amount of length implicit in such a segment can be assumed constant at all points in space, a hypothesis that we could call isometric, or the other logical option can be considered, that the same congruently rigid segment in other positions implicitly contains different amounts of length, depending on the physical nature of the space considered. And thus «dysmetric» spaces are born. In them, the «dysmetric» density of a magnitude can be defined as the dyadic quotient of two quantities of the same at two different points, expressed in the same congruent unit, which according to the new algebra is always a real number and, therefore, dimensionless. In this way, different physical areas can be conceived characterized by specific distributions of «dysmetric» densities of the fundamental magnitudes. Each space is thus characterized by convenient distributions of «dysmetric» densities, which are nothing more than real numbers associated with each point in the space in question. And this tool is by no means banal, but rather it changes everything. To demonstrate this, articles have been developed referring to the mathematization of «dysmetry», Newton's second «dysmetric» law, the observation that the number pi is not constant in a «dysmetric» space, nor is the speed. of light, and finally «dysmetric» gravitation. They must be examined with an open mind, not so much to understand the physical phenomenon exposed, but to glimpse the immense possibilities of «dysmetry» as a tool to represent physical phenomena of all kinds and what this can mean for the progress of science. Physical.

For example, with «dysmetry» it is evident that in a «dysmetric» space the physical constants do not have to be such. Specifically, the *relativistic hypothesis* of invariance of the speed of light, on which all *relativity* is based, is incompatible with the «dysmetry» of space, as explained in section XXXIV. Therefore, the algebra of magnitudes is not limited to resolving the «arithmetization» paradox of Physics and giving coherent meaning to operations with magnitudes without any practical change, but, on the contrary, it clears the way for us towards «dysmetric» spaces and is presented as the source of many other possible innovations. Overcoming the false hypothesis of the International System of Units through the algebra of magnitudes would in itself be a prodigious advance for science, because it gives full meaning to compound magnitudes, which are now nothing more than mere capricious symbolisms. But the fertility of the algebra created to correct it has led without seeking it to an even more radical and transcendent discovery: the «dysmetry» of space.

«Dysmetry» is a natural and epistemic product of physical dyads. Physics has tacitly assumed since its distant origins that magnitudes are rigid, that is, that they are not affected by any cause, what we call isometry and which is equivalent to admitting that the quantities of magnitudes only vary because their measurements change, remaining unchanged. units always constant.

But this is nothing more than a very crude simplification of reality, inherited from those times when modern algebra was not known. The complete truth is the «dysmetry» of space, because not only physical measurements can vary, but also the quantities of magnitudes implicit in any unit adopted as a standard.

In this way, «dysmetry» is discovered as a complete representation system of natural properties, and it is not a theory, but an eternal epistemic truth. To understand it with an example, it is as if to build or repair machines we were offered to choose a fixed spanner and an English wrench, which of them would we choose? The answer is obvious: the English one allows you to operate with a wide range of nut sizes, while the fixed one only allows one size. Well, something similar happens with isometry, the fixed key, and «dysmetry», the wrench. Physicists and technicians cannot avoid converting to «dysmetry» to formulate their theories and applications without the current isometric limitations. «Dysmetry» is a necessary option if we want to fully encompass the complexity of natural phenomena.

So «dysmetry» is a global phenomenon. Everyone must embrace it without remedy. And, since it is a complete representation system of natural properties, it will produce endless scientific advances and technical innovations, as other physical advances have produced in the past. Let us think, for example, about how thermodynamics led to the invention of heat engines, which mobilized the industrial revolution; or how the discovery of electricity and electromagnetism led to the electrification of industry, railway transport or homes; or how the physics of semiconductors has brought the immense development of digital and communication technologies today. «Dysmetry» is on par or one can even venture that it will surpass these colossal progress.

The «dysmetry» leads, in turn, to another revolutionary concept: the **«dysmetric» density** of each magnitude, which turns out to be dimensionless and allows the **flexibility or deformation laws of natural properties to be represented**, as they are affected by multiple causes. disturbing, today excluded by the rigidity of isometric simplification, stuffing Physics and dramatically trapping its capacity to represent the natural world.

In short, «dysmetry» implies that magnitudes are deformable, which means that due to the effects of multiple causes they can contract or dilate, and this property entails the notion of greater or lesser «dysmetric» density. Such a manifestation of density does not coincide with the classical form, which always relates different magnitudes, so heterogeneous density is dimensional, like the ratio between mass and volume, its composite unit is the dyadic ratio between the kilogram and the cubic meter. On the other hand, the «dysmetric» density is homogeneous, it refers only to the magnitude itself in relation to itself between two points in space. And with this, magnitude algebra teaches that the ratio between two quantities of the same magnitude is always a real number, not a dyad, so it has no dimension, and from this it follows that the «dysmetric» density is dimensionless.

How does the «dysmetric» phenomenon affect physics? Throughout. It is an advance in basic science, which has a root impact on everything we know. And as the distinguished Spanish scientist Margarita Salas Falgueras taught us, advances in basic science are the most fertile and spectacular. The «dysmetry» will cause the recasting of all theories since Newton. It will put an end to physical constants, as they are conceived today. It will force a complete review of the International System of Units. It will mean a flood of technical and social advances. Let us think, for example, of the possibility that opens up to cover large distances in very short times and at not very high speeds, simply by making the trajectories lengthless.

But even more: if a trajectory is reduced in length, with a slight impulse you can travel great distances with very little energy. A prodigious revolution in transportation, which for now can be described as more fiction than science, but which the episteme assures will materialize in the future.

«Dysmetry» is an autonomous and exclusive movement. It is unstoppable and inevitable. It has universal interest. It is inalienable. What can Physics do about the «dysmetric» phenomenon? Remain in the rigidity of the current isometric simplification, opting for the fixed wrench over the English wrench? Has no sense. That's not gonna happen. «Dysmetric» Physics welcomes isometry as a particular case, but it is much broader and richer than this. Therefore, what is intelligent for science? To remain trapped in the simplicity of childish silent isometry or to enrich itself with «dysmetry»? The answer is obvious and unobjectionable: Physics cannot be prevented from modernizing with «dysmetry». «Dysmetry» is an enduring legacy and will change the world rapidly. The fruits of «dysmetric» Physics are inexhaustible. The «dysmetric» future is spectacular, exciting, hopeful.

In reality, the «dysmetry» has always been there. You just had to discover it. And this work is already done. Now what remains is to disseminate it throughout the world as quickly and widely as possible, for the common good of humanity. And this movement will occur spontaneously or by promoting it sooner or later with greater or lesser speed of propagation. But it is inevitable that it will happen.

This summary is necessarily schematic to capture the fundamentals of the basic physics research reflected in the manual, which will conquer many minds. For the moment, the most didactic language possible has been chosen, as well as different forms of presentation, to make it accessible to the greatest number of intellects.

That is why it is submitted to public consideration, in order to create a group of followers, the broader the better, who found this new Physics. There are two very striking starting elements: the false hypothesis of the International System of Units and the sensational discovery of «dysmetric» spaces, which should attract the attention of even the most layman in the matter.

Section XXXI

HOW TO MATHEMATIZE THE
«DYSMETRY» OF MAGNITUDES
Culmination of the First Algebra of Magnitudes
and inexhaustible hotbed of physical innovations

We know that the fundamental elements of Physics are the quantities of magnitudes, whose representation is established by means of dyads with the form (μ, U), where the primary call μ represents the measure of the quantity of magnitude resulting from comparing it with the unit U or secondary. The affinity postulate allows any quantity of magnitude to be represented by a segment in which the unit of length associated with U is arbitrarily adopted.

The traditional way of classifying quantities is based on the mathematical nature of the primary μ, concluding that they can be scalars, vector or tensorial depending on whether μ is a real number, a vector or a tensor. However, Physics tacitly assumes that the units adopted in any measurement are invariant, that is, that they do not depend on the position in space or on any other agent that can act in it. Thus, the measurement would consist of comparing a certain quantity of the measured magnitude with a certain standard unit and the dyad thus formed would analytically represent the measurement carried out. However, the assumption that the units always contain the same quantity omits consideration of another plausible variant, which could be more realistic: the units could include different quantities of a certain magnitude due to various causes. To understand us, the traditional hypothesis of units with constant quantities could be called **isometry**, and the opposite general variant **«dysmetry»**.

To focus ideas, and given the affinity postulate, let's focus on the geometric segments. The geometric equality of segments is

295

established by **congruence**, so that two segments that are congruent or, in vulgar terms, that can be superimposed, are assumed equal and silently admitted to be the same quantity of length. These would be the invariant isometric conditions that have always been taken for granted. But this vulgar and puerile hypothesis admits the generalization that arises naturally from the concept of dyad (μ, U), since in these elements the unit U does not have to contain the same quantity of length in any circumstance. On the contrary, at least from a logical point of view, the idea that this is not the case, that it is «dysmetry», is very understandable.

The variation of the quantities of magnitude implicit in the same unit entails a very transcendent consequence: that **the geometric congruence of segments is not synonymous with equality**, but rather that two congruent segments can have different quantities of length as a result of different causes of a physical nature. not mathematical or axiomatic. With this, the equality of two dyads of the same magnitude does not have to refer to the mere congruence of their related segments, but to the equality of the quantities of magnitude implicit in them.

On the other hand, the measurement in a «dysmetric» space will consist of counting the number of congruent segments that fit in a certain length and this would be the **mathematical measure** by congruence; while the **physical measure**, which would determine the natural properties, would be established by the total quantity of length of those segments that, although congruent, would not be equal, as in an isometric space; and so it would be that the «dysmetric» measure does not have to be the same as the isometric. The «dysmetry» opens up an infinity of possibilities for new physical spaces, which is why it must constitute an inexhaustible hotbed of scientific innovations.

The omission of this transcendent observation has been caused by the traditional and simple «arithmetization» of Physics, forgetting that this science not only handles mathematical elements, such as real numbers, vectors or tensors, but also composes dyadic entities, which can vary not only because the primary ones do it, but also the secondary ones. This simple approach to tradition has been initially taken up with humility and self-denial in this investigation, described in the First Algebra of Magnitudes, which has limited itself to exposing, developing and solving the **paradox of «arithmetization» of Physics** by means of a dyadic algebra or not arithmetic, assuming in this first phase that the physical units are invariant or, what is the same, that the physical space is isometric. But, once this dyadic algebra has been accurately described, it is impossible to escape the temptation to postulate that the dyadic secondaries are not constant, a fact that is logically essential and that Physics has always ignored and cannot ignore now. However, fortunately the dyadic algebra presented here is easily generalizable to «dysmetric» spaces without more than putting into play the **density functions** δ that will be defined next.

So it is inevitable to culminate this work with an introduction to «dysmetric» spaces, which will be the subject of future and more complete research and publications. Let's start with a simple didactic formulation:

Imagine a magnitude such that, given its unit U_0 in a vacuum, it is affected by the influence of a mass M, so that in the neighborhood of any point in the space $P \in E^3$, E^3 being the ordinary affine point space, positioned by the vector with respect to the point mass, the unit in P, designated $U(P)$ or $U(\overline{r})$, congruent with U_0, contains implicitly a certain quantity of magnitude determined by a certain function $\delta(M, \overline{r})$ of $R \times E^3$ in R, such that $U(P) = \delta(M, \overline{r}) \circ U_0$, so that en $\overline{r} = \overline{0}$ let $\delta(M, \overline{0}) = 0$ and with tending to infinity let $\delta(M, \overline{r}) = 1$, where r is the modulus of. A magnitude like this, to understand us, could be said to be elastic; at a point far enough from the mass, it would act as in a vacuum, but in the environment of the mass it would be as if it were emptied of quantity. Note that the multiplication «$\circ$» is the one in section IX of the First Algebra of Magnitudes or product of a magnitude and a scalar, briefly described on page 124 here.

297

The «dysmetry» is manifested as well as the phenomenon that congruent segments appear to have different **density of length**. This concept would clash with the ordinary notion of density; but that in this type of space it acquires specific meaning. In **«dysmetric» spaces** the density of length would not be constant and the variation could be indicated by functions of the type δ, hence this Greek letter has been chosen to distinguish them. And nothing prevents us from suspecting that magnitudes of this nature could be length, mass, time or others such as temperature, electric charge, electromagnetic fields, and other physical magnitudes; hence the brand new innovation announced by these structures.

What characterizes «dysmetric» spaces is that, given a dyadic form (μ, U) indicative of the measurement of any magnitude, the **«dysmetric» equality criterion** would depart from the ordinary one, since two dyads could be equal with different μ measures even though the units are congruent, unlike what happens in isometry. Indeed, given two dyads (μ_1, U_1) and (μ_2, U_2), where $\delta_1 \circ U_0$ and $\delta_2 \circ U_0$ are the quantities implicit in two congruent units U_1 and U_2 and U0 the reference standard to establish the «dysmetric» density δ, the equality $(\mu_1, U_1) = (\mu_2, U_2)$ requires by dyadic algebra that $U_1 = (\mu_2/\mu_1) \circ U_2$. In turn, the ratio of the congruent segments $U_1 /\!/ U_2$ would have to be δ_1/δ_2, and thus it turns out that $\delta_1/\delta_2 = \mu_2/\mu_1$. On the other hand, in isometry, since U_1 and U_2 are congruent, as the «dysmetric» density δ is always unity, it would have that $U_1 /\!/ U_2 = 1$, and thus it would result $\mu_2/\mu_1 = 1$, with $\mu_2 = \mu_1$, concluding that it is impossible that congruent segments can produce different μ measures. Isometry is, therefore, a particular case of «asymmetry», when $\delta = 1$ in any case.

The previous hypothesis that the masses are the cause of the physical «dysmetry», so that the dyadic secondaries vary at each point in space while maintaining their congruence, which would mean that this space, which we could call mathematical or abstract, would remain unchanged, It can be contrasted with another hypothesis that attributes the «dysmetry» to empty space itself. Under these conditions the length or «dysmetric» density of the congruent segments could be represented by a function $\delta(\overline{r})$ of E^n in R, where the Euclidean space has been generalized to n dimensions, such that $U(P) = \delta(\overline{r}) \circ U_0$, where U_0 could represent the unit of magnitude at the origin of coordinates or any other certain point.

Under these conditions, the algebra of magnitudes seems to bet on a different principle of relativity than the established one, since relativity here would consist of the positional difference of the quantities of magnitudes in a mathematical space that maintains congruence in any case.

Another warning that the algebra of magnitudes gives us is the possible incorrectness of certain myths such as the constancy of the speed of light in all inertial systems, established on the basis of the «arithmetization» of Physics, which only considers the possible invariance of certain measurements, but tacitly or expressly admitting that the various units used in the measurement remain unchanged in any physical environment. And precisely this assumption has a good chance of not being true. This would occur in a «dysmetric» universe in which the quantities of congruent unit magnitudes are influenced by acting fields or position, as outlined in the preceding assumptions.

The algebra of magnitudes shows that the quantities of physical magnitudes are not reduced to a simple arithmetic number, which indicates the measure in relation to a certain unit, but must be described with physical dyads in which the second element is the corresponding unit. Therefore, physical environments could be found in which, the measurement being constant, what varies is the unit or second dyadic element, which would indicate variation in the quantity of magnitude described by the dyad.

Obviously, as with the speed of light, all physical constants are exposed to the risk of not being invariant, so the algebra of magnitudes could be warning us that the current physical constants do not reflect true invariant properties of the material world. And well thought, since physical measurements are made in a very small human environment of space, it is natural that some may seem invariable, but this observation does not guarantee at all that even the arithmetic part of such dyads will not remain constant in all spatial position. So the current claim of the International System of Units to refer the patterns to physical constants, seeking their invariance, could well be a chimerical aspiration, as the logic of the algebra of magnitudes shows without great difficulty.

This opens a new debate, very similar to the one that once pitted geocentrism against heliocentrism with dramatic noise. Now the algebra of magnitudes seems to be opposed to Einsteinian relativism, offering his innovative thesis that the mathematical or abstract space maintains its congruence and rigidity, the classical hypothesis, and that relativity consists of the variation by multiple causes of the congruent dyadic secondaries, according to to the «dysmetric» density functions δ, such that each magnitude can have its own, inevitably engaging us in the investigation of its different possible forms.

In short, up to now Physics, having «arithmetized», has only paid attention to the dyadic primaries, assuming that the units are imperturbable. On the contrary, the great contribution of the algebra of magnitudes is to direct the investigations to the secondary ones, warning that **mathematical congruence** does not have to be synonymous with **physical equality**, which would entail variations of the geometrically congruent units in the material space.

Recapitulating, in the process of measuring a length and, by affinity, of any magnitude, what you do is count the number of congruent segments that fit between two points whose distance you want to establish. Thus, a real number q is obtained that indicates the distance, understood as the number of times that the unit of measurement U comprises. In an isometric space the dyad resulting from the measurement (q, U) will indicate the quantity of length of the distance as multiple of that implicit in the unit U, which is tacitly assumed to be invariant or constant at all points in space; but in a «dysmetryc» one it would not be so, because each segment congruent with the standard unit U will have a different quantity of length depending on its spatial position. In other words, the classical measurement procedure conceals a hidden hypothesis, which is the assumption that the congruence of segments is equivalent to the equality of length quantities, and thus limits itself to establishing the number of segments congruent with the standard unit, but nothing is indicated about the true physical quantity of the measured magnitude.

Therefore, it is imperative to consider the «dysmetry», hitherto illogically ignored, and admit the more than **plausible distinction between congruence and equal quantity of length**. Thus, in the radial direction from the origin, the quantity of infinitesimal magnitude will be represented by the dyad $dQ=[dq, U(\overline{r})]$, where dq is the infinitesimal measure in the unit $U(\overline{r})$ congruent with the unit U_0 in the origin. The fundamental «dysmetric» relationship between congruent units will have the form $U(\overline{r})=\delta(\overline{r}) \circ U_0$, simply activating the **«dysmetric» density function** $\delta(\overline{r})$, where $\overline{r}$ it indicates the position by mere congruence of the infinitesimal element measured with respect to a given origin in which the unit of length is U_0. Thus we quickly arrive at the expression $dQ=[dq, U(\overline{r})]=dq \times \delta(\overline{r}) \circ U_0$ and conclude with this other equivalent:

$$\frac{dQ}{dq} = \delta(\overline{r}) \circ U_0 \qquad [31.1]$$

Note that the indicated «dysmetric» ratio is symbolized by two horizontal lines because it is not an arithmetic ratio, since the element dQ indicates a quantity of magnitude and, therefore, represents a dyad, specifically the $[dq, U(\overline{r})]$. This is the division derived from multiplication by a scalar in section XI of the First Algebra of Magnitudes or the division described on page 126 of this volume.

Insisting on the distinctive characteristic of «dysmetric» spaces, that is, that congruence does not equal segment equality, see how a distance can be established in them with the following scheme:

Let us be an «dysmetric» space and take two points of the same A and B. Suppose that the «dysmetry» of this space is characterized by the «dysmetric» density function $\delta(\overline{r})$, wher $\overline{r}$ it represents the position vector of any point. Let $d\overline{r}$ be the generic differential variation in the direction of segment AB. The quantity of length that this segment includes can be indicated by S_{AB} and will be given by the integral expression that adds all the differential elements included between A and B, that is:

$$ S_{AB} = \int_A^B \left[dr \circ U\left(\overline{r}\right) \right] $$

$U(\overline{r})$ indicates the unit of length congruent with the unit U_0 for the $\overline{r}$ position and dr refers to the modulus of the differential vector $d\overline{r}$. Therefore, U_0 and $U(\overline{r})$ are congruent unit segments, the first located in the surroundings of the origin of coordinates and the second in the surroundings of the position; but, being a «dysmetric» space, these segments do not have the same quantity of length, but their respective lengths are related by a certain «dysmetric» density function $U(\overline{r}) = \delta(\overline{r}) \circ U_0$.

Under these conditions, the quantity of length S_{AB} of segment AB is described by the expression:

$$ S_{AB} = \int_A^B \left[dr \circ U\left(\overline{r}\right) \right] = \left[\int_A^B dr \circ \delta\left(\overline{r}\right) \right] \circ U_0 $$

The «dysmetrc» density function of length $\delta(\overline{r})$ reproduces the physical fact that two geometrically congruent segments turn out to have different lengths or, what is the same, different distances between their end points. This quality, it is emphasized once again, is what characterizes «dysmetric» spaces and differentiates them from isometric ones. In these, congruence is synonymous with equal quantity of magnitude, while in the «dysmetric» it is not. For example, by identifying U_0 with the standard meter at the origin, in ordinary physics it will be possible to find congruent segments of 6 m, and all segments congruent with another 6 m will also measure 6 m, regardless of their location. On the other hand, in a «dysmetric» space, given a 6 m segment at the origin, others positioned at different points and congruent with it may have measures of 5 m, 3 m, 2 m or any other that is compatible with the function of «dysmetric» density $\delta(\overline{r})$.

Having established the distance between the points of an «dysmetric» space, the resulting mechanics of magnitudes should then be thought about. To do this, let us analyze the case of a radial movement with respect to a point mass M within a space in which the mass affects the density of length and not time. The kinematics of this problem would be like this:

Let $\overline{r}$ the position vector of the point whose radial motion is analyzed in relation to the position of the point mass M, which is taken as the origin. Let dS be the quantity of vector length in unit L for the $\overline{r}$ position of an equipment-lens vector (synonymous with congruence in the case of vector quantities) with $(d\overline{r}, L_0)$ or with the equivalent notation $d\overline{r}\, L_0$, where L0 is the unit of length in the void. The «dysmetry» of space will be indicated by a certain «dysmetric» density function $\delta(M,\overline{r})$ such that, as we have seen, $L(M,\overline{r}) = \delta(M,\overline{r}) \circ L_0$. The differential «dysmetric» ratio [31.1] allows us to establish the following:

$$\frac{dS}{d\overline{r}} = \delta\left(M,\overline{r}\right) \circ L_0$$

Suppose that the time magnitude is isometric, so it will not be affected by the mass and its unit in vacuum T_0 will also serve in the presence of the mass M, and thus the quantity of time in any case would be given by the dyad (t, T_0) or its equivalent notation $t\, T_0$. Given that the dyadic operations are isomorphic with those of R, the laws of differentiation would allow us to reach the conclusion indicated by the following analytical expression, which links several equalities, multiplying and dividing by $d\overline{r}$:

$$\frac{dS}{dt\, T_0} = \frac{dS}{d\overline{r}} \circ \frac{d\overline{r}}{dt\, T_0} = \left[\delta\left(M,\overline{r}\right) \circ L_0\right] \circ \frac{d\overline{r}}{dt\, T_0}$$

Taking into account that $dS = d\overline{r}\, L$, or what is equal, dS indicates the dyad $(d\overline{r}, L)$, the first member represents the physical velocity $\overline{v}$ at the point $\overline{r}$ at time $t\, T_0$ and that the last factor $d\overline{r}/dt$ is the measure of the speed in vacuum $\overline{v}_0$, we will have:

$$\overline{v}\, \frac{L}{T_0} = \left[\delta\left(M,\overline{r}\right) \bullet \overline{v}_0\right] \frac{L_0}{T_0} \qquad [31.2]$$

Thus it turns out that the density function of length δ relates not only the quantities of vector length, but also the physical and vacuum velocities. On the other hand, in the particular case of an isometric space, they will be $L = L_0$ y $\delta(M, \overline{r}) = 1$ at all points and the physical speed will coincide with that of a vacuum, also being $\overline{v} = \overline{v}_0$. It could also be understood $\overline{v}$ to indicate the «dysmetric» speed and $\overline{v}_0$ the isometric. But this expression [31.2] still hides a singular consequence. Observing it, it is not difficult to reach the conclusion that it must always be $\overline{v} = \overline{v}_0$, a result that could have been established directly in this way:

$$\overline{v}\, \frac{L}{T_0} = \left[\delta\left(M,\overline{r}\right) \bullet \overline{v}\right] \frac{L_0}{T_0} \Rightarrow \overline{v}\, \frac{L}{T_0} \neq \overline{v}\, \frac{L_0}{T_0} \quad si \quad \delta\left(M,\overline{r}\right) \neq 1$$

Remembering that L and L_0 are congruent lengths that, however, implicitly carry different quantities of length, in common parlance this means that in a «dysmetric» space the same measure of velocity will correspond to different quantities of velocity depending on the situation, since the unit of measure contains different quantities depending on the position.

The logical and mathematical confirmation of this result poses a very serious threat to the survival of the Theory of Relativity. This mythical theory is based on the constancy of the speed of light, a postulate that is based on measurements made in experiments in the terrestrial environment. And it is not necessary to have many lights to warn with the «dysmetric» algebra that the same measurements can indicate different speeds, since the secondaries of the dyads are not constant, despite the fact that they are the same standard unit, since their quantity of implicit length varies in the ratio that marks the density function, which in turn is the parameter indicated by a real number that relates the dyadic quantities to each other.

With all this it is observed that the algebra of magnitudes teaches that the «dysmetric» variations affect the kinematics of the movement, which opens up great possibilities to describe new physical phenomena until now unexplained. It even allows us to glimpse that the still pending unification of natural forces could be resolved in this way. We thus arrive at a more than probable Copernican turn of science, which could be summarized in these terms:

Given any two points in space, they would not change, they would be mere abstract references and congruent in any case with each other; what could vary would be the quantityies of magnitudes between the two, such as the quantity of length that comprises the distance that separates them. So an «dysmetric» space will be geometrically isometric and Euclidean, if the number of congruent segments between two points are simply counted in it, which would be equivalent to the classical measurement; but physically, for the purposes of natural phenomena, congruent segments can incorporate different quantities of length, thereby changing the laws that govern these spaces. With this background, some basic principles could be established that would characterize «dysmetric» spaces:

Empty space would be an abstract mathematical entity. It would not be a physical entity. It would be represented, by definition, by the three-dimensional Euclidean affine point space, which arises from the application of the hypothetico-deductive method of mathematics. The points in this space would be mere invariant positions of a non-physical nature and, therefore,

305

independent of any material disturbance. Given any two points, their positions would be fixed, they would not change, they would be simple spatial references of that immaterial, abstract and immutable sphere; however, the quantities of magnitudes present at these points and their variations between them could be sensitive to physical phenomena. For example, the quantity of length that encompasses the distance between any two points could depend on gravitational or electromagnetic fields or other physical actions. Admitting this perfectly plausible possibility, the effects of the different actions present in nature could be compounded by the principle of superposition of effects. Although the mathematical space did not change and remained congruent, the quantities of length and other magnitudes between its points could change due to the action of different physical disturbances. This immutable mathematical space could be the one observed when making measurements by congruence, so it could be called apparent space. On the other hand, the physical space, related to the mathematical one by means of the «dysmetric» density functions δ, could be taken for the real space; although it could also be seen the other way around: the mathematical space could be the real one and the physical one could appear as apparent. In any case, it would be necessary to know how to distinguish between reality and appearance, and to investigate which δ functions would be appropriate to represent the correspondences between the two.

In this way, the algebra of magnitudes takes a significant leap, because from being a mere logical instrument to give consistency to current Physics, revealing and solving the paradox of «arithmetization» of Physics, it becomes a fertile hotbed of physical innovations, which is a Copernican turn in the formulation of scientific laws that exceeds the limits of this manual on the First Algebra of Magnitudes, so this is limited with this article to provide some basic and general ideas of the tremendous novelty found, giving I move on to new research and publications that will deal with the nascent algebra of «dysmetric» spaces, characterized by their specific equality criterion, which departs from mere geometric congruence and focuses on the quantity of variable length for various reasons, so that «dysmetric» equality admits that congruent segments have different quantities of length, leading to the concept, something shocking and unusual, of density of length and their corresponding «dysmetric» density functions δ.

Continuing with these preliminary ideas, see how «dysmetric» spaces affect the classical classification of magnitudes. Traditionally, as the units are tacitly assumed to be constant, a hypothesis that here has been called isometry, the magnitudes are classified as scalars, vector or tensor, attending only to the mathematical element that the measure represents. On the other hand, under «dysmetric» conditions the same unit or affin segment will present different quantities of magnitudes depending on its position and depending on the various causes that determine such variation.

Without the intention of being exhaustive, simply to show the phenomenon, the «dysmetric» magnitudes can be classified as rigid and elastic. Rigid ones can be defined as those insensitive to position and any physical disturbance, in short they are isometric. The elastic ones would be those that change by position or by other acting agents. Recalling the definition of «dysmetry», given two congruent units of a certain magnitude U_0 and U, relative to any different positions, they will be represented by superimposable segments when transported as rigid bodies, as the canons of geometry command; but the quantities of magnitude they represent will not be equal, so their dyadic ratio will not be unitary, but a number other than one, which will express the density of length relative to U_0. Doing this operation at each point in the space positioned by the vector $\overline{r}$ with respect to the position of U_0, we will have the «dysmetric» density function $\delta(\overline{r})$, such that the dyadic ratio is $U/\!/U_0 = \delta(\overline{r})$. Obviously, in an isometric space $\delta(\overline{r}) = 1$ at all points, it will always be $U = U_0$ and the congruence will also mean equal quantity of length or in general of affine magnitude.

307

Another question to clarify is the concept of **simultaneity**. Look at figure 21. It assumes the production of a flash of light in all directions from a point *O*. In an isometric space, considering that such characteristic also includes the property of isotropy, the light would propagate at the same speed in all directions and the surfaces of temporal simultaneity from the moment of the flash would be perfectly symmetrical concentric spheres in *O*. This situation is the one that should occur in empty space in the

SIMULTANEITY
Difference between equitemporal surfaces in an
isometric space and in another dysmetric one

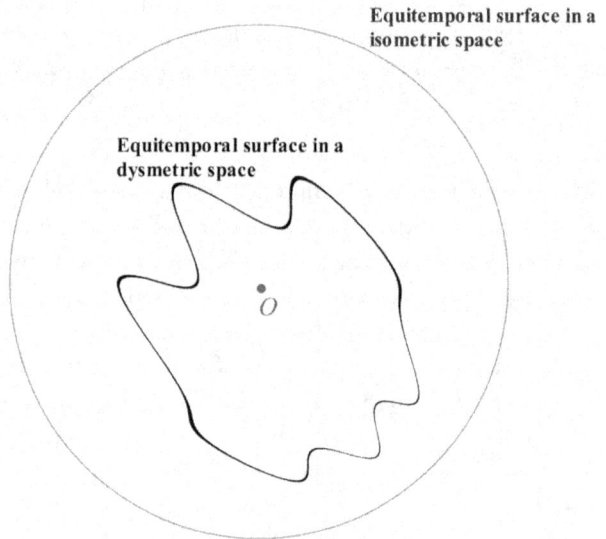

Equitemporal surface in a
isometric space

Equitemporal surface in a
dysmetric space

O

In an isometric space, which could be empty space, the equitemporal surfaces of simultaneity of the light rays originating from a point *O* would be concentric spheres centered at the point of the flash.

In an «dysmetric» space the speed of light varies from one point to another, even though its measure *c* could be kept constant, and equitemporal surfaces would be irregular or asymmetric surfaces.

Figure 21

absence of any disturbance. However, in a space that presents «dysmetry» for any reason, either by its own nature or by the presence of fields of different forces, the speed of light would vary from one point to another and the equitemporal surfaces would be irregular or asymmetric. Well, nothing is opposed to the consideration that the universe presents «dysmetries» due to multiple causes, which opens inexhaustible horizons of research and innovation for Physics.

The experiment carried out during an eclipse on May 29, 1919, conceived by Sir Frank Watson Dyson, is believed to confirm Einsteinian relativity versus Newton's classical gravitation. Well, the «dysmetric» spaces would allow the same phenomenon to be explained, that is, the curvature of the rays of light and, therefore, the visibility of a star hidden by the Sun. To verify this, let us consider that material space can refer to abstract to the three-dimensional Euclidean affine space, which would become its mathematical representation, assuming the criterion of segment congruence as the basis for measurements in this space. It can be associated biunivocally with a certain «dysmetric» density function $\delta(M,\overline{r})$ its transformed space in which physical phenomena should develop. Thus, the dual «dysmetry» is reached, based on correspondences such as the one indicated. Figure 22 describes the situation of the eclipse and its explanation by means of «dysmetric» dual spaces. Just as before, an «dysmetric» density function $\delta(M,\overline{r})$ of $R \times E^3$ in R was considered, such that en $\overline{r} = \overline{0}$ is $\delta(M,\overline{0})=0$ and that for $r \to \infty$ it is $\delta(M,\overline{r})=1$, and it was said that it would be as if the mass empties the length of the segments; in this case, nothing prevents formulating the opposite hypothesis, that is, that the mass fills the segments with more length than in a vacuum. Such a phenomenon would be represented by a density function $\delta(M,\overline{r})$ such that en $\overline{r} = \overline{0}$ is, for example, $\delta(M,\overline{0})=\infty$ and with $\overline{r}$ tending to infinity let $\delta(M,\overline{r})=1$, as in a vacuum. Thus we would find that the influence of a mass would increase the quantity of length of congruent segments, the more the closer the position is to the mass. As a consequence of this, the situation in Figure 22 could be

THE DUAL «DISMETRIC» SPACES
An example of dual «dysmetry» that fits to the eclipse experiment a hundred years ago

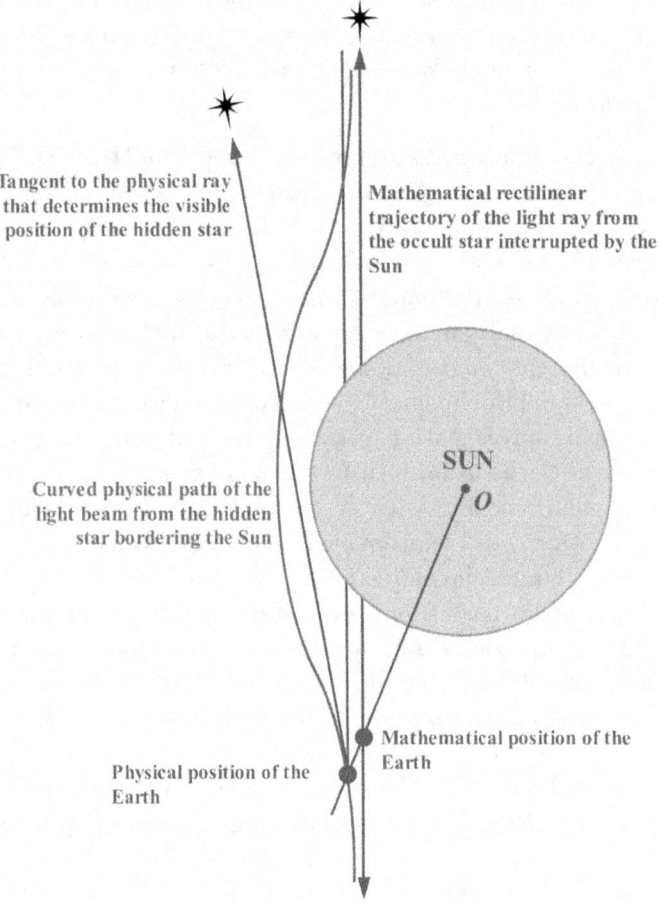

Tangent to the physical ray that determines the visible position of the hidden star

Mathematical rectilinear trajectory of the light ray from the occult star interrupted by the Sun

SUN
O

Curved physical path of the light beam from the hidden star bordering the Sun

Mathematical position of the Earth

Physical position of the Earth

Dual «dysmetric» spaces associate biunivocally the three-dimensional Euclidean affine mathematical space with the physical space in which natural phenomena would develop.

Figure 22

310

conceived in which the Sun hides a star with respect to the relative position of the Earth, which would present a congruent or isometric mathematical position in space, and a different physical position in the dual «dysmetric» space. In turn, given the assumed shape for the density function $\delta(M, \overline{r})$, the mathematical distance or by mere congruence to the origin O at the center of the Sun would always be less than the physical or «dysmetric» distance. In the same way, a line or mathematical trajectory of a light ray, would be transformed into a curve in the «dysmetric» space that would border the Sun. Under these conditions, the tangent to the physical trajectory of the light ray from the hidden star and in the physical position of the Earth it would make the star behind the solar sphere visible.

It is clear, therefore, that dual «dysmteric» spaces make it possible to establish physical models in which real and apparent space are associated, distinguishing between what is seen through the observation and measurement processes, and what is actually be the material space.

Mathematics unknowingly admits **dual « dysmetric» spaces**, so their existence is assured by this observation. Let's look at some simple cases. In Figure 23 an arc of circumference and a line tangent to it are represented. The central projection of the points of the arc on the line is of a «dysmetric» nature, because the quantity of length of the segments on the line presents a density of variable length, if the quantity of length of these segments is associated with their corresponding arcs, as happens, for example, in the known cartographic representations, in which the distances on the chart depend on the latitude, so that, at higher latitude, the same meridian arcs on the ground will appear represented by larger segments; or conversely, congruent segments on the chart will indicate different arcs on the ground. It is not necessary for one of the spaces to be curved, since the central projection between two non-parallel lines constitutes a dual «dysmteric» space, as shown in Figure 24. Congruent segments on one of the lines are transformed into different segments on the other, so that, without more than defining the corresponding amount of its

THE CENTRAL PROJECTION
Example of a dual «dysmetric» space

By projecting the points of a circle from its center onto a tangent line, a «dysmetric» correspondence is obtained. Different segments on the line correspond to equal arcs on the circumference.

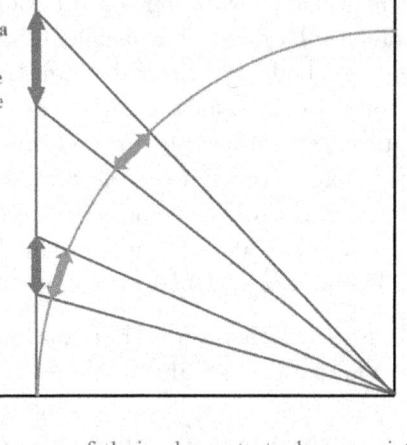

Dual «ysmetric» spaces allow any of their elements to be associated with mathematical or physical space. Thus, in this case, the line could be the physical one and the arc the mathematical one, or vice versa.

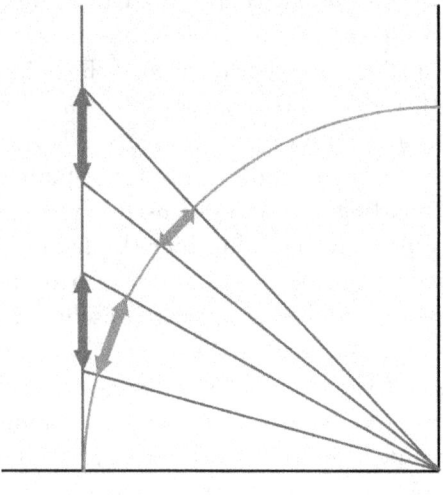

In the same way, in this «dysmetric» dual space, different arcs correspond to equal segments on the line.

Figure 23

312

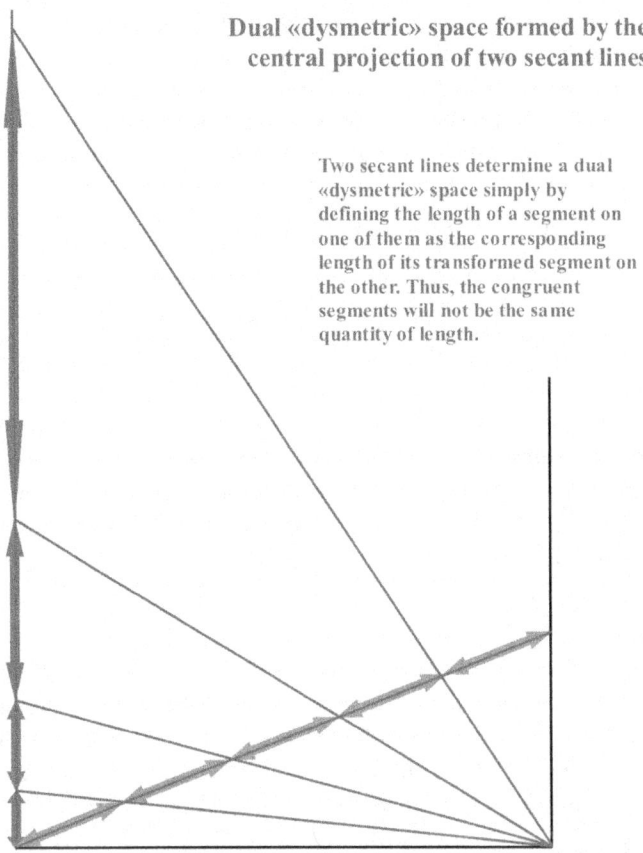

Dual «dysmetric» space formed by the central projection of two secant lines

Two secant lines determine a dual «dysmetric» space simply by defining the length of a segment on one of them as the corresponding length of its transformed segment on the other. Thus, the congruent segments will not be the same quantity of length.

Figure 24

transform as the quantity of length of a segment, each line will be transformed into a «dysmetric» space in which the length quantities of congruent segments will not be constant.

This phenomenon will not occur between parallel lines, because Thales' Theorem determines that congruent segments correspond to congruent segments by central projection.

In general, every non-linear function $y=f(x)$ represented by a Cartesian coordinate system defines a dual « dysmetric» space, provided that the quantity of length of its segments is defined in

313

either axis by its corresponding in the other axis. Thus, for example, the equality of the segments into which the abscissa axis is divided will be broken by establishing the difference between the ordinate of their extremes as the quantity of length of each one of them, given by the function $f(x)$, resulting in congruent segments on the abscissa axis will not be equal, since they will have different length quantities associated with them.

In mathematics, pure «dysmetric» spaces can be identified verywhere, in the sense of non-duals. For example, a logarithmic scale like the one in Figure 25 is a «dysmetric» space constructed with congruent segments such that the coordinate is not proportional but instead marks the antilogarithm of the distance to the origin. Thus, the congruent segments S_1, S_2, S_3, etc., when juxtaposed, mark respectively the abscissa 0, 1, 2, 3, etc. If in those same abscissa the antilogarithms of 0, 1, 2, 3, etc. are indicated, that is, 1, 10, 100, 1.000, etc., and it is considered that each segment has the quantity of length that determines the difference between the antilogarithms of its extremes, a «dysmetric» space will result in which the lengths associated with the segments S_1, S_2, S_3, etc., do not forget, are congruent with each other due to the scale construction itself, will be respectively $S_1=9$, $S_2=90$, $S_3=900$, etc. A single scale can be viewed as a one-dimensional «dysmetric» space. Two Cartesianly arranged scales will mark a two-dimensional «dysmetric» space. And three

THE LOGARITHMIC SCALE
Example of a pure «dysmetric» space

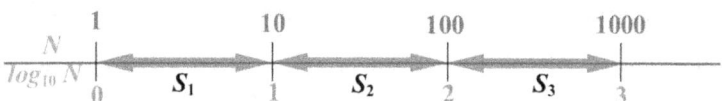

Defining the amount of length of each segment S_i as the difference of the antilogarithms of their extremes, a «dysmetric» space results in which the congruent segments S_i they have different length amounts. So they turn out $S_1=9$, $S_2=90$, $S_3=900$, etc.

Figure 25

314

scales another of dimension three. The common characteristic of all these spaces is that, although the coordinate axes are divided into Si segments congruent with each other, nevertheless each one of them will contain a different quantity of length according to the position it occupies in the space.

The operations with the slide rule and itself are based on the properties of the logarithmic scales, which are outlined in section XVIII, pages 167 and following of the repeated volume I.

In summary, the First Algebra of Magnitudes pays attention to the basic elements that Physics uses in its formulations, the dyads, which represent quantities of magnitudes. Discover and solve the «arithmetization» paradox of Physics and develop a specific dyadic algebra. Well, these physical quantities can obviously change not only because the first dyadic element, which is its mathematical component, varies, but also because the second, the one that represents the standard unit used as a comparison in the measurement, indicates quantities of different magnitudes depending on different causes. Note that we are not referring to mere changes of multiples or submultiples in the unit of measurement, but to intrinsic mutations in the quantity included in the same unit, which has been described as dissonance between congruence and equality, that is, the congruence of Two segments do not mean that they always have the same quantity of length, but that this will vary depending on the circumstances, which leads us to the innovative concept of density of length, which makes full sense when defining the «dysmetric» functions.

Thus we have the phenomenon that we have called «dysmetry» compared to classical isometry, which assumes invariant units in all its aspects. We are, therefore, before a sensational observation, until now omitted by science, which must be taken into account for the development of new scientific models. The «dysmetry» is inescapable, because it is mathematically and logically unobjectionable, and because Physics cannot do without the seedbed of innovations revealed by this inevitable revolution, which involves a Copernican turn similar to the famous dilemma between geocentrism and heliocentrism.

A dual «dysmetry» space can be mathematized in various ways. One would be by means of a tensor that serves as an operator to relate the mathematical and physical spaces, or real and apparent, transforming one into the other. The non-dual «dysmetry» spaces would be mathematized by means of a «dysmetric» density function that relates the quantity of length that the geometrically congruent segments accept as a function of their spatial or spatio-temporal position, without forgetting that this method it would serve any magnitude, not just length.

Mathematics and Physics have always been based on an almost invisible tacit hypothesis, it is the assumption that all congruent segments have the same quantity of length and that is why they are said to be equal. This criterion of equality admits what has been called here isometry of space or the hypothesis that the «dysmetric» density is constant regardless of position. It is recalled again that in the measurement process what is done is to count the number of congruent segments that fit between two points whose distance is to be established, and thus a real number q is obtained that indicates such distance, expressed in the unit U taken for reference, representing the measurement with the dyadic pair (q, U). In an isometric space the length quantity of the standard unit U is tacitly assumed to be invariable. Therefore, the classical measurement procedure has a fundamental defect, because it limits itself to counting the number of segments congruent with a standard unit, unintentionally neglecting the variability of the quantities that the units adopted as reference accept. In contrast to this, it has been seen and insists on the crucial observation that the quantity of length of the congruent segments depends on the circumstances as a function of various causes, or in other words, that the dimensional element U of the dyads, the so-called secondary, while maintaining its congruence, is variable in each space-time position, which leads us to «dysmetric» spaces, a Copernican innovation that is called to

316

revolutionize scientific models, because by the affinity postulate the «dysmetry» it can be extended to all magnitudes.

This article has outlined the fundamental lines and potential of «dysmetric» spaces, so that any attentive reader will be able to appreciate their significance for science with relative ease. Anyone can understand that for physical units there are only two possibilities: that their quantity of magnitude is the same in all space or that such quantities vary according to the space-time position, which is the **«dysmetric» observation** that arises out of pure logic. Currently this option is tacitly omitted and, therefore, the opposite hypothesis is not explicitly formulated, which is the belief in the intrinsic imperturbability of physical units, which here has been called the **silent hypothesis of isometry**. However, sooner or later and for the progress of Physics, the «dysmetry» will find its way without remedy, because it is logically unobjectionable and materially plausible, colliding head-on with the very existence of physical constants and everything that this would entail, from the recasting of the current International System of Units, based on fixed and immutable patterns, up to the more than suspicious invariance of the speed of light in vacuum for all inertial systems, which is a fundamental principle of Einsteinian relativity, in such a way that, if this invariance were not real, all relativity would fall to its foundations, being reduced to a mere very well-constructed mathematical model, but without a material reference. And, as we have seen in this article, mathematically and logically the constant measurement of the speed of light in the terrestrial environment does not guarantee at all that it remains unchanged throughout space.

The «dysmetry» is called to revolutionize Physics, or better still, **it creates a new Physics**, overcoming the more than childish «arithmetization» of this science, which is mutilated by the prevailing elementary, arbitrary and invisible isometry, preventing the development of creative and realistic innovations. More than a revolution, it would be a **new Physics**, much richer in its potential for representation than the current one.

317

Section XXXII

«DYSMETRIC» FORMULATION
OF PHYSICAL LAWS
Second law of Newton

In order to illustrate how physical laws are transformed with «dysmetry», this section is developed, which tries to guide scholars and researchers about the notable influence of the flexibility of magnitudes, to the point of unfailingly committing ourselves to a new Physical. The previous evidence presented to us is the distinction between observing space as a continuous whole or, on the contrary, as a set of discrete and independent environments. The first perspective, the most rigorous and complex, leads to what we might call the complete formulation of the «dysmetric» physical laws; the second, less elaborate and much simpler, could be called a discrete formulation of those same laws.

We chose Newton's second law as the first element of «dysmteric» analysis. We know that for physical algebra this law relates three magnitudes: force, mass and acceleration. Force and acceleration are vector magnitudes, while mass is scalar. In article 16 of section XXVIII in the First Algebra of Magnitudes the «dyadic form of physical equations» was explained. One of the cases studied there is precisely Newton's second law. It was said that it is about the multiplication of a scalar physical quantity by another vector, so that, given a mass expressed with the scalar dyad $(M\ kg)$, the acceleration indicated by the vector dyad $(\overline{a}\ m /\!/ s^2)$ and the acting force symbolized by the vector dyad $(\overline{F}\ N)$, Newton's second law should be written in terms of the dyadic algebra $\mathscr{D}$ with the form $(\overline{F}\ N) = (M\ kg) \odot (\overline{a}\ m /\!/ s^2)$, where the multiplication «$\odot$» corresponds to the composition law of section XX or 11 of XXVIII. The definition of this operation allows us to write the same previous equation in the form

$(\overline{F}\ N)=[(M\bullet\overline{a})\ (kg*m/\!/s^2)]$, in which three composition laws appear : the multiplication «•» of a real number by a vector, the scalar dyadic multiplication «*» and the scalar dyadic division «/\!/». For simplicity, the case of an «dysmetric» space such that the only flexible fundamental magnitud is length can be assumed. Let the primary measure be the vector measure of the velocity at a point P in space. By hypothesis, the quantity of length of the standard meter m in the reference environment and the congruent meter m_P at point P will be related by the «dysmetric» density δ_P with $m_P = \delta_P \circ m$; while time is rigid or isometric, so the quantity of time a second welcomes will always be the same. Thus, the velocity in P will be:

$$\overline{v}\ \frac{m_P}{s} = \overline{v}\ \frac{\delta_P \circ m}{s} = \left[\delta_P \bullet \overline{v}\right]\ \frac{m}{s} \qquad [32.1]$$

Instead of abstract units, we are using terrestrial units as reference elements, and we designate the corresponding ones at the generic point P indicating it with a subscript, so that Newton's second law at any point P would be written like this:

$$(\overline{F}\ N_P)=[(M\bullet\overline{a})\ (kg*m_P/\!/s^2)] \qquad [32.2]$$

Mass and time are magnitudes that by hypothesis in the case studied are rigid, hence their units have not been indicated with the subscript P, which indicates the point in space where this physical law applies, a notation reserved for flexible magnitudes, in this case only length and force, which is a composite magnitude.

We define the «dysmetric» acceleration at point P as the differential dyadic quotient between the dyads quantity of velocity and quantity of time:

$$\overline{a}\ \frac{m_P}{s^2} = \frac{d\left(\overline{v}\ \dfrac{m_P}{s}\right)}{d(t\,s)}$$

Therefore, by differentiating the second member of equation [32.1] with respect to time, we will have the «asymmetric» acceleration in P. Since the meter m is the reference length standard it is invariable with time. In turn, the second, unit of time, by hypothesis, always contains the same quantity of time. Therefore m and s are constant with respect to time, and thus we have the following derivative:

$$-\overline{a}\,\frac{m_P}{s^2}=\frac{d\left(\overline{v}\,\dfrac{m_P}{s}\right)}{d(t\,s)}=\frac{d\left(\delta_P\bullet\overline{v}\right)\dfrac{m}{s}}{dt\,s}=\left(\frac{d\delta_P}{dt}\bullet\overline{v}+\delta_P\bullet\frac{d\overline{v}}{dt}\right)\frac{m}{s^2}$$

Since $m_P=\delta_P\circ m$, it will be $m=m_P/\!/\delta_P$ and the second member of the previous equation can be written like this:

$$-\overline{a}\,\frac{m_P}{s^2}=\frac{\left(\dfrac{d\delta_P}{dt}\bullet\overline{v}+\delta_P\bullet\dfrac{d\overline{v}}{dt}\right)}{\delta_P}\,\frac{m_P}{s^2}\qquad[32.3]$$

Since the secondaries of these two dyads are equal, the primaries must also be equal, so by virtue of the criterion of dyadic equality, the vector measure of the acceleration in P will be given by

$$-\overline{a}=\frac{\left(\dfrac{d\delta_P}{dt}\bullet\overline{v}+\delta_P\bullet\dfrac{d\overline{v}}{dt}\right)}{\delta_P}\qquad[32.4]$$

Substituting the previous measure [32.4] of the acceleration vector $\overline{a}$ in [32.2], the «dysmetric» expression of Newton's second law results, whose measure does not simply correspond to the classical result of the measure of the derivative of the speed, but a more complex primary appears:

321

$$\left(\overline{F}\,N_P\right) = M \bullet \frac{\left(\dfrac{d\delta_P}{dt} \bullet \overline{v} + \delta_P \bullet \dfrac{d\overline{v}}{dt}\right)}{\delta_P}\; \frac{kg * m_P}{s^2}$$

Newton's second law when only length is «dysmetric»   [32.5]

This result accepts the general case in which the density function varies both through space and time. If δ_P did not depend on time, its derivative would be null and would result:

$$\left(\overline{F}\,N_P\right) = M \bullet \frac{d\overline{v}}{dt}\; \frac{kg * m_P}{s^2} \qquad [32.6]$$

Just as [32.5] is the general «dysmetric» form of Newton's second law in a space in which the only flexible fundamental magnitude is length, [32.6] corresponds to a particular case of flexibility in which the density function of length does not depend on time, but only on position, resulting in that the form of Newton's second law remains uniform at all points in space without more than contemplating at each point the unit of length inherent in it, that is, the meter m_P, congruent with the reference standard meter m at a given fixed point, since the derivative $d\overline{v}/dt$ is the vector measure of acceleration, which is independent of the unit of length that corresponds to each point P; and thus the existence of «dysmetric» spaces is mathematically verified in which the measures of physical properties, such as length, velocity, acceleration and force, remain constant even though the quantities associated with these measures are not constant, such as a consequence of the variation of the implicit quantities in the corresponding congruent units, differences reflected in the «dysmetric» densities.

In definition [32.3] it is observed that acceleration, understood as the dyadic quotient of the differentials of velocity and time, has the vector $\overline{a}$ as measure, and this vector does not coincide

322

with the arithmetic derivative of the measure of velocity with respect to the time $d\bar{v}/dt$, as in classical mechanics, the effect of the density function of length δ_P also intervenes, as reflected in equation [32.4].

Only if δ_P is independent of time, its derivative being null, will the result [32.4] coincide with the classic one and we will have $\bar{a} = d\bar{v}/dt$. Otherwise, the acceleration measure will have the general form [32.4]. In turn, since $m_P = \delta_P \circ m$, equation [32.6] can obviously be written in this way:

$$\left(\overline{F}\,N_P\right) = \delta_P \times M \bullet \frac{d\bar{v}}{dt}\ \frac{kg*m}{s^{2}}$$

It thus follows that, if the density δ_P is constant at each point P, the greater its value, the greater must be the force to be applied to a mass M to impart a certain fixed acceleration $\bar{a} = d\bar{v}/dt$; and on the contrary, the smaller is δ_P, the smaller will be that force. This result shows the influence of the «dysmetric» densities on physical phenomena and clearly indicates the representative qualities of this new Physics in those cases in which the «dysmetric» densities are affected by various causes. Thus, the same laws can be adapted to different environments without more than considering the «dysmetric» densities associated with each of them. When we speak of environments we refer to the terrestrial, atomic or cosmic spheres, as well as any other pertinent to the object of study.

For all these reasons, the assumption of classical physics that physical equations can be replaced by arithmetic relationships between the measures of the various magnitudes is nothing more than a crude simplification of the general case of «dysmetry». The «dysmetric» space examined here, one of the simplest that can be conceived, already reveals that «arithmetization» engenders a vital atrophy of physical models, severely limiting them in their capacity to represent natural phenomena.

At this point, it is time to move on to the second «dysmetric» analysis scheme, the **discrete formulation** of Newton's second law.

To do this, suppose the space divided into various environments separated from each other by enormous distances. Let us admit that the separation between them is such that the effects they produce on each other are negligible for observers located in each of these places. Let us take one of them as a reference environment, for example, the terrestrial sphere, and indicate the physical units of it, omitting any subscript. In the field of another environment E, the quantities of the units of the various magnitudes congruent with those of the reference field will not remain constant and the relationship between them will be established by the corresponding density. Thus, the fundamental units in both systems of length L, mass M and time T, will be related in this way:

$$m_E = \delta_L \circ m; \; kg_E = \delta_M \circ kg; \; s_E = \delta_T \circ s \qquad [32.7]$$

Obviously, δ_L, δ_M and δ_T indicate the densities of the fundamental magnitudes of length, mass and time in the environment E with respect to the terrestrial reference, which are the fixed patterns adopted.

Suppose Newton's second law is valid in all such environments. In E this law can be written with the following notation:

$$(\overline{F} \; N_E) = (M \; kg_E) \odot (\overline{a} \; m_E /\!/ s_E^{\,2}) \qquad [32.8]$$

Referring equation [32.8] to the terrestrial reference environment through the relations [32.7], it will result:

$$(\overline{F} \; N_E) = [M \; (\delta_M \circ kg)] \odot [\overline{a} \; (\delta_L \circ m /\!/ \delta_T^{\,2} \circ s^{\,2})]$$

Operating with dyadic algebra, we easily arrive at the following expression:

$$\left(\overline{F} \; N_E\right) = \left(\frac{\delta_M \times \delta_L}{\delta_T^2} \times M \bullet \overline{a}\right) \frac{kg * m}{s^{\,2}} \qquad [32.9]$$

The dyadic formula [32.9] is the discrete «dysmetric» expression of Newton's second law in the environment E and linked to the terrestrial sphere of reference. It is easily observed

324

that the factor in which the «dysmetric» densities appear, the subscripts formally reproduce the dimensional equation of the force quantity. Indeed, in accordance with classical dimensional analysis, it can be written:

$$[F] = \frac{M \times L}{T^2}$$

Therefore, the«dysmetry» gives full meaning to dimensional analysis, which was void of content due to the vice of «arithmetization», saved by dyadic algebra, knowing that the previous dimensional expression, in mathematical purity, should be written from this other way:

$$[F] = \frac{M * L}{T * T}$$

Where, in the absence of symbolic simplifications, the asterisk designates the algebraic operation that multiplies quantities of magnitudes and the double line shows the dyadic quotient.

On the other hand, repeating the analysis we made of the physical law [32.6], now regarding the corresponding [32.9], we can write it with the following notation:

$$\Delta_E = \frac{\delta_M \times \delta_L}{\delta_T^2} \Rightarrow \left(\overline{F} \, N_E \right) = \left(\Delta_E \times M \bullet \overline{a} \right) \frac{kg * m}{s^2}$$

We will say that Δ_E is the composite «dysmetric» density. It is easily observed that in an environment E in which Δ_E is very large, the force necessary to impart a certain acceleration $\overline{a}$ to a mass M will also be very large, in relation to another environment in which Δ_E is very small and for the same mass and acceleration. Therefore, the force corresponding to Newton's second law differs in each environment E as a function of the quantity of its composite «dysmetric» density Δ_E.

The preceding «dysmetric» analysis of Newton's second law, with this double approach, shows how classical physics childishly simplifies this law by means of an equation of vector algebra, the well-known $\overline{F} = M \cdot \overline{a}$ que, which uses only the measures of the related magnitudes, force, mass and acceleration, regardless of how the relationship between the units of these magnitudes may affect the phenomena represented. This is how the nefarious «arithmetization» of Physics tacitly dispenses with something so essential, such as physical or non-arithmetic algebras. Quantities of quantities and their relationships are the core of physical laws. Ignoring them and maintaining «arithmetization» is a guilty anachronism that keeps this science from flying. Here it has been verified how the specific dyad algebra of Physics transforms that essential law of nature into configurations never seen before, predicting a horizon of innovations never imagined from the simplicity of arithmetic algebra.

Apart from this, given the verification that «dysmetric» densities affect the measurements of physical phenomena, as we have observed, so that the higher the density, the greater the force for equality of mass and acceleration, it makes it clear that the same laws apply they can be applied to different physical environments without more than taking into account the appropriate «dysmetric» densities for each of them.

hus, any attentive reader who has followed the preceding «dysmetric» analysis of Newton's second law will easily glimpse where the new Physics is going and will have no objection to joining the «dysmetric» movement. You will agree that this tool is necessary and inalienable, and that it should be advocated, disseminated and generalized. Perhaps there are those who are intimidated by its apparent mathematical complexity, but these must be encouraged with the well-known adage that great conquests were never simple or lazy, and it turns out that there are very advanced mathematical tools, such as tensor calculus, which are suitable for accommodating «dysmetric» shapes. In any case, it would not be wise to give up any advance in our sight simply because of its complexity. The only limit of any researcher should not be other than the impossibility of something, never the effort or intelligence necessary to achieve it. Aristotle said that «There is only happiness where there is virtue and serious effort, because life is not a game». Or as Seneca sentenced: «Nothing would ever be discovered, if we considered ourselves satisfied with the things discovered». Ultimately, in the end, when everything is over, the only thing that remains and matters is what you have done. And it turns out that the new Physics, the «dysmetric» Physics, has yet to be done. It is an infinite and exciting task that can fill many lives. Nor can we imagine the surprises that this trip can bring us. The only sure thing is that it will be worth it.

Section XXXIII

THE «DYSMETRIC» PI NUMBER
In a «dysmetric» space, neither
number π remains constant

The purpose of this section is to analyze the number π from a «dysmetric» point of view. To do this, consider a flat space with a reference point O where there is the standard meter m, which will serve as a unit of universal length to which the congruent meters will be related at other points in space. The problem is described in Figure 26:

The starting point is a circle with a center at O and a mathematical radius r m, a quantity of length measured by mere juxtaposition and congruence with the standard unit m. The «dysmetric» space is assumed to be isotropic, that is, the «dysmetric» density is distributed in the same way over any radius OP. Let X be a generic interior point of the radius OP. Its mathematical distance to O will be given x m, where x represents the number of congruences of the measure of OX with the meter m. A differential segment in X of measure dx will be represented by the dyad dx m_X, with the meaning (dx, m_X), different from $d(x\ m_X) = d(x, m_X)$, where m_X is the unit of length or meter in X congruent with the meter m in O. The relationship between the two units will be given by the «dysmetric» density δ_X in X, and it can be written in the form $m_X = \delta_X \circ m$.

The quantity of mathematical length by mere congruence of the radius OP we have postulated to be r m. Let's now calculate the quantity of physical length of the same radius OP. To do this, all the differential segments dx m_X between O and P must be added, which could be indicated with the following integral:

$$\int_0^r \left[dx\ m_X \right] = \int_0^r dx \circ \left(\delta_X \circ m \right) = \left[\int_0^r \delta_X \times dx \right] m \qquad [33.1]$$

In this calculation, the properties of operations with magnitudes from section IX and 6 of XXVIII of the First Algebra of Magnitudes have been used. Thus, we have the quantity of physical length of the radius OP reduced to the reference meter m in O. Since what is sought is the uniform dyadic ratio between two physical lengths, that of the circumference and that of the diameter, it is necessary to determine next the first. To do this, the differential arc segments $ds\ m_r = (\delta_r \times dr)\ m$ must be added along the entire circumference. The mathematical measure ds can be put in the form $ds = r \times d\theta$, where θ is the angle in radians of the

**Pi number for a circle
of a flat «dysmetric» space and
isotropic about a central point**

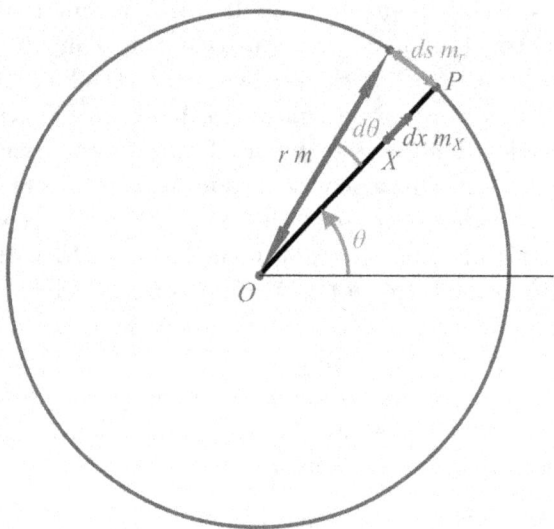

Representation of a circle in a plane «dysmetric» space with center at point O, with respect to which the space is isotropic for the density of length. The number pi is dependent on the radius and the «dysmetric» density.

Figure 26

330

generic radius OP with respect to another reference. With this, it is easy to see that the differential element to be integrated is ds $m_r = (\delta_r \times r \times d\theta)$ m, in accordance with the provisions of section IX and 6 of XXVIII, already mentioned. The density of length dr along the circumference is constant, given the isotropy of the postulated space. Therefore, the amount of physical length of the circumference will be given by the following integral:

$$\int_0^{2\pi} \left(ds\ m_r \right) = \left[\int_0^{2\pi} \delta_r \times r \times d\theta \right] m = \left(2 \times \pi \times r \times \delta_r \right) m \quad [33.2]$$

The uniform dyadic quotient between the quantities [33.2] and twice [33.1], both being two lengths, will be the ratio sought, which is nothing more than the physical number pi in the «dysmetric» space considered for the circumference of radius r and center O. Such quotient is defined in section XI or 7 of XXVIII of the First Algebra of Magnitudes, where it was justified that the dyadic ratio of two scalar concrete quantities referred to the same unit is the equal real number to the arithmetic ratio of its primaries. Therefore, said number pi, which could be noted π_r, since it has been obtained for a specific circle of radius r, will be given by the following uniform dyadic quotient:

$$\pi_r = \frac{\left(2 \times \pi \times r \times \delta_r \right) m}{2 \times \left[\int_0^r \delta_X \times dx \right] m} = \frac{\pi \times r \times \delta_r}{\int_0^r \delta_X \times dx}$$

In conclusion, in the «dysmetric» space considered, the ratio between the physical lengths of the circumference and its diameter is not constant or equal to the number pi, but depends on the radius r and the «dysmetric» density δ_X along the radius, as well as its dr value at any point on the circumference, according to the previous formula.

In particular, if the «asymmetric» density were constant and equal to unity at all points, it would result that $\delta_X = \delta_r = 1$, and the integral of the denominator would be equal to ar, with which the result would coincide with the classic one and we would have

331

$\pi_r = \pi$. This checks the coherence of the «dysmetric» tool, which always has to conclude the coincidence with classical geometry and physics in the particular case that the «dysmetric» density is constant at all points in space and equal to unity.

This experimentation is of the utmost importance, because it shows that «dysmetry» by nature excludes the existence of physical constants even at the geometric level for the mythical number pi. So, if real space obeyed that same nature, the physical constants should be considered an unrealistic simplification, which would only have to be taken into account on a purely theoretical plane.

Section XXXIV

«DYSMETRIC» ANALISYS OF THE SPEED OF LIGHT
In a «dysmetric» space the speed of light
it doesn't have to be constant

The speed of light acquired an iconic role with the publication in 1905 of Einstein's Theory of Special Relativity or Special Relativity. The relativistic postulate about the constancy of its speed, as well as the supposed impossibility of exceeding it, have fascinated everyone, provoking endless fables about time travel and other fantasies more typical of literary fiction than of scientific works.

Einstein based all his Theory of Relativity on a single basic postulate, the supposed invariability of the speed of light with independence of the movement of the source in all inertial systems, cementing all the mathematical development on this hypothesis, which he established assuming the famous Michelson and Morley experiment of 1887. At that time the theory of ether was still believed. The Theory of Special Relativity had its origin precisely in the negative result of this experiment. Various experimental facts had led to admit the existence of an ether at absolute rest, which would not participate in the movement of matter and would constitute the basis for the propagation of electromagnetic waves. Today we believe that electromagnetic waves propagate in a vacuum. From this concept of immobile aether, it would seem to inevitably follow that the value of the speed of light, measured by an observer in motion with respect to the ether, would depend on said movement and, in particular, on the direction of its speed. If c is the speed of light with respect to the immobile ether and v is the speed of the observer, the observer should measure, according to classical kinematics, a speed of light $c-v$ or $c+v$, in absolute value, as it moves in the same direction and sense that the light or in the opposite direction. An observer who

333

was initially unaware of its movement with respect to the ether could experimentally appreciate it by emitting a light signal in all directions and measuring the times it took for said signal to reach the points of a sphere in motion with the observer and centered on the emitting point. Of the signal. If there were movement with respect to the ether, the ether wind would have to blow the signal, so that it would first reach the point of the sphere directly opposite the direction of movement and lastly the point corresponding to the direction and sense of movement.

This was the reason for the famous experience of Michelson and Morley, through which they tried to determine the state of motion of the Earth with respect to the ether. In relation to the Copernicus axes, which have their origin in the center of gravity of the solar system, the speed of the center of gravity of the Earth on its trajectory is approximately 30 km/s, and in six months this vector becomes another sensibly opposite. It could happen that at a certain moment the unknown movement of the axes of Copernicus with respect to the ether annulled the absolute movement of the Earth, but such a coincidence could not exist for a year or at any two opposite points in the Earth's orbit.

Thanks to a well-known interferential device, Michelson was able to reveal an aether wind equal «only» at 1,5 km/s. Which is equivalent to measuring the quantities of 28,5 km/s and 31,5 km/s as the speed of light with respect to the observer, depending on whether the movement is in the same direction as the light or in the opposite direction. But, given this «so small» difference, it was considered negligible and attributable to the error of the measuring instruments. Thus, instead of admitting that there would be variation in the speed of light, it was agreed that there was not. This induced artificially and approximately, not exactly, the conclusion that there would be no dependence of the speed of light on the state of motion of the observer. Under these conditions, Lorentz and Einstein took as a true starting point the illusory result of Michelson's negative experiment and formulated the following theoretical principle: the speed of light c is constant in all inertial reference frames, which implies admitting that the speed of light is the same for all systems moving relative to each other at constant speed, without any acceleration. The pure mathematization of this hypothesis leads to the Lorentz transformations and these to Einstein's Relativity.

Therefore, all Relativity has as its Achilles heel the only theoretical principle on which its entire logical scheme is based, that of constant speed of light in all inertial systems. All the other supposed experimental verifications in his favor, adduced a posteriori by his unconditional loyal followers, would not have the least value of scientific or logical proof, if that fundamental postulate were refuted in some way, even if it was not conclusive. And precisely the aforementioned principle is based on the already old Michelson experiment, whose conclusion is highly doubtful and imprecise. Relativity has an amazing seductive effect, because with a single basic principle it produces a prodigious mathematical apparatus, but it hangs from that single thread, which could break at any moment. It is very plausible to suspect that this fine support, already very weak in itself, given the fragility of Michelson's experiment, is definitively broken by the « dysmetry» of space, and perhaps it will not take long to admit that the Theory of Relativity it is nothing more than a magnificent mathematical speculation based on a false principle, something that will not happen with Newton and his mechanical laws.

As has been observed with the analysis of the geometric number pi, the «dysmetry» has a nature such that it is incompatible with the existence of physical constants, which would also determine the impossibility of the speed of light c being invariant. Below is the concrete analysis for this current universal constant, which is dangerously serving current metrology even to define certain fundamental standard units such as the meter, the second and even the kilogram. Consider an «dysmetric» space in which length and time are flexible. Let O be the reference point with respect to which the congruent units of all other points P in space will be associated. The standard units of length and time in O are assumed to be the meter m and the second s. At any other point P the units of length and time congruent with the previous ones will be designated, as usual, with the notation m_P for the meter and s_P for the second, being related to those of the origin O as a function of the «dysmetric» densities of length and time in P, that is, respectively δ_{LP} and δ_{TP}, with which we will have the relationships $m_P = \delta_{LP} \circ m$ y $s_P = \delta_{TP} \circ s$.

Obviously, Michelson's experiment refers to the measure c of the speed of light as a constant parameter, since at that time it was not imagined at all that the standard units could vary in their quantity of implicit magnitude from one point to another in space. If the measure of the speed of light c were to be constant at all points, it would also be constant at P, where the speed quantity would be the dyad $c\, m_P /\!/ s_P$. This quantity of velocity can be referred to the units in O simply by operating with the «dysmetric» densities and the transformation into its uniform quantity will be easily obtained with the following logical sequence of dyadic algebra:

$$c\ \frac{m_P}{s_P} = c\ \frac{\delta_{LP} \circ m}{\delta_{TP} \circ s} = c \times \frac{\delta_{LP}}{\delta_{TP}}\ \frac{m}{s}$$

Consequently, the measurement c_P of the speed of light at any point P referred to the standard units in O is given by the expression:

$$c_P = c \times \frac{\delta_{LP}}{\delta_{TP}} \qquad [34.1]$$

Thus, if c were the measure with $m_P /\!/ s_P$ of the speed of light in P, its measure referred to O with $m /\!/ s$ would have to be $c \times \delta_{LP} / \delta_{TP}$. It is thus concluded that the measurement of the speed of light in O and in P cannot coincide, since in general $c \neq c \times \delta_{LP} / \delta_{TP}$, except in the particular and strange case that the equality $\delta_{LP} = \delta_{TP}$ was given in all point P. In a «dysmetric» space where this is not the case, that is, where there is some point P where $\delta_{LP} \neq \delta_{TP}$, the invariance of the speed of light would not be verified. It is clear that this conclusion would be absurd, because as an initial hypothesis the absolute invariance of the speed of light had been imposed, according to Michelson's experiment. Therefore, this hypothesis could not be verified if $\delta_{LP} \neq \delta_{TP}$ in some P, and in such a general «dysmetric» space the measure of the speed of light could not be invariant. That is to say, in a «dysmetric» field that is widely variable in terms of its densities of length and time the uniform measure of the speed of light and, therefore, its quantity of speed cannot be kept constant. On the other hand, the ratio δ_{LP}/δ_{TP} determines that if its value in P tends to zero, the measure and with it the amount of speed of light tend to zero and, if that ratio tends to infinity, both tend to infinity, making Einstein's postulate impossible. In turn, the speed of light has no limit in a general «dysmetric» space contradicts the Einsteinian conclusion that speeds greater than the current constant c cannot occur in nature, just as it would not have a limit or the speed of propagation of electromagnetic waves would be constant.

Let's next do a physical-mathematical exercise beyond Michelson's experiment and see how «dysmetry» affects the amount of speed. At point O such physical quantity will be represented by the dyad $c \; m /\!/ s$. In turn, at a generic point P and, therefore, with «dysmetric» densities δ_{LP} and δ_{TP}, there will be the quantity indicated by $c_P \, m_P /\!/ s_P$. Developing this last quantity through dyadic algebra, we have:

$$ c_P \frac{m_P}{s_P} = c_P \frac{\delta_{LP} \circ m}{\delta_{TP} \circ s} = c_P \times \frac{\delta_{LP}}{\delta_{TP}} \frac{m}{s} $$

At P the measurement c_P should be observed materially and would have a given value. Since the quantity $m /\!/ s$ is finite and invariant, if δ_{LP}/δ_{TP} tended to zero, the last term of the previous equality, which represents the amount of velocity at P, would tend to zero, and if δ_{LP}/δ_{TP} tended to infinity, said speed would also tend to infinity. Therefore, «dysmetry» generally prevents

337

the amount of velocity from being the same at all points in space and has a range of variation between zero and infinity.

Let us now see what conditions should be met for the invariance of the quantity of velocity to be verified at every point P. Such a premise is reflected in the following dyadic algebra reasoning:

$$c \, \frac{m}{s} = c_P \, \frac{m_P}{s_P} = c_P \, \frac{\delta_{LP} \circ m}{\delta_{TP} \circ s} = c_P \times \frac{\delta_{LP}}{\delta_{TP}} \, \frac{m}{s}$$

The first and last terms of the previous chain have the same secondary, the compound unit $m/\!/s$, then, applying the criterion of dyadic equality, their numerical primaries must be equal, and it can be ensured that:

$$c = c_P \times \frac{\delta_{LP}}{\delta_{TP}} \implies c_P = c \times \frac{\delta_{TP}}{\delta_{LP}} \qquad [34.2]$$

For this condition to be met and the amount of speed of light to be constant, the relationship between the measurements of the speed c in O and c_P at any point P could not be any, but rather the ratio δ_{LP}/δ_{TP} between the densities «dysmetric» of length and time would have to be proportional to the ratio c/c_P, according to expression [34.2], which would mean establishing an unproven physical law, which cannot be arbitrarily admitted and which obviously does not have to be satisfied at all priori throughout the widely variable «dysmetric» space.

In conclusion, whether law [34.1], constant measurement c of the speed of light, or its complementary law [34.2], constant quantity of speed $c_P \, m_P$, were admitted, in no case is said invariance compatible with the «dysmetry», which It has very important implications in relation, for example, to the determination of the age of the universe or the estimation of distances, which are currently calculated with the constant c, to which the definitions of the standard units of fundamental magnitudes, formulated by the International System of Units, also refer with great probability of error.

To illustrate this fact let us consider a numerical example. Let O be the reference point of space, which could represent the terrestrial environment, with respect to which the «dysmetric» densities are established at any other point P. The «dysmetric» densities of the magnitudes length and time in O will both be unity, with which we will have $\delta_{LO} = \delta_{TO} = 1$. Suppose that at point P the length has a density equal to 2 and time is isometric, so its density will be unity, so that $\delta_{LP} = 2$ and $\delta_{TP} = 1$. Under these conditions, the measurement of the speed of light in P, which we denote c_P, admitting *Einstein's postulate* and the current criterion of the International System of Units, both framed in law [34.1], will be given by the following calculation:

$$c_P = c \times \frac{\delta_{LP}}{\delta_{TP}} = c \times \frac{2}{1} = 2 \times c$$

Therefore, under the conditions of the example the measured speed of light at P is twice that at O and c would not be constant.

In assumption law [34.2], applied to the example, the «dysmetric» densities of length and time would have to satisfy the following condition:

$$\frac{c}{c_P} = \frac{\delta_{LP}}{\delta_{TP}} = \frac{2}{1} = 2$$

It would thus result that the length and time densities could not be any, as would be expected in a generic «dysmetric» space, but that the ratios c/c_P and δ_{LP}/δ_{TP} would both be equal to 2, which would be as much as admitting without foundation a strange restriction, which in any case should be tested experimentally.

In this section we have briefly described the meaning of Michelson's experiment according to the explanation of André Lichnerowicz in his text *Elements of Tensor Calculation*. However, we must observe that in our reasoning exposed in the first place, the result of said experiment or the relativistic hypothesis is not denied, but rather, in order not to contradict them, what we do is the hypothesis that they are correct, resulting that, if the measurement of the speed of light or its quantity in the terrestrial environment, which is the object of that isometric experiment, were constant, the quantity of speed does not have to be so in all points of «dysmetric» space. Therefore, what we are questioning here is the extension of the result to all points in space. What's more, with dyadic algebra we prove that the relativistic hypothesis that the speed of light is invariant in all space is generally incompatible with «dysmetry», except for very singular cases.

On the other hand, «dysmetry» does not refute classical Newtonian mechanics at all, as was verified when analyzing *Newton's second law* from the «dysmetric» point of view in section **XXXII**, as well as it is also verified in the following section **XXXV** dedicated to «dysmetric» gravitation.

Section XXXV

THE «DYSMETRIC» GRAVITATION
Alternative rational explanation to dark
matter for gravitational anomalies

Let's begin by giving a brief review of classical gravitation with the help of Figure 27. From a mechanical point of view, it is considered that the forces observed in nature can be at a distance, such as gravitation, or contact, when bodies seem to come together and support or collide with each other. Today the atomic model seems to reveal that matter would never come into contact, although for all practical purposes our theories assume otherwise. In mechanics the actions that are considered are alien to other phenomena such as electricity and magnetism, which have their own descriptions and explanations. Precisely those branches of Physics were born because mechanics was not capable of representing certain experiences, giving rise to other specific models. Therefore, mechanics is limited to the study of actions at a distance or in contact, including in these the interiors of linkage or linking that hold some bodies together, that material points are exerted between them, considered as minuscule groupings of matter. The mechanical laws are applied in such a way that, both at a distance and in contact, separated or linked, two material points influence each other by means of two forces applied to each of them with the direction of the line that passes through both, opposite direction and equal modulus. This is the fundamental basis of mechanical theory, which is completed with the fundamental law that relates forces, inertial masses and accelerations. The current theory of actions at a distance is due to Isaac Newton, who formulated the law of universal gravitation by determining the quantity of the forces at a distance with which two material points P_1 and P_2 are influenced. Following the notation of the third postulate, such forces are identified with the

341

Laws of mechanics
Newton's postulates

I. Inertia postulate

There is an absolute reference system in relation to which every isolated material point has zero acceleration, so its motion is rectilinear and uniform.

II. Vector relationship between force, mass and acceleration

Since $\Sigma \overline{F}_i(t)$ is the resultant of all the forces acting on a material point of mass M and acceleration $\overline{a}(t)$ at instant t, the law represented by the vector equation is admitted:

$$\sum \overline{F}_i(t) = M \cdot \overline{a}(t)$$

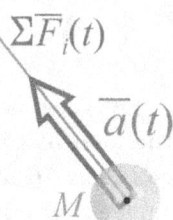

III. Equality of action and reaction (at a distance or in contact)

In any state of motion or rest, contact or distance, every material point P_1 exerts on another P_2 an action called force, which is assimilated to a fixed vector $\overline{F}_{21}$ applied to P_2, whose direction is that of the line that passes through P_1 and P_2; and reciprocally, the point P_2 exerts on the P_1 a force given by the fixed vector $\overline{F}_{12}$ applied in P_1 and with the same direction as the previous one; these two forces are admitted to have the same modulus and opposite direction, which is given vectorly by $\overline{F}_{21} = -\overline{F}_{12}$.

Superposition principle: For any system of material points $\{P_i\}$ with i taking values of 1 to n, each material point P_j is subjected to $n-1$ forces $\overline{F}_{jk}$ with $j \neq k$, because the points do not exert action on themselves. These forces are assumed to be applied at P_j and their resultant j also applied at P_j is admitted that it can be calculated by vector algebra and that it will produce the same effect on P_j. As the relationship between force and acceleration is the inertial mass, constant for each P_j, analytically the following vector equations will be obtained:

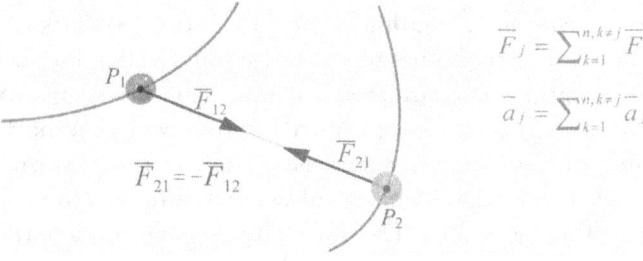

$$\overline{F}_j = \sum_{k=1}^{n, k \neq j} \overline{F}_{jk}$$

$$\overline{a}_j = \sum_{k=1}^{n, k \neq j} \overline{a}_{jk}$$

Figure 27

fixed vectors $\overline{F}_{21}$ for the action of P_1 on P_2, and $\overline{F}_{12}$ for that of P_2 on P_1, applied $\overline{F}_{21}$ on P_2 and $\overline{F}_{12}$ on P_1, they must have the direction of the line that passes through P_1 and P_2, in the opposite direction, which means that $\overline{F}_{21} = -\overline{F}_{12}$. The principle of action and reaction does not pay attention to whether these forces are attractive or repulsive or to the extent of the modulus of these

342

reciprocal actions, because the principle affects all types of mechanical forces. Well, the law of gravitation establishes those of a gravitational nature with the formulation of the experimental hypothesis that determines that the gravitational forces at a distance $\overline{F}_{21}$ and $\overline{F}_{12}$ are attraction and their equal modules are directly proportional to two positive numbers designated μ_1 and μ_2 such that they do not depend on time and represent a certain characteristic and invariant of each material point P_1 and P_2, or any other, as well as these modules are observed inversely proportional to the square of the distance r that separates the two material points considered, everything which is analytically expressed by the well-known equation that is written below:

$$\left|\overline{F}_{21}\right| = \left|\overline{F}_{12}\right| = \frac{\mu_1 \times \mu_2}{r^2}$$

[35.1]

The positive real numbers μ_1 and μ_2 are considered to measure a certain quality of the material points P_1 and P_2, which is called **gravitational mass** and refers to a kind of capacity of matter to exert attraction at a distance. So at every material point there are two characteristics present that are described by the mass of inertia M or measure of the resistance to be accelerated according to the second postulate of mechanics, and by the gravitational mass μ or measure of the ability to exert attraction on other material points by virtue of the law of gravitation. But there is still more, because it turns out that the quotient between the gravitational mass and that of inertia is a number that seems to be the same for all matter, it seems to be the same for all material points, so it is considered a universal constant; so representing it with the symbol, so that when multiplied by itself it gives $\sqrt{G}$, the law of universal gravitation can be expressed as a function of the **inertial masses** M_1 and M_2 of the material points and the universal gravitation constant $G = 6,67 \times 10^{-11}$ $N*m^2 /\!/kg^2$, easily resulting in the equalities:

$$\left|\overline{F}_{21}\right| = \left|\overline{F}_{12}\right| = G \times \frac{M_1 \times M_2}{r^2}$$

For this reason in Physics when we speak of mass, no distinction is made between inertial and gravitational mass, understanding mass for both purposes as mass, so that to quantify the gravitational attraction between two given inertial masses, the constant must be considered G. The law of gravitation describes the only known forces at a distance of mechanical origin and allows the power to describe analytically the interactions at a distance between the points of any system of material points; so that, for a system of material points $\{P_i\}$ with inertial masses M_i, with i taking values of 1 to n, given another material point P_0 with inertial mass M_0, the total resulting force $\overline{F}_0$ of the forces $\overline{F}_{0i}$ exerted by the points of the system $\{P_i\}$ on this other material point P_0, where $\overline{u}_i$ is the versor of the sense of the vector with origin at P_0 and end at P_i and r_i the distance between P_0 and P_i, as the validity of the algebra of mathematical vector spaces is admitted, the resultant $\overline{F}_0$ is given by the expression:

$$\overline{F}_0 = \sum_{i=1}^{n} \overline{F}_{0i} = \sum_{i=1}^{n} G \times \frac{M_0 \times M_i}{r_i^2} \bullet \overline{u}_i$$

If the versor $\overline{u}_i$ were set in the opposite direction, that is, from P_i to P_0, a minus sign would have to be placed before the second member, and a minus sign would also correspond if instead of adding the forces $\overline{F}_{0i}$, the $\overline{F}_{i0}$ exerted on each P_i by P_0, because, given the principle of action and reaction, it will be $\overline{F}_{0i} = -\overline{F}_{i0}$ for all i. On the other hand, in the previous scheme a different criterion from the second postulate has been used, in which the analyzed point was considered included in the system, so it was necessary to differentiate it with the condition $j \neq k$; instead, here the points $\{P_i\}$ have been differentiated from P_0, making this distinction unnecessary. It is clear that the physical results are not influenced at all, except by the notation used in each case. Knowing the total force 0 exerted by the set of points $\{P_i\}$ on another P_0, it is possible to calculate its acceleration 0 with respect to an inertial reference system, simply dividing the previous force by the mass M_0 to obtain it:

$$\overline{a}_0 = \frac{\overline{F}_0}{M_0} = \sum_{i=1}^{n} G \times \frac{M_i}{r_i^2} \bullet \overline{u}_i$$

It is observed that the acceleration that the set of points $\{P_i\}$ communicates to any other point does not depend on its mass, but only on its position in space, as reflected by the distances ri between the positions of the points $\{P_i\}$ and that of another foreign point P_0. This means that two different material points, with different masses, if placed at the same point and with the same speed, under the influence of the point system $\{P_i\}$ would move with the same trajectory. If these two points are considered next to each other, they will form another material point such that their movement due to the set of material points $\{P_i\}$ will not depend on the joint mass of the two joined points either, so they will also move with the same trajectory with which they would do it individually, assuming that they were at the same point and with the same initial speed in all cases. This result allows defining a mathematical concept called **gravitational field,** since the action of a system of material points $\{P_i\}$ only depends on the position in space, so that any mass located in generic coordinates (x,y,z) will experience the same acceleration, hence the gravitational **field strength** of a given system of material points $\{P_i\}$ in the coordinates (x,y,z) is defined as the vector $\overline{E}(x,y,z)$ representing the acceleration that the system would communicate to any other material point located in that position, so its expression coincides with that of the acceleration without more than modifying the notation of the first member:

$$\overline{E}(x,y,z) = \sum_{i=1}^{n} G \times \frac{M_i}{r_i^2} \bullet \overline{u}_i$$

The previous summation is composed of addends that represent the gravitational field induced by each isolated material point P_i, so that it can be stated that the field strength of a system at each position in space is the vector addition of the fields that correspond to each one of its points in that same position. A field

345

is nothing more than a mathematical abstraction that consists in establishing an application between the geometric space E^3 and R, if each point in space is assigned a real number, or V^3, if each point in space is assigned a three-dimensional geometric vector; in the first case there would be a scalar field and in the second a vector field, which could ultimately be described as an application of R^3 to R^3, if the geometric vectors are replaced by their three components or abstract vectors. Obviously, the notion of field is included in the general concept of a vector function as an application of R^n in R^m, so by means of the names of gravitational field as a set of field intensities, the general mathematical concept of a vector function is being endowed with the specific meaning which refers to the action of a system of material points on any other that is located at a point in space, with the effect of transmitting a certain acceleration. On the other hand, it must be observed that in all the previous expressions, if it could be considered that matter is distributed in a continuous way, the summations would become integrals, since the material points could be assimilated to volumetric infinitesimal differential elements. The notion of gravitational field, applied to the system of material points that constitute any star, means the acceleration that would be communicated to any material point located in a certain place; assimilating that star to a material point with mass of inertia M and defining a versor $\overline{u}$ with direction and sense of the center of the star towards a point located at a distance r, the intensity of the gravitational field of M, noted $\overline{E}_M(r)$ at that distance r, will be given by the vector equation:

$$\overline{E}_M(r) = -\frac{G \times M}{r^2} \bullet \overline{u}$$

Let us apply the above to the case of a binary system formed by a very massive body and another much less heavy one, such that, for practical purposes, it can be considered that the largest remains immobile at a point O and that the smallest rotates around it in a circular orbit. The mass of the smallest body does not matter, as we have just verified, so the smallest element of the system can be any with negligible mass compared to the largest. Under these conditions, each orbit, defined by its radius r, will be associated with a unique value of the field, that is, an acceleration, and therefore an orbital period τ_r. As we are postulating the existence of a uniform circular motion, the tangential acceleration will be zero and we will only have a normal acceleration, which will be precisely the value of the gravitational field in the generic orbit. By designating ω_r the angular velocity for the orbit of radius r, we know from kinematics that the normal acceleration, also called centripetal, is $\omega_r^2 \times r$. In turn, the period and the angular velocity are related by the equation $\omega_r \times \tau_r = 2 \times \pi$. By Newton's second law, this acceleration must be the opposite of the field, given the defined sense for the versor. Therefore, the following reasoning can be spun:

$$\frac{G \times M}{r^2} = \omega_r^2 \times r \implies \tau_r = \frac{2 \times \pi}{\omega_r} = 2 \times \pi \times \sqrt{\frac{r^3}{G \times M}} \quad [35.2]$$

This result allows us to conclude that the period associated with each orbit of radius r is a function of r and M. So, given a mass M, each orbit r will have its specific period τ_r, according to the previous equation.

So far a brief exposition of classical gravitation, in which, as is conventional custom, one operates only with mathematical entities without explicitly taking into account the units of physical quantities. We will now extend this theory with the algebra of magnitudes to the «dysmetric» view of the phenomenon. For this, the first step must be to reformulate the classical gravitation of the law [35.1] applying the algebra of magnitudes, so that, taking two material points P_1 and P_2, being the versor with the direction of the line that joins the two points and their sense is from P_1 to P_2, assuming that their respective gravitational masses are μ_1 and μ_2, recalling the expression [33.1], which gives the «dysmetric» measure of the segment P_1P_2, referring the units to the congruent of P_1, in which the integral with the factor λ_r can be named, results:

$$\int_0^r [dx\, m_X] = \int_0^r dx \circ (\delta_X \circ m_1) = \left[\int_0^r \delta_X \times dx\right] m_1 = \lambda_r\, m_1$$

Admitting the Newtonian criterion that the measure of the action that these two material points exert on each other coincides in absolute value and that it is given by the expression [35.1], with all this, we have that the «dysmetric» expression of Actions $\overline{F}_{21}$ and $\overline{F}_{12}$, in their corresponding units of force in newton N_1 and N_2, which are exerted on each other at a distance from both points are:

$$\overline{F}_{21}\, N_2 = -\frac{\mu_1 \times \mu_2}{\lambda_r^2} \bullet \overline{u}\ \frac{kg_2 * m_2}{s_2^2}$$

$$\overline{F}_{12}\, N_1 = \frac{\mu_1 \times \mu_2}{\lambda_r^2} \bullet \overline{u}\ \frac{kg_1 * m_1}{s_1^2}$$

Referring the units of the points P_1 and P_2 to the congruent ones in another reference O, where the standard units m, kg and s exist, for length, mass and time, we can write $m_1 = \delta_{L1} \circ m$, $kg_1 = \delta_{M1} \circ kg$, $s_1 = \delta_{T1} \circ s$ y $m_2 = \delta_{L2} \circ m$, $kg_2 = \delta_{M2} \circ kg$, $s_2 = \delta_{T2} \circ s$. Operating with the algebra of magnitudes, the «dysmetric» expressions of the new law of gravitation are easily reached:

$$\overline{F}_{21}\, N_2 = -\frac{\delta_{M2} \times \delta_{L2}}{\delta_{T2}^2} \times \frac{\mu_1 \times \mu_2}{\lambda_r^2} \bullet \overline{u}\ \frac{kg * m}{s^2}$$

$$\overline{F}_{12}\, N_1 = \frac{\delta_{M1} \times \delta_{L1}}{\delta_{T1}^2} \times \frac{\mu_1 \times \mu_2}{\lambda_r^2} \bullet \overline{u}\ \frac{kg * m}{s^2}$$

348

Assuming the classical fact that the proportionality between the gravitational and inertial masses for all material points is assumed as an inherent quality of matter, we will have $\mu_1 = \sqrt{G} \times M_1$ and $\mu_2 = \sqrt{G} \times M_2$, and the previous formulas become these others:

$$\overline{F}_{21}\, N_2 = -G \times \frac{\delta_{M2} \times \delta_{L2}}{\delta_{T2}^2} \times \frac{M_1 \times M_2}{\lambda_r^{\,2}} \bullet \overline{u} \quad \frac{kg*m}{s^2}$$

$$\overline{F}_{12}\, N_1 = G \times \frac{\delta_{M1} \times \delta_{L1}}{\delta_{T1}^2} \times \frac{M_1 \times M_2}{\lambda_r^{\,2}} \bullet \overline{u} \quad \frac{kg*m}{s^2}$$

LAWS OF «DYSMETRIC» GRAVITATION   [35.3]

These are the two «dysmetric» expressions of gravitation. In them, it is observed that in a «dysmetric» space, the universal constant of gravitation does not strictly exist. First, because G is altered by the factor formed by the densities δ of the three fundamental magnitudes; and second, because the denominator of the product of the masses is not the square of the mathematical distance between the material points, but the quantity of physical length between them λ_r.

On the other hand, these «dysmetric» laws of gravitation, in general, if the factors with the densities δ in each of them are different, the reciprocal forces at distance $\overline{F}_{21}$ and $\overline{F}_{12}$ exerted on each other will also be different in modulus. two material points. And this result is the alternative «dysmetric» explanation for the gravitational anomalies observed today that are attributed to the mysterious dark matter. The «dysmetry» shows that the forces with which the bodies interact do not necessarily correspond to classical gravitation, hence some observed orbits deviate from those expected. To make this phenomenon clearer, let us calculate the «dysmetric» periods of the gravitational orbits.

In order to facilitate didactic understanding, let us consider the binary system already described and suppose that the space presents an isotropic «dysmetry» with respect to the central point

O. This assumes that the «dysmetric» density of any magnitude is distributed in the same way along the along any line that passes through *O*, or what is the same, that the «dysmetric» density of all magnitudes will be constant at all points of any sphere centered on *O*. The problem to be solved is to calculate the period of rotation of the body smallest P_2 in circular orbit with respect to the largest P_1 located at *O*. Since the «dysmetric» densities remain constant along every circumference centered at *O*, we can conclude that the classical kinematic laws apply to these cases. Which would not necessarily hold for other circles centered on points other than *O*.

The normal or centripetal acceleration of P_2 in an orbit of radius *r* will be $-\omega_r^2 \times \lambda_r \bullet \overline{u}$ $m_2 /\!/ s_2^2$, which must be equal to $\overline{F}_{21}$ $N_2 /\!/ M_2$; which brings us to the following «dysmetric» equation:

$$\overline{F}_{12} \, N_1 = G \times \frac{\delta_{M1} \times \delta_{L1}}{\delta_{T1}^2} \times \frac{M_1 \times M_2}{\lambda_r^2} \bullet \overline{u} \quad \frac{kg*m}{s^2}$$

To reduce both members to the same composite unit, it must be taken into account that $m_2 = \delta_{L2} \circ m$ and $s_2 = \delta_{T2} \circ s$, with which it results:

$$\frac{\delta_{L2}}{\delta_{T2}^2} \times \omega_r^2 \times \lambda_r \quad \frac{m}{s^2} = G \times \frac{\delta_{M2} \times \delta_{L2}}{\delta_{T2}^2} \times \frac{M_1}{\lambda_r^2} \quad \frac{m}{s^2}$$

As the secondaries are equal, the dyadic equality criterion allows us to identify the primaries, and thus we have:

$$\omega_r^2 = \frac{\delta_{M2}}{\lambda_r^3} \times G \times M_1$$

Considering the «dysmetric» π_r number π_r, calculated for this same space in the corresponding article, knowing that the relationship between the angular velocity and the period of rotation is given by $2 \times \pi_r \ rad = \omega_r \ rad /\!/ s_2 * \tau_r \ s_2$, that is to say, $2 \times \pi_r = \omega_r \times \tau_r$, where τ_r is said period, the expression we were looking for easily results:

$$\tau_r = \frac{2 \times \pi_r}{\omega_r} = 2 \times \pi_r \times \sqrt{\frac{\lambda_r^{\,3}}{\delta_{M2} \times G \times M_1}}$$

Comparing this result with the classical one, described in equation [35.2], it can be easily verified in which situation the classical and «dysmetric» orbital periods would coincide, as well as when one would be greater than the other, producing the orbital anomalies observed and explained with the existence of a supposed dark matter, unnecessary for the «dysmetry». So we could simplify the notation and call Φ_C and Φ_D the respective multiplicative factors of the classical and «dysmetric» periods that do not include the constant G and the mass $M_1 = M$ in this case. That is to say:

$$\Phi_C = 2 \times \pi \times \sqrt{r^3} \quad ; \quad \Phi_D = 2 \times \pi_r \times \sqrt{\frac{\lambda_r^{\,3}}{\delta_{M2}}}$$

Under these conditions, only if $\Phi_C = \Phi_D$ will the classical and the «dysmetric» periods of the same orbit of radius r coincide around the same mass. If it were $\Phi_C < \Phi_D$, the classical orbital period will be less than the «asymmetric» and vice versa if $\Phi_C > \Phi_D$. Thus, «dysmetry» is able to explain gravitational anomalies without the need for dark matter.

We must note that to arrive at this result the Newtonian criterion has been applied in order that the measure of gravitational action is constant. And so, we have overlooked a striking consequence, relative to the fact that the reciprocal forces $\overline{F}_{21} \ N_2$ and $\overline{F}_{12} \ N_1$ do not have to indicate the same quantity of force, which would seem to contradict Newton's third postulate on equality in modulus of action and reaction.

However, this result is not contradictory at all, because «dysmetric» gravitation perfectly quantifies the two acting forces, which allows the independent motion of points P_1 and P_2 to be calculated. What would really happen is that Newton's third postulate would be a specific and restricted case of «dysmetric» space in which this law would be verified; but, in general, the principle of equality of action and reaction need not always hold, as the two «dysmetric» laws of gravitation show.

Finally, it is enough to observe the «dysmetric» laws [35.3] to verify that the gravitation in these spaces does not have to be constant, that is, there is generally no single value of G, since its measurement depends on the position as a function of the corresponding densities and λ_r/r, so gravitation can be referred to the points P_1 and P_2 with the notations G_1 and G_2 using the following expressions:

$$G_2 = G \times \frac{\delta_{M2} \times \delta_{L2}}{\delta_{T2}^2} \times \frac{r^2}{\lambda_r^{\,2}}$$

$$G_1 = G \times \frac{\delta_{M1} \times \delta_{L1}}{\delta_{T1}^2} \times \frac{r^2}{\lambda_r^{\,2}}$$

Obviously, in the case that the «dysmetric» densities are unity at all points, we will have $\lambda_r = r$ and it will result that $G_1 = G_2 = G$, coinciding with classical gravitation. But, in general, it is clear that «dysmetry» makes it possible to represent spaces in which gravitation is not constant.

Section XXXVI

LAWS OF EMPTY SPACE
Tensor formulation of «dysmetric»
properties of physical space

This subject is developed more fully in section XXXVI of *The new Physics of «dysmetric» spaces*, published in Spanish with the title *La nueva Física de los espacios «dismétricos»*, by the same author. Here we will summarize the most basic things, but first a very important observation must be made. In the development of the laws of empty space and the «dysmetric» tensors we apply the dyadic property described in section XI and in number 7 of section XXVIII, which also has to do with what is explained in section XXIX, all of them sections of the *First algebra of magnitudes*. Said dyadic property is the following: **we have on the one hand a mathematical quantity of length, obtained by measurement with the standard meter, and on the other hand the quantity of physical length implicit in it and given by the «dysmetry» of space; both quantities are represented by two dyads, which are homogeneous, because they refer to the same magnitude, the length; and under these conditions we know that the dyadic ratio of homogeneous quantities is a real number equal to the arithmetic ratio of the primary quantities when the secondary quantities are expressed with the same unit of the corresponding magnitude. This property fully justifies operating only with the primaries of the dyads to base the determination of the «dysmetric» tensors that characterize the empty space.**

As a consequence of the above, a dyadic equality in which both members are indicated in the same unit, in this case a unit of length, becomes an ordinary equality of the primaries of both members.

353

A **simple «dysmetric» space** is defined as a set of three elements, a mathematical space $\mathcal{M}$, a physical space $\mathcal{F}$, both with an affine point space structure of the same dimension n, and a map between both $\mathcal{D}$, which transforms $\mathcal{M}$ in $\mathcal{F}$. The mathematical space is the one in which the measurements take place, it can also be considered as the apparent or perceived and visible space, and the physical space is where the phenomena take place, it is the real and invisible space.

The application or transformation $\mathcal{D}$ is called the **spatial deformation tensor** and describes the difference and relationship between mathematical and physical spaces. Its components d_i^j are arranged in matrix form of order $n \times n$.

In physical applications it is practical to identify $\mathcal{M}$ and $\mathcal{F}$ with the ordinary three-dimensional space R^3. Thus, being $\overline{v}$ the vector of the transformed physical space of $\overline{u}$, a vector of the mathematical space, the transformation of in through the tensor $\mathcal{D}$ can be indicated as follows:

$$\overline{v} = \mathcal{D}\left(\overline{u}\right)$$

Let $\{\overline{e}_i\}$ be any base of R^3. Let u^i be the contravariant coordinates of $\overline{u}$ and v^j those of the vector $\overline{v}$. We will have $\overline{u} = u^i \, \overline{e}_i$ and $\overline{v} = v^j \, \overline{e}_j$. Each vector of the base $\{\overline{e}_i\}$ will have a transform by $\mathcal{D}$ that can be written $\mathcal{D}\left(\overline{e}_i\right) = d_i^j \, \overline{e}_j$, where d_i^j are the contravariant coordinates of the vector $\overline{e}_i$ with respect to the same base $\{\overline{e}_j\}$. Thus it turns out that the transformation $\overline{v} = \mathcal{D}\left(\overline{u}\right)$ can also be written $v^j \, \overline{e}_j = \mathcal{D}(u^i \, \overline{e}_i)$. Assuming that $\mathcal{D}$ is linear, $v^j \, \overline{e}_j = u^i \, \mathcal{D}\left(\overline{e}_i\right)$ and, therefore, $v^j \, \overline{e}_j = u^i \, d_i^j \, \overline{e}_j$. We thus conclude the equality of coordinates of both members, which we reflect $v^j = u^i \, d_i^j$. We call this expression the *law of deformation of empty space* and the set of 3×3 elements d_i^j, or in general $n \times n$, we will call it the space **deformation tensor**. Therefore, we can conclude that the relationship between mathematical and physical space is established by the following physical-mathematical truth:

$$v^j = u^i \ d^j_i$$

Matrix-wise we can write this law with the row matrices $[\![u]\!]$ and $[\![v]\!]$ for the coordinates u^i and v^j, and with the square matrix $[d]$ for the tensor d^j_i, of generic order $n \times n$, indicating i row and j column. With which we arrive at this formulation:

$$[\![v]\!] = [\![u]\!] \ [d]$$

Next let us reflect on the need to assign to each point in space a real number that represents the corresponding «dysmetric» density. To do this, a linear application $\triangle$ of the tensor product $\mathscr{M} \otimes \mathscr{F}$ in R can be conceived, such that the tensor product $\overline{u} \otimes \overline{v}$ of every vector $\overline{u}$ with its transform $\overline{v}$ has as an image the real number $\delta(P) \in$ R that represents the «dysmetric» density in the point P affine to the associated vectors $\overline{u}$ and $\overline{v}$. In analytical terms this application can be described as follows:

$$\triangle(\ \overline{u} \otimes \overline{v}) = \delta(P) \in R$$

Substituting $\overline{u}$ and $\overline{v}$ for their expressions as a function of the basis vectors $\{\overline{e}_i\}$, that is, $\overline{u} = u^i \ \overline{e}_i$ and $\overline{v} = v^j \ \overline{e}_j$, we easily arrive at the following law:

$$\triangle(\overline{u} \otimes \overline{v}) = \triangle(u^i \ \overline{e}_i \otimes v^j \ \overline{e}_j) = u^i \ v^j \ \triangle(\overline{e}_i \otimes \overline{e}_j)$$

The $n \times n$ elements $\triangle(\overline{e}_i \otimes \overline{e}_j)$ can be designated with the notation Δ_{ij}. Each value Δ_{ij} represents a real number associated with the image by $\triangle$ of the tensor products of the base vectors $\overline{e}_i \otimes \overline{e}_j$. In this way, we arrive at the following law:

$$\triangle(\overline{u} \otimes \overline{v}) = u^i \ v^j \ \Delta_{ij}$$

Replacing the contravariant coordinates v^j with those provided by the deformation tensor through $v^j = u^k \ d^j_k$, we are finally left with the following tensor formulation:

$$\triangle(\overline{u} \otimes \overline{v}) = u^i \ u^k \ d^j_k \ \Delta_{ij} = \delta(P) \in R$$

355

We call this expression the *«dysmetric» law of empty space*, which in matrix notation can be written in this other way:

$$\triangle\left(\overline{u}\otimes\overline{v}\right)=[u][d][\triangle]^{T}\{u\}=\delta(P)\in R$$

We call the $n\times n$ elements Δ_{ij} or its associated matrix $[\triangle]$ the **«dysmetric» density tensor of empty space**.

The $n\times n$ elements d_{i}^{j} of the deformation tensor and the Δ_{ij} of the density tensor can be indicated in general as functions of the coordinates ui and the magnitude time. This is explained in more detail in the aforementioned section XXXVI of *La nueva Física de los espacios «dismétrios»* in Spanish, by formulating the laws of deformation and density in general curvilinear coordinates.

In concise mathematical language a **general «dysmetric» space** can be conceived as a set of four elements notated by $\{\mathcal{M},\mathcal{F},\mathcal{D},\triangle\}$: a mathematical space $\mathcal{M}$, a physical space $\mathcal{F}$, both with affine point space structure of the same dimension n, a map between both $\mathcal{D}$, which transforms $\mathcal{M}$ into $\mathcal{F}$ ($\mathcal{D}$: $\mathcal{M}\rightarrow\mathcal{F}$) and a map $\triangle$ of the tensor product $\mathcal{M}\otimes\mathcal{F}$ into R ($\triangle$: $\mathcal{M}\otimes\mathcal{F}\rightarrow$ R). The application $\mathcal{D}$ relates the spaces $\mathcal{M}$ and $\mathcal{F}$, we call it the **deformation tensor**. The $\triangle$ application assigns a real number to each two homologous points in the spaces $\mathcal{M}$ and $\mathcal{F}$, we call it a **«dysmetric» density tensor**.

It must be noted that the formulation presented here of the «dysmetric» tensors and the associated laws is nothing more than one of the ways that can be mathematically devised among the infinite possible ones to establish the field of «dysmetric» densities, which does nothing but confirm the wealth of tools that «dysmetr» provides to express the physical-mathematical truths that wait to be observed by eyes that can understand them. For this reason, «dysmetric» spaces should not be ignored by any creative mathematician or physicist, because they extend to infinity the possibilities of representation of general physical phenomena, not only isometric ones, which constitute a very particular and restricted case of «dysmetry», which is

356

theoretically reduced to isometry when in a very singular way the «dysmetric» densities of all magnitudes are the numerical unit at all times and places, as we physicists have tacitly assumed until now. So «dysmetry» is the most generic character of space, which is reduced to isometry, currently considered exclusively, only in very special cases.

Section XXXVII

LAW OF DYADIC VARIATION

The physical-mathematical truth that proves
the fact that «dysmetry» is natural

In section XXXVII of *The New Physics of «dysmetric» spaces,*
published in Spanish with the title *La nueva Física de los espacios*
«dismétricos», by the same author, differential «dysmetry» is
discussed more extensively. Here we will limit ourselves to
describing the fundamental mathematical truth that refers to the
elementary expression of the differential variation of any dyad,
representative of a certain quantity of a magnitude.

To do this, let's analyze the variation of any dyad (q, U). Let us
remember that every dyad represents a quantity of magnitude
that can be symbolized in multiple indistinct ways: (q, U), $(q\ U)$,
$q\ U$, $q\ \circ U$, $q\ \circ(1, U)$ or similar. A dyad can vary because its
primary q changes or because its secondary U changes.

In classical Physics unfortunately «arithmetized» only the first
option is contemplated. On the other hand, the generality of
«dysmetry» also admits the second variant. The general case of
infinitesimal variation of a dyad can be represented by $d(q, U)$,
which must be the difference between the quantities $(q+dq, U \oplus dU)$
and (q, U). Obviously, the addition of the term $U \oplus dU$ is the
dyadic addition and the marked difference as well.

We observe in $U \oplus dU$ that it is a homogeneous sum, because the
quantities of the addends refer to the same magnitude; but the
addends are not uniform, because they are different units. At the
end of sections V and XXVIII-3 from Part I dedicated to the
First Algebra of Magnitudes, the analytical form of these singular
cases of dyadic addition is established, based on the postulate of
affinity with the geometric algebra of segments. This is a typical
exception to the axiom of uniformity when the primaries coincide

359

and the secondary ones are homogeneous but not uniform. It follows that, although the units of the addends do not coincide, nothing prevents analytically formulating the addition of homogeneous and non-uniform quantities. So, given two dyads (q, U_1) and (q, U_2) of the same magnitude, with U_1 and U_2 homogeneous, the following additive law can be described:

$$(q, U_1) \oplus (q, U_2) = (q, U_1 \oplus U_2)$$

Considering the previous property of dyadic addition, with $U_1 = U$ and $U_2 = dU$, from $d(q, U) = (q + dq, U \oplus dU) \ominus (q, U)$, we have the following reasoning:

$$d(q, U) = (q + dq, U \oplus dU) \ominus (q, U) =$$

$$= (q + dq, U) \oplus (q + dq, dU) \ominus (q, U) =$$

$$= (q, U) \oplus (dq, U) \oplus (q, dU) \oplus (dq, dU) \ominus (q, U) =$$

$$= (dq, U) \oplus (q, dU) \oplus (dq, dU)$$

The term (dq, dU) is an infinitesimal of the second order, so it can be neglected with respect to the other two, which are of the first order, resulting in:

$$d(q, U) = (dq, U) \oplus (q, dU)$$

We could call this conclusion the law of dyadic variation. The addend (dq, U) represents the modification of the dyad (q, U) as a consequence of the change in the primary, which could be called **metric variation** and describes the conventionalism used to analyze variations in quantities of magnitudes since always. In turn, the innovative term (q, dU) could be called **«dysmetric» variation** and determines the component attributable to the homonymous effect, which refers to the change experienced by the quantity of magnitude implicit in every standard unit U_0, a transcendent phenomenon ignored until now.

The *law of dyadic variation* has a very relevant meaning, it is proof that «dysmetry» is natural, so «dysmetric» phenomena cannot be ignored, except at the cost of severely curtailing Physics. So «dysmetry» is an eternal truth that will be implanted sooner rather than later and that all creative mathematicians and physicists will take advantage of in multiple ways to develop their theories and innovations.

To implement «dysmetry» it is not necessary to start Physics from scratch, but only to consider that isometry is a local observation that cannot be generically attributed to all space, given its true «dysmetry» nature. The new horizons that this reveals are so broad that it is unimaginable to conceive where the application of that **fundamental physical-mathematical truth** that is «dysmetry» can take us, as soon as mathematicians and physicists focus on it and understand it with full understanding. help of dyadic algebra.

BREVIARY

REVELATIONS OF DIADIC ALGEBRA
*Discovery of remarkable truths
emerged from mathematical certainty*

In this section we are going to briefly present the most important physical-mathematical truths that dyadic algebra and «dysmetry» reveal to us without great effort, for the benefit of all present and future physicists and mathematicians, multiplying by many the potential of their knowledge.

As a preamble, the first observation to make is the importance of understanding the difference between magnitude and quantity (section I, page 27). Such a distinction is not easily discernible in the English language, which seems to view both concepts as synonyms. In English, magnitude refers to measurement or extension (section II, page 31), so those who think in this language have to make more efforts than Hispanics to differentiate these two terms. Language is essential to represent reality and even more so to communicate it. It is almost impossible for the mind to transmit a reality outside of itself that is not expressly defined by linguistic concepts shared with others.

By magnitude we understand any physical phenomenon affine to length and by quantity any portion of a magnitude. Which means that the set of quantities of a given magnitude can be put into one-to-one correspondence with the set of all lengths, or what is the same, with the set of all geometric segments. In this way we make it possible to see quantities of different magnitudes through their homologous geometric segments.

Once the quantities of magnitude are made visible through abstract rectilinear segments, it is easy to realize that the addition of segments is very simple, it is enough to juxtapose one addend after another to reach a new segment that we call the sum of the

363

given segments. So adding segments or physical units of the same magnitude is not a difficult problem to solve or to structure in an additive algebra, because the addends and the sum belong to the same magnitude. Hence the axiom of uniformity (page 43), which implicitly implies that the addition of magnitudes is an internal operation.

It is important to remember the notation assigned to the quantities of magnitudes through dyads (section III): every quantity of magnitude or dyad is indicated by a pair q and U, where q is a numerical or vector value and U an arbitrary quantity adopted as a standard and represented with a certain symbol. **These pairs can be designated interchangeably with the forms $q\ U$, $(q\ U)$, (q,U)** or any other form. The choice of description in each case must be made so **that the dyad is well described and there is no ambiguity in the context of the corresponding study**.

Focusing on the affine addition of segments (sections V to XI, page 43; section XXVIII, articles 4 to 7, page 197), we observe the external operation that consists of multiplying a number by a dyad or affine segment (section IX, page 57; section XXVIII, article 6, page 201). It is immediate to appreciate that this multiplication is based on homogeneous addition, since it is enough to take the number as a multiplier and the given affine segment as a multiplicand, to add it with itself as many times as the numerical multiplier indicates. The result of multiplying a segment or quantity of magnitude by a number is another segment or quantity of the same magnitude, so it is enough to take the product as a dividend and the multiplicand as a divisor to arrive at an important property: the ratio of two homogeneous quantities it is a real number.

And by the affinity postulate (pages 122, 190, 197, 225, among others) we can admit that the ratio between any two quantities of the same magnitude is a real number. This property of additive multiplication will be very useful to define important concepts such as the **«dysmetric» density of magnitudes** (pages 133, 290, 292, among others).

Said multiplicative operation of additive origin should not be confused with the multiplication of quantities of any magnitude, as is a common error at all academic and scientific levels, including the International System of Units. The multiplications of magnitudes are **external generating laws,** they produce new magnitudes. For example, the product of two lengths is an area, the product of three lengths is a volume (sections XII to XVIII, pages 69 et seq.; section XXVIII, articles 8 to 13, pages 219 et seq.). Without prejudice to the above, if we want to go deeper into the formulation and visualization of magnitude quantities, we can resort to abstract algebra and conceive of concrete numbers or physical dyads as equivalence classes defined in dyadic sets. Thus the form of the hyperbolic surface is elegantly arrived at as the geometric meaning of all possible quantities of any magnitude (article 7, pages 212 and 215). In this way we fully enter into abstract dyadic algebra without needing to resort to affinity with geometric segments. Both approaches are perfectly compatible and lead to the same result.

Generative multiplicative operations, not being internal laws of composition, cannot have unitary or inverse elements, as is assumed in the regulations of the International System and in all scientific uses at any level. Therefore, it does not make mathematical sense to talk about the inverses of a second or a meter or any other physical unit or any quantity of magnitude, in the sense that we have established for numbers (section XIV, page 79; and XXVIII, article 9, page 225 and article 13, page 233).

Once the generative nature of operations with magnitudes has been appreciated, all the multiplicative laws of composition and their derived operations such as division make sense. So that we come to formulate them all with an isomorphic appearance, but without identifying them with the numerical ones, because the elements that are composed of them are not numbers or vectors,

365

but scalar or vector dyads. With these premises, the foundations are laid to build a totally coherent dyadic algebra, which certainly does not provide the dyadic sets with an algebraic structure of an abelian multiplicative group, as was assumed until now. Hence we say that the algebra of magnitudes has its own and original structure that we designate with the name dyadic algebra, noted with the letter $\mathscr{D}$ (section XXVIII, article 13, page 233).

The observation of the dyadic elements suggests the analysis of their form of variation. We immediately realize that the quantity of magnitude represented in a dyad can vary in two ways: first, because its primary or numerical part varies; and second, because the secondary or dimensional part varies. The first option is obvious and refers to the observation of variation through measurement. What is not so evident is the variation due to the quantity implicit in the physical unit. This variant has been ignored until now and constitutes the most general provision that should not be omitted at all, because it has never been proven that the quantity of magnitude implicit in a physical unit is indifferent to position in space or time or material disturbances. We call such indifference isometry.

In other words, isometry is not proven, so its opposite, «dysmetry» which is the generic variant, must be the forecast to be considered unless proven otherwise. Furthermore, as we explain in section XXXVII, pages 359 et seq., the *law of differential dyadic variation* constitutes the mathematical proof that «dysmetry» is natural. In short, we have no proof of isometry, which is logical, because what can be proven is «dysmetry», so that «dysmetric» spaces are not a mere invention of mathematical abstraction, but a physical reality. .

That is why in section XXXVI, pages 353 et seq., we analyze the **«dysmetric» properties of empty space**, based only on the «dysmetry» of length and, where appropriate, time, without material disturbances associated with masses or electromagnetic fields. We thus aim to characterize these properties and we achieve this through the spatial deformation and **«dysmetric»**

density tensors. Without great difficulty it is established that a ray of light does not follow a straight path in «dysmetrical» empty space nor does its speed remain constant, which clashes head-on with the postulates that are currently assumed to be dogmatic truths.

In general, we have defined the «dysmetric» space as a set of four notated elements $\{\mathcal{M},\mathcal{F},\mathcal{D},\triangle\}$: a mathematical space $\mathcal{M}$, a physical space $\mathcal{F}$, both with affine point space structure of the same dimension n, a map between both $\mathcal{D}$, which transforms $\mathcal{M}$ into $\mathcal{F}$ ($\mathcal{D}: \mathcal{M}\rightarrow\mathcal{F}$) and a map $\triangle$ of the tensor product $\mathcal{M}\otimes\mathcal{F}$ into R ($\triangle: \mathcal{M}\otimes\mathcal{F}\rightarrow$R). The mathematical space is the one in which the measurements take place, it can also be considered as the apparent or perceived and visible space, and the physical space is where the phenomena take place, it is the real and invisible space. The application $\mathcal{D}$ or **spatial deformation tensor** describes the difference and relationship between mathematical and physical spaces. The application $\triangle$ or **«dysmetric» density tensor** reflects the «dysmetry» of the space, associating a real number with each two homologous points of $\mathcal{M}$ and $\mathcal{F}$. The components d_i^j and Δ_{ij} of both tensors are arranged in matrix form of order $n\times n$. In physical applications it is practical to identify $\mathcal{M}$ and $\mathcal{F}$ with the ordinary three-dimensional space R^3.

The expression $v^j=u^i d_i^j$ (page 355) represents the transformation $\overline{v}=\mathcal{D}(\overline{u})$, being any vector $\overline{u}$ of the mathematical space $\mathcal{M}$ and the vector $\overline{v}$ its image in the physical or deformed space $\mathcal{F}$. In matrix form this relation can be written $[v]=[u][d]$, where $[d]$ is the matrix symbol for the deformation tensor $\mathcal{D}$. The $n\times n$ elements d_i^j of this tensor can be conceived in general as functions of u^i coordinates and time.

The law $\triangle(\overline{u}\otimes\overline{v})=u^i u^k d_k^j \Delta_{ij}=\delta(P)\in$R (page 355), in matrix form it can be put $\triangle(\overline{u}\otimes\overline{v})=[u][d][\triangle]^T\{u\}=\delta(P)\in$R, where $\delta(P)$ represents the **«dysmetric» density** at each point P. Any set of $n\times n$ values ordered in the matrix $[\triangle]$ is called the **«dysmetric» density tensor** and is such that at every affine point P of the vector

367

determines the «dysmetric» density of the space at that point, given by $[u][d\,][\triangle]^T\{u\} = \delta(P) \in \mathrm{R}$. The $n \times n$ elements Δ_{ij} of this tensor can be indicated in general as functions of the coordinates u^i and the magnitude time.

It should be noted that the formulation presented here of the «dysmetric» tensors is nothing more than one of the ways that can be mathematically devised among the infinite possible ones to establish the field of «dysmetric» densities, which only confirms the wealth of tools that «dysmetry» provides to express the physical-mathematical truths that wait to be observed by eyes that can understand them. For this reason, «dysmetric» spaces should not be ignored by any creative mathematician or physicist, because they extend to infinity the possibilities of representation of general physical phenomena, not only isometric ones, which constitute a very particular and restricted case of «dysmetry», which is theoretically reduced to isometry when in a very singular way the «dysmetric» densities of all magnitudes are the numerical unit at all times and places, as we physicists have tacitly assumed until now. So «dysmetry» is the most generic character of space, which is reduced to isometry, currently considered exclusively, only in very special cases (pages 359 et seq., among others).

In sections XXXIII, page 329, and XXXIV, page 333, it is clear that in a «dysmetric» space neither the mathematical number pi nor the physical speed of light are constant. And we conclude it not by questioning geometry or Michelson's experiment or *Einstein's hypothesis* or any other fundamental knowledge, but by admitting and completing them with dyadic algebra, which leads us by pure mathematics to the fact that these isometric assumptions are not compatible with the «dysmetr». This result warns of the dangerous drift of the International System that seeks to refer all physical units to certain universal constants, which can only be such in a strangely isometric space. This approach completely excludes the «dysmetric» variant, preventing students and researchers from contemplating reality in all its extension and variability.

In section XXXII, page 319, we explain how physical laws can be reformulated using the example of *Newton's second law*. At the same time, in section XXXV, page 339, «dysmetric» gravitation is generalized. In the same section XXXV, pages 351 and following, we analyze the orbital perturbations that are currently explained with the capricious invention of the mysterious and unobservable dark matter, and we verify that these apparent anomalies are subsumed in the nature of a «dysmetric» space. In this area it does seem pertinent and necessary to refound cosmological theories.

In the same way, other contradictions of current cosmology, which only thinks in isometric terms, are also embraced by the effects of «dysmetr». For example, we talk about the age of the universe, and it is assigned an official age of about 13,787 million years. But this estimate is not free of contradictions, as is the case of the star Methuselah or HD 140283, located about 190 light years from Earth and whose age dates back to 14.46 billion years. It is evident that no star can be older than the universe, so one of the two estimates must be wrong, or both, which is most likely. So, instead of thinking that isometric physical models are wrong, what is done is to distort the contradictory data and modify them appropriately, concluding that Methuselah is younger or that the universe is older. Thus, for example, there are those who maintain that the universe dates back 21 billion years. Others claim that the error in measuring Methuselah's age is 800 million years. In both cases the discrepancy would disappear and that star could be younger than the universe. But the problem is not resolved, because at the same time other studies of the age of the universe put it at about 11.4 billion years, resurfacing the paradox.

In our opinion, contradictions of this type prove the obsolescence and closure of isometric theories, which do not work for a «dysmetric» space, although the reciprocal would be fulfilled, that is, the «dysmetric» laws take isometric ones as a particular case. The current nineteenth-century models are based on constants such as the speed of light, which we have already proven are not such in «dysmetrical» environments, resulting in the fact that making calculations with a simple rule of three proportional to the invariant speed of light leads to conclusions that They do not correspond to reality. The correct thing to do is that both distances and times are established taking into account precisely the «dysmetry» of length and time. Only in this way can we reach cosmic models consistent with the material reality of space. What does not seem sensible is to continue trusting in the kindergarten method of the rule of three as the only means of calculating very complex phenomena. It is obvious that this crude simplicity cannot give reliable results, as experience shows.

We cannot exclude from this breviary the experiences with the measurement of time on GPS satellites, which seem to us to constitute physical proof of the «dysmetry» of temporal magnitude. It is proven that time passes differently depending on the height of the orbits with respect to the surface of the Earth. At 20,200 km altitude, time would pass slightly more slowly than on the Earth's surface; specifically, it seems that clocks move back 4.53×10^{-10} seconds every second. Therefore, one second at 20,200 km would implicitly contain a quantity of $1 + 4.53 \times 10^{-10}$ pattern seconds on the surface. It seems to have been proven that at 3,200 km altitude time passes the same as on the surface and that below that altitude the clocks move forward. Assuming that these experiences are correct, we would have time be have «dysmetrically», so that it would be observed that the «dysmetric» density of this magnitude δ_T would be a function of the height above the Earth's surface. At zero km and at 3,200 km we would obviously have $\delta_T(0) = \delta_T(3,200) = 1$. Between zero and 3,200 km altitude the «dysmetric» density of time would be less than unity, time would appear less dense than on the surface, so it would run faster than at zero level. And at 20,200 km the opposite would happen, the «dysmetric» density would have the value $\delta_T(20,200) = 1 + 4.53 \times 10^{-10}$, a quantity that we remember is pure numerical or dimensionless, time at that height would be denser than at the surface and would pass more slowly relative to it.

To implement «dysmetry» it is not necessary to start physics from scratch, but only to consider that isometry is a local observation that cannot be generically attributed to all space, given its true «dysmetry» nature. The new horizons that this reveals are so broad that it is unimaginable to conceive where the application of that **fundamental physical-mathematical truth** that is «dysmetry» can take us, as soon as mathematicians and physicists focus on it and understand it with full understanding. help of dyadic algebra.

371

ANNEX

THE DYADIC INVERSES
The logical formal sense for the notation
of unitary and inverse magnitudes

In sections XIV and XXVIII, article 9, we noted that multiplicative operations on magnitudes are external generating laws, so they cannot accommodate unitary elements or inverses in the sense of the internal laws of classical algebra. When two quantities of homogeneous or non-homogeneous magnitudes are multiplied, the law of composition involved is external generating, so the product can never be homogeneous with the factors, even if these are. For example, the product of two lengths expressed in meters results in an area in square meters; hence, we cannot find any quantity of length that, when multiplied by any other, gives a product expressed in meters. For this reason, homogeneous unitary elements do not exist for this multiplicative operation. Likewise, we cannot find a quantity of length that is the inverse of another given length, because their product will not appear expressed in meters, but in square meters. Since the unitary element does not exist, the product could not equal a nonexistent unit. However, what we can attempt is to give formal meaning to the inverse notations, to harmonize multiplicative operations on magnitudes with common algebraic expressions. This is what we attempt in the following, seeking a logical meaning for the nomenclature of the inverses of magnitudes, but adapted to the algebraic meaning associated with the external composition laws specific to magnitudes.

The difference with ordinary algebra is that, for example, the inverse of the number 2 is the number 0.5, such that multiplied together they give the real unit 1, all three belonging to the same numerical set. Well, as we have reiterated, this cannot be done with quantities of magnitudes, because their product is external and generative, so it is not homogeneous with the factors. So, if we want to simulate an isomorphism between ordinary algebra and dyadic algebra, we have to invent something that is algebraically valid.

The multiplication of magnitudes is developed in section XII and in article 9 of XXVIII. Among the notations that we have accepted for the dyads, $q\ U$, $(q\ U)$ or (q, U), we opt here for the most explicit one, which is (q, U). According to the definition established in those sections, given any two dyads (q_1, U_1) and (q_2, U_2), their geometric affine product can be written analytically in the following way:

$$(q_1, U_1) * (q_2, U_2) = (q_1 \times q_2, U_1 * U_2)$$

Taking $(q_1 \times q_2, U_1 * U_2)$ as dividend, (q_2, U_2) as divisor, not zero, and (q_1, U_1) as quotient, the ratio of two quantities of homogeneous or non-homogeneous magnitudes has been defined by the expression:

$$(q_1, U_1) = \frac{\left(q_1 \times q_2, U_1 * U_2\right)}{\left(q_2, U_2\right)}$$

We define the dyad (q_2^{-1}, U_2^{-1}) and the inverse magnitude with unit U_2^{-1} as those that satisfy the following condition:

$$\frac{\left(q_1 \times q_2, U_1 * U_2\right)}{\left(q_2, U_2\right)} = \left(q_1 \times q_2, U_1 * U_2\right) * \left(q_2^{-1}, U_2^{-1}\right)$$

With this notation we can write the initial dyadic ratio with the following formulation:

$$(q_1, U_1) = (q_1 \times q_2, U_1 * U_2) * (q_2^{-1}, U_2^{-1}) = (q_1 \times q_2 \times q_2^{-1}, U_1 * U_2 * U_2^{-1})$$

Since in R $q_1 \times q_2 \times q_2^{-1} = q_1$, dyadic equality requires that the dyadic product $U_1 * U_2 * U_2^{-1}$ must be identical to U_1. That is, the magnitude defining the product $U_1 * U_2 * U_2^{-1}$ must be the same as the one corresponding to U_1. Therefore, for the inverse notation U_2^{-1} to make sense in the case of an external generating law, the condition $U_1 * U_2 * U_2^{-1} = U_1$ must be fulfilled for any quantity U1 of any magnitude. This means that $U_1 * U_2 * U_2^{-1}$ corresponds to an affine volume and thus, on the one hand, U_1 is affine to a length on the right side, and in turn is affine to a volume on the right side. This ambivalence clearly contradicts the affinity postulate. However, in mathematics it is common to resolve such singularities with axioms. And in this case, that is what can be done; therefore, we admit in principle that the unitary and inverse elements are special, that they do not satisfy the general affinity, and that the expression $U_1 * U_2 * U_2^{-1} = U_1$ is valid.

Some unruly people might object that the above is obvious, because $U_2 * U_2^{-1} = 1$ and so $U_1 \times 1 = U_1$. But this reasoning would be entirely erroneous and would mean falling right into the «arithmetization» trap that we have warned about so often in this work. Indeed, $U_2 * U_2^{-1}$ is the product of two quantities of magnitude and can never result in a real number, because said dyadic product is an affine surface, that is, it is another quantity of the magnitude generated by U_2 and U_2^{-1}, which are also quantities of different magnitudes, that is, not homogeneous.

However, what we do observe in $U_1 * U_2 * U_2^{-1} = U_1$ is that the dyadic product $U_2 * U_2^{-1}$ is such that it leaves any quantity U_1 invariant when both are multiplied dyadic, and this property reflects what can be expected of a unitary element. Therefore, we are authorized to define the unitary and inverse elements of the structure of multiplicative dyadic algebra as follows: **given any quantity of a certain generic magnitude, represented by U_2, we will say that the quantity $U_2 * U_2^{-1}$ is a unitary element of the multiplicative operation of magnitudes if and only if the condition $U_1 * U_2 * U_2^{-1} = U_1$ is fulfilled for any other magnitude and quantity U_1; and in turn, we will say that the inverse magnitude and quantity of U_2 are determined by U_2^{-1}.**

So, on the one hand, we have the **unitary magnitude** to which the unit element $U_2 * U_2^{-1} = U$ belongs; and on the other, the

inverse magnitude to which U_2^{-1} refers, for any other quantity indicated by the quantity U_2. The unitary magnitude and the inverse magnitude are not homogeneous. By definition, U_2 and U_2^{-1} are quantities of magnitudes that are inversely correlated with each other.

We insist that we should not be fooled by the common inverse notation U_2^{-1}. Its numerical meaning would be $U_2 \times U_2^{-1} = 1$ for the internal law «×» of multiplication of real numbers; but here it means a quantity of a certain magnitude such that it satisfies the condition $U_1 * U_2 * U_2^{-1} = U_1$ for any U_1, where «*» is the external multiplicative law generating magnitudes. Here the unitary element is not the number one, but the quantity U of the unitary magnitude such that $U_2 * U_2^{-1} = U$.

We must now ask whether these unitary and inverse elements exist, and if so, whether they are unique. Let us first examine the existence of unitary elements. Let $U = U_2 * U_2^{-1}$ be the unitary element of the magnitude defined by the quantity U_2. It is evident that U denotes the affine surface related to the product of the affine lengths U_2 and U_2^{-1}. Therefore, since this product is unique, by definition of dyadic multiplication, the unitary element U exists and is unique. We can reach the same conclusion by supposing that two unitary elements U and U' exist. Then, since U is a unitary element, $U' * U = U'$ is required, and since U' is a unitary element, $U * U' = U$ is required. Therefore, given the commutative property, $U = U'$ is required. So, in effect, the unitary element of any magnitude exists and is unique in the sense established here.

Let us now examine the existence and uniqueness of the inverse elements. According to the definition of inverse established above, given the quantity U_2, its inverse quantity U_2^{-1} will be the one that satisfies $U_1 * U_2 * U_2^{-1} = U_1$ for any other magnitude U_1. It is evident that the abstract affine parallelepiped defined by the dyadic equation $U_1 * U_2 * U_2^{-1} = U_1$ has three dimensions U_1, U_2 and U_2^{-1}, with a volume U_1, then, the edge U_2^{-1}, which is the inverse magnitude, is like the other two an affine length that exists and is unique for a given volume. Another way of seeing uniqueness is to suppose that there exist two inverse quantities U_{21}^{-1} and U_{22}^{-1}. Since U_{21}^{-1} is the inverse of U_2, we will have $U_1 * U_2 * U_{21}^{-1} = U_1$. And since U_{22}^{-1} is the inverse of U_2, it will also be verified that $U_1 * U_2 * U_{22}^{-1} = U_1$. With which we have $U_1 * U_2 * U_{21}^{-1} = U_1 * U_2 * U_{22}^{-1}$. By definition of dyadic product, the terms $U_1 * U_2 * U_{21}^{-1}$ and $U_1 * U_2 * U_{22}^{-1}$ define two right parallelepipeds of equal volume with two equal edges, U_1 and U_2, so the third edge, U_{21}^{-1} and U_{22}^{-1}, must be equal and thus $U_{21}^{-1} = U_{22}^{-1}$. Therefore, the inverses exist and are unique.

Let us apply the definition of unit elements and inverses to dyadic fractions. Let U_1 and U_2 be two quantities of magnitudes represented by their units. Multiplication allows us to take U_1 as the affine surface, U_2 as the affine length, and $U_1 /\!/ U_2$ also as the affine length, which gives $U_1 = U_2 * U_1 /\!/ U_2$. It also allows us to consider U_2 as the affine surface, U_1 as the affine length, and $U_2 /\!/ U_1$ as the affine length, giving $U_2 = U_1 * U_2 /\!/ U_1$. Substituting U_2 in $U_1 = U_2 * U_1 /\!/ U_2$ by its value, given by $U_2 = U_1 * U_2 /\!/ U_1$, gives:

$$U_1 = U_2 * \frac{U_1}{U_2} = U_1 * \frac{U_2}{U_1} * \frac{U_1}{U_2} \Rightarrow \frac{U_2}{U_1} * \frac{U_1}{U_2} = U$$

This means that the product of two reciprocal fractions, that is, those in which their numerators and denominators are interchanged, is the unitary element of magnitude U, because it leaves any quantity U_1 unchanged. In turn, each fraction is the inverse element of the other.

Finally, given any dyad (q_2, U_2), let us check that its inverse quantity is (q_2^{-1}, U_2^{-1}). Indeed, for any quantity U_1 we will have the product $U_1 * (q_2, U_2) * (q_2^{-1}, U_2^{-1})$, operating on this product, it is $(q_2 \times q_2^{-1}, U_1 * U_2 * U_2^{-1})$, and as $U_1 * U_2 * U_2^{-1} = U_1$ and in R is $q_2 \times q_2^{-1} = 1$, we obtain:

$$U_1 * (q_2, U_2) * (q_2^{-1}, U_2^{-1}) = U_1 * U_2 * U_2^{-1} = U_1$$

In conclusion, $(q_2, U_2) * (q_2^{-1}, U_2^{-1})$ is a unit element, so (q_2^{-1}, U_2^{-1}) is the inverse dyad of (q_2, U_2), and vice versa.

To summarize in practical terms, when a quantity multiplied by its inverse appears in a dyadic equation, this product must be understood as a unitary element that keeps the rest of the factors invariant. However, if the inverse element appears alone, multiplying other quantities, it must be interpreted as its divisor or fractional denominator. Thus, inverse dyadic elements cannot appear in isolation. This is a significant peculiarity of dyadic inverses. For example, the speed unit written in inverse notation $m * s^{-1}$ means the ratio or quotient $m /\!/ s$; expressing $s * s^{-1}$ indicates the unit magnitude of time and its unit $U_T = s * s^{-1}$; or for length $U_L = m * m^{-1}$; and isolated notations s^{-1} or m^{-1} are meaningless for magnitudes. The International System establishes s^{-1} and el m^{-1} as units of frequency and wavenumber, but dyadic algebra teaches that these magnitudes must be formulated in the compound units $cycle * s^{-1} = cycle /\!/ s$ and $cycle * m^{-1} = cycle /\!/ m$. Isolated inverse units are an undesirable aberration. Therefore, the dyadic algebraic structure does not have unitary elements or inverses in the same sense attributed to the structures of ordinary algebra. However, respecting the external generative laws of dyadic algebra, we can define unitary elements and inverses, understood as quantities of singular magnitudes with some of the formal and symbolic qualities commonly expected of such algebraic elements, as we have established in this annex. We have endeavored to show with tenacious reiteration throughout the text by various means that it is not correct to attribute to the multiplicative operations of magnitudes the internal properties of the group structure, a presumption that underlies the substitute of symbolic and «arithmetized» algebra that standardizes the International System of Units, because the group structure is not possible for the external laws of composition that generate the magnitudes themselves, except in the form described in this annex or another that could be imagined according to the laws of algebra.

APPENDIX

«PSYCHOFUNCTIONAL» ANALYSIS
OF *QUANTUM THEORY*
Why a theory that works contradicts
common sense and it is paradoxical

In this section we are going to examine *quantum theory* with an epistemological vision, without going into depth to analyze the physical-mathematical details that make it up. In reality, the examination could be done on any other theory, but we have chosen the *quantum theory* for its clearly anti-realistic nature, which facilitates the understanding of what we want to show, which is nothing more than the mental character of every theory, which in no case is identified with the true essence of extramental reality.

The analysis method will be that described in the same author's publication *Psychofunctional Theory* [23], whose basic principle is the **difference between mental reality and extramental reality**, which would be connected through perception. From these connections sensory data would arise, from these abstract data and by conveniently grouping them, concepts, common languages and scientific theories expressed in their specific languages would be born. In such a way that all theory or language belongs to the domain of mental reality and, if they are well constructed, that is, if they are connected with extramental reality through perception, they seem to replace it, although it is always an illusory identity, because mind and reality are totally different things that can never coincide at all. And this without forgetting

---

[23] Although we try to make the «psychofunctional» explanation as intuitive as possible here, it is recommended that you read this work to accurately understand the terminology used.

that the mind can also imagine with total freedom and fantasize to its liking, which is why it is capable of creating fictional universes totally unrelated to the outside world.

It is enough to observe that thoughts settle in the brains and that the world is outside of them. If we are able to understand this simple fact, that the brain is not the entire world, we will properly differentiate thought and what is foreign to it, which is everything that exists and is not thinking.

We will thus understand that, from a «psychofunctional» point of view, it is perfectly feasible that the mental model can be counterintuitive and yet work, in the sense that it seems to adjust to reality and can make us believe that the world is the model, a very common sensation that can never be true, so we must always reject it.

Thus, what we think when speaking, writing or reading is the sensory meaning that we give to those signs that make up all language, whether ordinary or scientific. And it is clear that language is not extramental reality, so arguing about whether a theory or language is the world itself makes no sense, because neither can be. The only thing we can establish is whether a model conforms to what can be expected of it as it does not contradict the facts as they are perceived.

In short, no theory replaces extramental reality. Every theory is merely a kind of more or less detailed map of reality or similar to it. The better the correspondence between the map and reality, the more accurate the mental image of the world will be. But the map will always lack details that will distance it more or less from reality, depending on its quality, so it is a gross mistake to identify them.

So *quantum theory*, like any other, is specified in specific principles and language, which result in a system of thought with which it is intended to describe or create a «map» of extramental reality that, in the case of *quantum*, corresponds to the physical universe of the very small. This map does not have to be intuitive or obey common sense, it simply has to reflect what it was created for, without being the same, because it is impossible.

Precisely this inclination to confuse the mental model with the extramental reality is what has been generating controversial debates among scientists since the birth of *quantum theory*, named after the influence of the principle formulated by Max Planck in 1901, which introduced the postulate based on the observation that hot objects appear to emit discontinuous range radiation using small discrete amounts of energy called quanta. Thus was born the name *quantum theory*, which was extended to all forms of electromagnetic radiation, including light. Planck even formulated his famous law that relates the emitted energy E to its frequency v and the constant h that bears his name, through the relationship $E = h \times v$. The value of Planck's constant is set to the measure $6.62607015 \times 10^{-34}$ when expressed in the product joule per second.

In 1905 Albert Einstein published an article in which he explained the photoelectric effect related to the emission of electrons from a metal surface by projecting a beam of light on it. He observed that the brighter or more energetic the light was, the greater the number of electrons the metal radiated, but the emission of electrons did not occur for any value of light energy, but only when it exceeded a certain threshold energy, coinciding with the experience of Planck's *theory of quanta*.

In a very reduced synthesis and only for the purposes of this analysis, *quantum theory* assumes that the states of *quantum* systems are defined by what is called a wave function or state vector, postulating that these entities are specific mathematical elements, specifically vectors of a Hilbert space with certain characteristics, which is a variant of the vector spaces on the field of complex numbers, called Hermitian. The important thing is to understand that this *quantum* postulate does not attribute to its mathematical elements the representation of physical states, but rather it is a probabilistic density distribution of each of the possible variants of a given system.

Thus, if we think about the trajectory of an electron or other particle, the wave function or state vector reflects the probabilities that the electron fits each of the eventual trajectories. The fundamental *quantum* postulate admits that probabilistic states can be operated algebraically as if they were vectors of a Hilbert space. With this arises the notion of *quantum superposition*, which translates into the idea, not extramental, but merely algebraic, that the electron in the example can follow all possible trajectories at the same time, each one with an associated level of probability. And this is solely due to the algebraic structure accepted to operate mathematically with the different mental states and their probability levels. Obviously, it has nothing to do with extramental reality, because nothing is proven about the true physical nature of the electron or the system examined.

Under these conditions, it seems clear that *quantum* construction is nothing more than a mental mathematical work, which bridges the void of sensory information that we all suffer with respect to the extramental behavior of matter at the very small level, so that intuition is left out of *quantum* phenomena, because no one can immediately perceive them sensorially.

To overcome this difficulty of absence of direct sensory information, *quantum theory* uses the imagination and formulates its **completely arbitrary postulates** to check what happens next in its material application and make the appropriate corrections. To do this, it assigns its probability levels to the different possible states of a *quantum* system and the observations would have to agree with the corresponding wave functions or state vectors. So, by experimentally determining the states of a system, they would have to conform to their mathematical images.

What *quantum theory* does is similar to a cartographer who drew up a map of a territory by drawing it only with his imagination and then checking whether or not his drawing conforms to reality. It is evident that *quantum* postulates do just this, which is why we should not try to understand or examine them materially, but rather we must believe them and take them as abstract and mental entities, operate with them and check whether the mathematical predictions agree with the observed *quantum* reality.

In our opinion, the general incomprehension that *quantum theory* arouses at all levels is caused by the need to understand its arbitrary postulates, something impossible to achieve, because they are born from the physical-mathematical whim of its creators. But above all the confusion arises from the natural inclination present in all of us to identify the map, the theory, with the extramental reality represented or associated. And it is extremely easy to get carried away by the unhealthy need to believe that what we think has a real entity, as demonstrated by the objections that the most eminent scientists have been raising on this issue.

Thus, for example, in 1926, Albert Einstein was intransigent with the probabilistic formulation of the world established by *quantum mechanics*. For Einstein, matter must always obey the laws of physics, which are essentially deterministic. He admitted nothing against this belief. This is how Einstein expressed it with crystal clarity when responding to a letter from Max Born in which he said: «*Quantum mechanics* is very impressive. But an inner voice tells me that it is not yet the real thing. The theory produces a good deal but hardly brings us closer to the secret of the Old One. I am at all events convinced that He does not play dice». This phrase summarizes his thoughts against *quantum theory*: «God does not play dice with the universe».

Richard Feynman revealed his lack of understanding of the difference between mind and world when he said: «I think I can safely say that nobody understands *quantum mechanics*». It is evident that Feynman considered *quantum theory* incomprehensible because he ideally attributed to it an entity of extramental reality. However, it is enough to refer only to the abstract mathematical formulation that this theory postulates to

383

verify that it is perfectly understandable by anyone, like any other algebraic structure.

Niels Bohr was the promoter of the new Physics inspired by the philosophy of radical antirealism. Bohr believed that a particle has no definite position until it is measured. He believed that the exact location is not in a specific place, but depends on probabilities. Niels Bohr used *quantum theory* to describe his *atomic model*. He said in this regard: «If any of this seems disconcerting to you, it's because you haven't understood it».

Bohr also said: «The movement of particles follows laws of probability». It seems obvious that Bohr also fell into the error of confusing mental reality with extramental reality, since he attributed true existence to *quantum* phenomena, he did not consider them mental entities.

Although *quantum theory* is not about cats, it is perhaps Erwin Schrödinger's cat thought experiment that best describes the confusion generated by the probabilistic wave function that he himself contributed substantially to defining. This experiment, properly described as mental, consists of the following: let us imagine an opaque box and place inside it a live cat as well as a radioactive atom with a fifty percent probability of emission; let's put a detector connected to a device that activates a hammer if radiation is measured; finally, suppose that, if the hammer is operated, it falls on a glass container containing a poison and breaks it. Under these conditions there would be a fifty percent chance that the cat would be poisoned once the box was closed. As long as the box is not opened, that is, as long as what is happening inside is not observed, the cat can be alive or dead with a probability level of fifty percent for both options.

What *quantum theory* says is that, if the box remains closed, the *quantum* state is an algebraic combination of the two possible states, which are that the cat is alive and that it is dead, with a probability level each of fifty percent. The Hilbert sum of these two states of the considered system is another Hilbert state that has an associated probability equal to unity. However, when the box is opened, what is called the collapse of the *quantum* wave will occur and the real state will manifest, that is, the cat will be alive or dead. This experiment is also known as Schrödinger's paradox, which for us is not such, because it has its origin, as we have been repeating, in childishly confusing mental reality with extramental reality. The Hilbert space used by *quantum theory* is a mental entity that is used to assess the possibilities that the cat is alive or dead as long as it is not known what happened inside the box. Now, when it is opened and what is inside is observed, the extramental reality takes control and imposes on us what truly exists. The so-called collapse of the *quantum* wave is simply this transition from mental reality to extramental reality. *Quantum theory* is only valid in the interval prior to observation.

Furthermore, this theory can never fail, as long as the probability distribution for the different states of a system is well constructed. It is as if someone claimed that the *quantum* state of a lottery was the sum of the states of any number being awarded. Obviously the probability of this mental state prior to the drawing would be one, but if the game is carried out a result will occur and there will be some extra-mentally favored number, although it must have a *quantum* probability less than unity. Hence, the myth about the supposed infallibility of *quantum theory* does not have the meaning that is commonly attributed to it. It would not be physical infallibility, but rather abstract mathematical truth.

We take advantage of this section to perform the «dysmetric» analysis of Planck's constant. To do this, let us express the dyad that indicates its value in the original reference point O, which could be the terrestrial environment:

$$(6.62607015 \times 10^{-34}, J*s)$$

The composite unit $J*s$ will have the following fundamental expression:

$$J*s = \frac{kg*m^2}{s^2}*s = \frac{kg*m^2}{s}$$

As we have done in other cases with the number pi and the speed of light in sections **XXXIII** and **XXXIV**, let us take any other point P and suppose that the «dysmetric» densities of the magnitudes length, mass and time in this position are respectively δ_{LP}, δ_{MP} and δ_{TP}. The magnitude quantities of the units m_P, kg_P and s_P in P congruent with those of O, noted m, kg and s, will be related, given the definition of «dysmetric» density, by $m_P = \delta_{LP} \circ m$, $kg_P = \delta_{LP} \circ kg$ and $s_P = \delta_{LP} \circ s$. The International System defines Planck's constant as the quantity of magnitude $h = 6.62607015 \times 10^{-34} \, J * s$. Therefore, although it is indicated with the only letter h, it is actually a dyad composed of a numerical part, the value $h = 6.62607015 \times 10^{-34}$, and another dimensional part, the composite unit $J * s$.

Let us remember that dyadic notation is multiple with the only condition that there is no ambiguity. Thus the following notations are equivalent: $(h, J*s) = (h \, J*s) = h \, J*s$. We will use the first one here, although we have used the other two in other calculations.

In «dysmetry» saying that the Planck quantity is constant necessarily includes two possibilities: that the primary of the dyad remains constant, the number $6.62607015 \times 10^{-34}$, or that the quantity of magnitude h is constant.

In the first case, the Planck constant quantity in P, which is a dyad that we can note h_P, is easily calculated from the corresponding dyad and assuming that the numerical primary is the same at all points in space:

$$h_P = \left(6{,}62607015 \times 10^{-34}, \frac{kg_P * m_P^2}{s_P} \right) =$$

$$= \left(6{,}62607015 \times 10^{-34}, \frac{(\delta_{MP} \circ kg) * (\delta_{LP}^2 \circ m^2)}{\delta_{TP} \circ s} \right) =$$

$$= \frac{\delta_{MP} \times \delta_{LP}^2}{\delta_{TP}} \circ \left(6{,}62607015 \times 10^{-34}, \frac{kg * m^2}{s} \right) = \frac{\delta_{MP} \times \delta_{LP}^2}{\delta_{TP}} \circ h$$

386

We observe that Planck's constant in P is related to its value in O through the «dysmetric» densities of the fundamental magnitudes, according to the following law:

$$h_P = \frac{\delta_{MP} \times \delta_{LP}^2}{\delta_{TP}} \circ h$$

Using the definition of dyadic division from sections XI and article 7 of the XXVIII, we can express the quotient $h_P/\!/h$ between these quantities of magnitude, which is always equal to the real number that intervenes as a multiplier in the second member, resulting:

$$\frac{h_P}{\,h\,} = \frac{\delta_{MP} \times \delta_{LP}^2}{\delta_{TP}}$$

Therefore, if we make the hypothesis that the Planck measure is constant at all points in space, we come to the conclusion that «dysmetry» is incompatible with said assumption, unless the second member of the previous expression were equal to the numerical unit at all points in space. Since this would be admitting a strong arbitrary constraint, we have to infer that the measure of the Planck quantity h cannot be kept constant in a generic «dysmetric» space. Furthermore, it is clear that the dyadic ratio $h_P/\!/h$ can take values from zero to infinity, depending on how much the densities δ_{LP}, δ_{MP} and δ_{TP} are in P. Obviously, in the particular case of an isometric space, in which the densities «dysmetric» of all magnitudes at all points are unity, as is currently assumed, we would always have $h_P=h$ for all P, thus observing what we have already repeatedly warned: that **isometry is a case particular of «dysmetry».**

In the second case, if the Planck quantity were admitted constant, the equivalence between the value at O and at any other point P would have to be given. Expressing this analytically, the dyadic equality would have to be verified $(\eta, J*s) = (\eta_P, J_P*s_P)$. In particular, note that at O is $\eta = 6.62607015 \times 10^{-34}$.

In this assumption, η and η_P represent the primary or numerical value of the Planck quantity, a quantity that would be constant by hypothesis. Developing the quantity (η_P, J_P*s_P) according to the laws of dyadic algebra, the following will easily result:

$$(\eta_P, J_P*s_P) = \left(\eta_P, \frac{kg_P * m_P^2}{s_P}\right) = \left(\eta_P, \frac{\delta_{MP} \circ kg * \delta_{LP}^2 \circ m^2}{\delta_{TP} \circ s}\right) =$$

$$= \frac{\delta_{MP} \times \delta_{LP}^2}{\delta_{TP}} \circ \left(\eta_P, \frac{kg * m^2}{s}\right) = \frac{\delta_{MP} \times \delta_{LP}^2}{\delta_{TP}} \times \eta_P \circ \left(1, \frac{kg * m^2}{s}\right)$$

On the other hand, it is elementary to algebraically transform the quantity expressed by the dyad $(\eta, J*s)$ in this way:

$$\left(\eta, \frac{kg * m^2}{s}\right) = \eta \circ \left(1, \frac{kg * m^2}{s}\right)$$

Since we are analyzing the case that both quantities are the same, the laws of dyadic algebra determine that the ratio of these two quantities of magnitude must be the unit of the real numbers and must coincide with the arithmetic ratio of the numerical multipliers, since the numerator and denominator are referred to the same uniform unit, which disappears when the dyadic reason is developed. Which in analytical terms is written like this:

$$\frac{\dfrac{\delta_{MP} \times \delta_{LP}^2}{\delta_{TP}} \times \eta_P \circ \left(1, \dfrac{kg * m^2}{s}\right)}{\eta \circ \left(1, \dfrac{kg * m^2}{s}\right)} = \frac{\dfrac{\delta_{MP} \times \delta_{LP}^2}{\delta_{TP}} \times \eta_P}{\eta} = \frac{\delta_{MP} \times \delta_{LP}^2}{\delta_{TP}} \times \frac{\eta_P}{\eta} = 1$$

In sum, the ratio of the measurements of the Planck quantity at O and at any other point P is proportional to the ratio of the «dysmetric» densities, according to the following arithmetic law:

$$\frac{\eta}{\eta_P} = \frac{\delta_{MP} \times \delta_{LP}^2}{\delta_{TP}}$$

Therefore, in the hypothesis of a constant Planck quantity, the measurement η_P can vary between zero and infinity, depending on the values adopted by the «dysmetric» densities of the second member of the previous equation. And, since the composite unit in P has been reduced to that of O, which represents a quantity of definite and finite magnitude, the Planck quantity in P can also vary between zero and infinity, which contradicts the invariance hypothesis, and so the incompatibility of «dysmetry» with this supposed constant is manifested.

Of course, as in the other cases, if space were isometric, the only current variant visible to physicists, all the «dysmetric» densities are equal to unity and the second member of the last formula as well, so $\eta_P = \eta$ in all P, the Planck quantity remaining constant and it is verified again that **isometry is a particular case of «dysmetry»**.

In view of the above, we must ask ourselves if *quantum theory* is incompatible with «dysmetry», and the answer must necessarily be negative, because it does not invalidate the *quantum* postulates. What's more, it is enough to implement the «dysmetric» effect in the *quantum* formulations to enrich their approaches in an absolutely parallel way to what we did with Newtonian mechanics in sections XXXII and XXXV.

In short, Planck's constant, like all the others, including dimensionless ones like the number pi, are not so constant and must vary, if the magnitudes are «dysmetric», depending on their densities at each point P with respect to the origin of reference O, and this assuming that physical laws remain isomorphic at all points in space.

Returning to the purpose of this section, in conclusion, *quantum theory* or any other theory is the expression of how extramental reality manifests itself to our brains. They are models that integrate mental realities connected with extramental reality through perception. Therefore, it is not correct to identify them with what exists in the world external to the ideas that are formed in the brain, an error that we are all very prone to make, since it seems natural to immaturity to attribute material entity to our thoughts. In fact, in this section we provide evidence of how very notable scientists fall into this trap and provoke absurd discussions for epistemology.

Schrödinger's cat cannot really be alive and dead at the same time, but this does not prevent *quantum theory* from allowing the prediction of states of a system to be assessed with probabilities and to operate algebraically with them through correct mathematical laws, simply by postulating the valid application. of an abstract and incontestable algebraic structure, such as Hilbert's Hermitian vector spaces. So from this point of view *quantum theory* is nothing special, its apparent irrationality is in our opinion just a mere misinterpretation of its postulates. **In no case is *quantum theory* allowed to violate the laws of formal logic.**

If we look, for example, at *classical rational mechanics*, the same thing is done with it. We represent velocities, accelerations or forces with geometric vectors and we postulate that it is allowed to operate with these mathematical entities as if they were elements of an algebraic structure that we call in this case Euclidean vector space on the body of real numbers. Here it would not be correct to identify the true phenomena of speed, acceleration or force with the mathematical entity chosen to form the model, to which we must only confer the property of mentally representing what happens in the extramental realm.

The difference between *classical* and *quantum mechanics* is that intuitively it is easy to associate measurements of masses or energies with scalars or measurements of velocities, accelerations and forces with vectors, because our sensory experience suggests it to us, resulting in the *classical* postulate of associating the magnitudes with scalars or vectors, the former members of the body of real numbers and the latter of three-dimensional Euclidean space, it is easy to assimilate it by any mind, since it is compatible with common sensory experience; On the other hand, for *quantum* phenomena we lack a sensory reference and we act blindly. Hence, the creators of *quantum theory* have chosen to choose a priori a pre-existing mathematical structure that serves to operate with *quantum* measurements, establishing as a fundamental postulate that this structure of abstract algebra describes its operations. Therefore, whoever tries to understand this choice will go crazy without succeeding, because it is nothing more than an unprecedented working method consisting of considering that a part of mathematics defined in the abstract, without any experimentation, is valid to precisely represent *quantum* phenomena, and this only because we have decided so without prior evidence. It is as if the scientific method were inverted and, instead of putting observation before the subsequent mathematical formulation of the phenomena, we proceeded in reverse: first we choose the mathematical apparatus and then we check and adjust it with observations, which in the case of *quantum* facts cannot be direct, but must necessarily be done with very sophisticated measuring instruments and inaccessible to most of us, so we cannot examine them and we have to settle for learning the mental mathematical model that is offered to us without the possibility of face it even with a minimal sensory experience of our own, which we cannot possess because *quantum* reality does not interact with our perception.

We have admitted that *classical mechanics* is more intuitive and, therefore, more understandable than *quantum mechanics*; however, this is not as obvious as it may seem. To observe it, let's take a significant case: a priori it seems that the free fall of objects depends on their mass, because that is what we observe when dropping, for example, a feather and a much heavier object from the same height, what we observe is that the feather falls more slowly. However, if we ignore air resistance, that is, in a vacuum, as Galileo demonstrated, the feather and the heavy element take the same time to reach the ground from the same height. The visual experiment carried out on the Moon by Apollo 15 is famous, showing that in the lunar vacuum a feather and a hammer touch the ground at the same time when the astronaut releases them in free fall. Therefore, *classical mechanics*, which seems so intuitive to us, is not so intuitive, and this is because, as we have been warning, no theory, no matter how simple and obvious it may seem to us, ever replaces extramental reality in any way. Every theory is a mental entity alien to extramental reality. Reflecting on a theory and analyzing its elements is a very different thing from the true qualities of nature. Physical magnitudes are nothing but physical-mathematical mental entities whose material condition we ignore and only slightly perceive through their impression on the senses. In short, thought and reality are not equal entities but related through perception. Taking this simple fact into account we are in a position to better understand what it means to understand the meaning and suitability of any physical theory or any other field.

In the realm of the very small, the *quantum* state is a representation of the physical state of a phenomenon under the perspective of *quantum mechanics*. In *classical rational mechanics* it is accepted that measuring a physical magnitud repeatedly produces the same value. However, in *quantum theory* it is assumed that the measurement of any physical magnitud can offer different quantities in different measurements on identical *quantum* states. That is, if the measurement could be repeated, a different observation may appear each time the magnitude is measured. Hence, *quantum physics* uses a probability distribution to express the result of a measurement.

In short, the main difference between *quantum* and *classical theories* is that intuition adapts well to the *classical model*, because we can easily conceive the measurements of magnitudes as real numbers and mathematical vectors. On the other hand, in *quantum theory* intuition does not work in the same way and seems incomprehensible, but no matter how irrational or anti-realistic it may appear, everything fits into it when it is seen for what it is: a simple map that draws in its own way the extramental world of the very small through the predetermined algebraic mental structure of Hilbert's Hermitian vector spaces, with its particular abstract character that may deviate from common sense, but for algebra it is a mathematical language as unappealable as the body of real numbers or Euclidean spaces.

So, in the same way that no one thinks of identifying the map of a territory with the territory itself, because the difference is obvious, no one should confuse any physical theory or its postulates with the real phenomena they purport to reflect. Therefore, it would be a gross error to pretend in *classical mechanics* that the masses, energies, velocities, accelerations or forces of the extramental world are effectively real numbers or geometric vectors, even if it seems plausible to us sensorially; and in the same way it would be childish to identify wave functions or state vectors with real *quantum* phenomena: **mental reality can in no case be the same as extramental reality.** Each brain can only know its own mental reality and the extramental reality can only have indirect reference through the sensory data received from perception, data that are also mental entities, since they are housed in the brain itself, but they are not even elements of external reality.

393

ADDENDUM

«DYSMETRY»
*Discovery of a new dimension
of physical magnitudes*

The preceding breviary has presented the compendium of the research and revelation of the physical-mathematical truths that have emerged within it, carried out by the author of this work. Here we are going to propose another summary with a somewhat different approach, focusing more on what «dysmetry» means. We prefer to be repetitive and err on the side of excessive argumentation rather than defect, given the difficulty of the subject, due above all to the blindness induced by the bad habits of archaic «arithmetization». We hope to alleviate the work necessary to become informed about what this book means and to overcome the resistance of the most conservative minds, converting them for their own good to support the «dysmetric» movement.

Until now, physicists have assumed, without thinking about it, that our units of measurement for physical phenomena would be constant, that is, that we were indifferent to the location in space-time and the material environments of our measurements, because the measurement standards would never change. We assumed that these standards would not be affected by the supposed impassive and immutable nature of empty space or by the existence of variable matter or energy in the different measurement environments.

Such a mentality is nothing but a primitive and clumsy simplification of the infinite variability observed in the universe from the smallest to the most immense, arbitrarily assuming, without conclusive proof, that everything that exists manifests itself in the same way as in our limited perceptible human environment. However, in this work we have found that such an assumption is illusory, rather childish, and certainly not at all justified. We now proceed to a succinct summary of the steps that led us to discover the «dysmetry» and to the conclusion that we are faced with an inescapable truth: the **«dysmetric» dimension of physical magnitudes**.

At the beginning of the investigation we recover the concern of the classical physicists of the late nineteenth and early twentieth centuries about the lack of foundation of the operations with magnitudes, summarized in the **mystery of the composite magnitudes**, which has produced so much controversy without ever deciphering the enigma, an unsolved mystery that has been shelved by the International System of Units by simply arbitrarily postulating a reckless pseudo-algebra consisting of the fictitious rule of operating with physical units with the same laws established for numerical sets, what we have called here the unhealthy «arithmetization» of Physics.

To address this major inconsistency, the first important observation is to differentiate between measure and quantity of a magnitude. Measure is a real number, as an expression of a quantity in relation to its reference unit or standard of the given magnitude. Quantity, on the other hand, is indicated by the binary set of measure and unit, which we have called a dyad here and which in classical mathematics is called a concrete number. Well, all quantities of any scalar magnitude are represented in relation to any unit U by the set of all dyads (q, U) where q is a real number. Hence we call the grouping of all real numbers R associated with the unit U, which we write $\{R, U\}$, a dyadic set. It is enough to observe this notation to understand that the set of real numbers R does not coincide with any dyadic set $\{R, U\}$ of any magnitude. Therefore, establishing without further ado that the algebra of dyadic sets is identified with that of real numbers is absurd and has no mathematical sense. **Dyadic sets are different from R and, as such, required a specific algebra**, pending development, which constituted the origin of this research and which has led to the fortunate chance discovery of «diysmetry», one more case of serendipity among the many that appear in the history of science.

The same thing happens with vector magnitudes. Dyadic sets can be described in this case in the form $\{R^3, U\}$ and, of course, the set R^3 is different from $\{R^3, U\}$. In the text we developed the algebra of dyadic vector sets and, as it is completely analogous to the algebra of scalar magnitudes, in this philosophical addendum we will limit ourselves exclusively to scalar magnitudes, so as not to tire the reader with superfluous reiterations.

Therefore, with the previous dissertation we have already established the inexorable need to develop a specific algebra for dyadic sets, representative of the quantities of each magnitude considered. The first operation to be defined is addition. This requires appreciating that in order to add quantities, they must refer to the same magnitude, an observation that we have described as the need for **homogeneity**. But there is more, in order to add two dyadic quantities, they must be referred to the same unit, which we call the **axiom of uniformity**. In short, it is possible to add, for example, kg with g, but to do so both addends must be expressed in *kg* or in *g*. In this way, when the addends are uniform, the sum is obtained by adding the measurements with the addition of R, maintaining the common unit of the addends. Once the addition of homogeneous quantities is defined in this way, it is easily demonstrated that this operation confers to each dyadic set the structure of an **abelian additive group**.

The addition of quantities does not, therefore, present too many algebraic problems. But the same does not occur with multiplication, because in this case the dyadic product is not reduced to that of R, since it is not possible to identify a multiplicand or a multiplier. And this phenomenon is vital to unravel and understand the mystery of composite magnitudes. Let us take length as an example of magnitude. When two lengths are multiplied, the product is not another length, but a surface, and it is a volume when there are three lengths multiplied. On the other hand, the multiplication of real numbers is reduced to

397

abbreviated sums and gives another real number as a product. The product of numbers is an internal law, while the product of lengths is an external law, because, as has already been said, when multiplying two lengths, another length is not obtained, but a surface, or a volume if three lengths are multiplied. We thus observe that the multiplication of lengths is a new law of composition that we have baptized with the name of **external generative law**. It is external, because the product is not a length, but a surface or a volume, which are different geometric magnitudes. And it is generative because the multiplication of lengths produces another different magnitude, the surface or the volume.

The product of lengths is visualized as the product of segments and this allows us to draw inspiration for this operation from the **geometric algebra of segments**. Then, considering that any quantity of other magnitudes can be represented by segments, it is easy to establish a biunivocal affinity or correspondence between quantities of diverse magnitudes and the set of geometric segments, as Newton did in his Principia, following in turn the criterion of Euclid's Elements. We call this maneuver the **affinity postulate**, which already allows us to conceive the multiplication of any quantities by means of segments, surfaces or affine volumes. And in this simple way, multiplicative operations of any magnitudes can be developed, making it clear **that composite magnitudes do not contain any mystery**, but rather present an irrefutable geometric simplicity.

The fact that multiplicative operations are generating external composition laws prevents dyadic sets from having the structure of an abelian multiplicative group, contrary to what happens with addition. And thus the false hypothesis of the International System of Units is exposed, which wrongly attributes this structure to the multiplication of magnitudes. The most immediate consequence of true algebra is that it proves that **there are no unitary or inverse multiplicative elements**. With this we verify the nonexistence of elements such as m^{-1}, kg^{-1} or s^{-1}, which supposes a clear incongruity of the current system for some magnitudes such as frequency, which the International System of Units measures with the fake isolated unit s^{-1}.

With these foundations we build a complete specific algebra for magnitudes that we call **dyadic algebra** and we identify the multiple operations that compose it, revealing that the famous mystery of the composite units is not such and that anyone can observe the underlying mathematical truths behind the operations with magnitudes.

Dyadic algebra not only solves the mystery of composite magnitudes, overcoming the erroneous «arithmetization» that the International System of Units arbitrarily normalizes, but it also gives us a splendid gift: «dysmetry». Let us look at the simple process that reveals this phenomenon to us in such an unexpected way.

Let us consider a generic scalar dyad (q, U). The quantity represented by this binary element can obviously vary by changing the measure q for any other. Now, is this the only form of dyadic variation? Until now we thought so, because we assumed that every unit U contained a quantity of constant magnitude and independent of any circumstance. Is this belief consistent from a scientific and logical point of view? Let us see: What prevents us from theorizing that the quantity implied in any unit U can vary in space, time or by the influence of physical actions? Nothing. Therefore, this assumption should not be excluded a priori and without any proof to refute it.

For the sake of clarity, the first condition, that is, that the quantity of magnitude associated with any unit U is constant in any context, is called isometry. The opposite property, that is, that any unit U can indicate quantities of magnitude that vary for various reasons, is called «dysmetry». Thus, we have only two variants of the same thing: isometry and «dysmetry». Nature is either isometric or «dysmetric». If we admit without proof that it is isometric, we will be excluding in one fell swoop all «dysmetric» phenomena that may have real existence. On the other hand, it is clear that the «dysmetric» variant is broader than the isometric one and that isometry is included in «dysmetry». because if the latter did not exist, isometry would manifest itself alone as an implacable phenomenon. In conclusion, philosophically and by pure logic, the principle to be established a priori on the nature of the universe must be the «dysmetric» variant, as long as there is no proof to the contrary. The search for such proof is extremely difficult, because it would require, for example, carrying out experiments very far from Earth. It would be more feasible to experiment with the «dysmetric» phenomenon in the atomic realm.

Perhaps the simplest proof we can conceive to refute isometry is the simple observation of what exists, which manifests itself in an infinite variety of forms and essences, because no two things are the same. Then, this simple experience common to all should lead us to establish that the obvious thing is «dysmetry». And, if we go further, doing a differential mathematical experiment and studying the mathematical variation of a dyad (section XXXVII), we arrive at the irrefutable mathematical conclusion that **the natural thing is «dysmetry»**

The complexity of dysmetry» seems to make it difficult to incorporate it into the description of physical phenomena. But fortunately, we have found a relatively simple way of representing it, taking advantage of the dyadic algebra of magnitudes. The method consists of finding the quotient between the magnitude quantities of the same physical unit in different environments. We know that the dyadic quotient of homogeneous quantities is in any case a real number and we call such quotients **«dysmetric» densities**. It follows that every «dysmetric» density, being a real number, is dimensionless. Thus we arrive at the discovery of the **«dysmetric» dimension of physical magnitudes**, characterized by a field of magnitude densities, given as the dyadic ratio between the quantity implied in a certain unit for each point in space-time in relation to the quantity contained in the same unit corresponding to a fixed point taken as a reference. This also allows us to confirm that isometry is a particular case of «dysmetry», because the latter is reduced to the former when all the «dysmetric» densities are equal to one, which constitutes a definitive argument for adopting the «dysmetric» variant as a **fundamental principle of Physics**, if we do not want to risk our equations and physical laws indicating very limited phenomena with respect to all those that really exist.

As the research in this work progressed, considering only the «dysmetric» dimension of length, we were able to characterize tensorially the physical properties of empty space (section XXXVI), which does not manifest itself as something inert, but rather produces by itself material effects as important as the variation of the speed of light (see also section XXXIV) and the curvature of its rays, without the need for any other perturbation. Or also the variation of the geometric number pi (section XXXIII). Likewise, we reformulated some classical laws such as Newton's second law or gravitation by simply incorporating into them the «dysmetric» dimension of the intervening magnitudes (sections XXXII and XXXV).

In short, this work develops the dyadic algebra of magnitudes, solves the mythical mystery of composite units, and provides Physics with the first specific algebra in history, correcting essential errors such as the nonexistence of unitary or inverse multiplicative elements and establishing a separate mathematical structure for physical phenomena based on the generative external composition laws. At the same time, this new algebra reveals a fundamental principle hidden until now, the **«dysmetric» dimension of magnitudes**, which reveals new astronomical, cosmological and physical laws, offering an infinite horizon of research and innovation.

THEMATIC PAPER I

DYADIC ALGEBRA OF MAGNITUDES
Natural structure for physical operations

Abstract

Contemporary physics rests on a tacit isomorphism between physical magnitudes and the field of real numbers. However, this *arithmetization* overlooks the fact that the multiplication of magnitudes is not an internal operation, but rather a generative function that produces new dimensions and breaks the classical field structure. In this work we present the Dyadic Algebra of Magnitudes, a mathematical framework in which quantities are not mere points on a numerical line, but entities with their own extent and internal degrees of freedom. We introduce the *law of dyadic variation*, which shows that the current paradigm of the International System of Units (SI) is only the particular case of *zero dysmetric variation*. By relaxing the rigidity of magnitudes, this approach makes it possible to model phenomena in non isometric regimes —where the dimensional structure itself is dynamic—, that remain invisible to conventional arithmetic. Our results suggest that adherence to numerical algebra limits the observation of essential properties of matter, and we argue for an ontological modernization suited to the challenges of advanced theoretical physics.

Introduction

This synthesis faithfully condenses the entire dyadic algebra formulated in the works *First Algebra of Magnitudes and Dysmetry* and *The New Physics of Dysmetric Spaces* by the same author, J. M. Arnaiz (2016), which are the reference texts. It omits the motivations that previously justify the definitions of the various concepts, in order to focus exclusively on the mathematical structure formed by operations with quantities of physical magnitudes, sacrificing all literary elegance so as to impose only the rigor of the necessary mathematical logic that allows the abstract algebra of physical operations to be configured with the greatest possible order, brevity, and coherence.

The origin of this work lies in the uncomfortable paradox that Physics operates with magnitudes using a numerical algebra, assuming that the behavior of magnitudes is identical to that of real numbers or, equivalently, admitting a tacit and never-proven isomorphism between the set of magnitudes and the field of real numbers. This *arithmetization* overlooks the fact that the multiplication of magnitudes is not an internal operation, but rather a function that governs the emergence of new physical dimensions, thereby breaking the field structure and the possibility of the isomorphism suggested by classical arithmetic. It can be observed that Newton, in his *Principia*, did not rely exclusively on arithmetic to operate with magnitudes. He reserved operations with abstract numbers solely for the measures of magnitudes, that is, for the numerical part of any measurement, abstracting from the physical unit. To compensate for this limitation, Newton employed classical geometry to operate with magnitudes, since at that time abstract algebra had not yet developed sufficiently. In any case, Newton recognized that operating with magnitudes using arithmetic alone was not correct. This was also noted by Fourier in his *Théorie Analytique de la Chaleur* (1888). Hence, Fourier developed a principle of an algebra of magnitudes, basing it on a simple symbolic imitation of arithmetic operations; thus dimensional analysis was born, and it has persisted to the present day. However, Fourier's method did not resolve the problem that such dimensional analysis is not a true algebra of magnitudes, since it is evident that it is not. This was confirmed by the numerous studies published by leading figures of twentieth-century Physics. We shall not reproduce here the interesting philosophical debates that took place at that time among Planck (1906), Bridgman (1922), Tolman (1917), or the different factions of the Vienna Circle.

Special mention should be made of the Spanish physicist Julio Palacios and his work *Análisis dimensional* (1964), internationally recognized, in which he explicitly refers to the lack of an algebra of magnitudes:

> A widely held view, dating back to Clerk Maxwell and shared by many physicists of my generation, is that these symbols, and therefore dimensional formulas, refer to units; thus, for example, one writes

$$1\,erg = \frac{1\,cm^2 \times 1g}{1\,s^2}$$

> without realizing that we would find ourselves in difficulty if an inquisitive student were to ask how one is supposed to multiply a square centimeter by a gram and divide the product by a second squared (p. 12).

Elsewhere in his work, Julio Palacios explicitly refers to *arithmetization*, emphasizing that physical formulas relate measures —that is, numbers— rather than quantities of magnitudes:

> Since this method requires excessively abstract speculations, it has seemed advisable not to attribute to equations any meaning other than that of relations between measures, that is, numbers. The verbal statement of the equation $f = ma$ would be: at every material point, the *measure of the quantity* of force acting on it is equal to the *measure of the quantity* of mass of the point multiplied by the *measure of the quantity* of acceleration. The relentless repetition of the underlined phrase [measure of the quantity] would turn physical language into an unbearable and pedantic singsong; it is therefore quite rightly omitted, it being understood that the same name serves to designate both the magnitude as an abstract entity and the measures of its quantities in each particular case (p. 32).

Another relevant warning concerning the same problem of the *arithmetization* of quantities can be found in the article of Giovanni Giorgi (Italy, 1871-1950, founder of the MKS system, origin of current International System of Units), titled *Sistemi e unita di misura* (Systems and units of measurement) in the Enciclopedia delle Matematiche Elementari (Encyclopedia of Elementary Mathematics), states that:

> The Theory of dimensions must be considered as an object of artificial linkages, of pure determining conventions of a certain particular metrological structure, and as a consequence, a special way of writing the equations of Physics; conventions theoretically arbitrary and practically dictated for reasons of mere opportunity and for this very reason deprived of all foundation, of all physical meaning.

One final example, among the many that exist, referring to the lack of an algebra of physical quantities. R. M. Cooke and J. Hilgevoord in *The Algebra of Physical Magnitudes* (1980, pp. 363 to 373), they summarize the debates of the classics thus:

> Philosophers have long been interested in the question of the physical presuppositions underlying the application of algebraic operations to physical magnitudes, and this interest has quickened as a result of the existence of hidden variables underlying quantum mechanics.

What is formalized in this work, precisely, is the need to modernize Physics by means of a proper algebra capable of overcoming the current *arithmetization,* noted by those authors and many others, but consolidated through theoretical metrology and classical dimensional analysis. These have crystallized in the International System of Units, which represents little more than the simplest case of total rigidity of magnitudes and ignores their dynamic qualities. This requires prior knowledge of abstract algebra, geometry, vector spaces, and tensor calculus, taking into account the foundations established by other authors, such as Hilbert (1899), Herstein (1980), Fraleigh (1982), Lichnerowicz (1972), Santaló (1970), Puig Adam (1970), and Catalá Moreno (1972). Consideration has likewise been given to the dimensional analysis of Bridgman (1922), Giorgi (1943), Cooke (1980), Palacios (1964), Kurth (1972), Catalán Chillerón (1973), as well as to the recommendations of the Bureau International des Poids et

Mesures, which greatly oversimplify the relations among physical magnitudes by reducing them to operations that merely imitate arithmetic ones, without any genuine algebraic foundation. From this survey, the reader will realize that the algebra of magnitudes and the *dismetry* derived from it in fact concern something much deeper: a proposal for an algebraic structure of magnitudes with distinct properties, differing from the arithmetic ones on which dimensional analysis is based, a discipline that represents only a particular case or, rather, a very limited expression of the reality of magnitudes. The absence of such a structure became evident during the development of the textbooks *Matematizar* 1-3, eventually leading to *La nueva Física de los espacios «dismétricos»* and *La Primera álgebra de magnitudes y «Dismetría»* (J. M. Arnaiz, 2016), which have made it possible to bring this article to completion.

In the analysis of addition we conclude here that each set of quantities corresponding to any given magnitude satisfies the properties of an abelian additive group. The same does not hold for the multiplication of magnitudes, since the structure of a multiplicative group, characteristic of the real numbers and of fields in general, is not valid for this operation. Rather than constituting an internal law of composition, it is an external, generative operation, lacking unit and inverse elements in the classical sense, as explained in Sections 32 and 33. Consequently, the multiplication of magnitudes cannot satisfy the axioms of a multiplicative group, contrary to what is currently assumed. The true isomorphism is found between the set of magnitudes and the set of geometric segments, as described in Section 36, once the dyadic algebra of magnitudes has been established, thereby recovering the view of magnitude as an entity possessing its own extension rather than as a mere point on the numerical line.

Finally, the last three sections introduce the notion of *dysmetric* mathematics, emphasizing that persistence in classical arithmetic limits predictive capacity in non-isometric regimes. By reducing magnitudes to a mere numerical coefficient subordinated to the standard of the International System, an implicit ontological

407

reduction of reality takes place, in which the complexity of magnitudes is emptied in order to force their accommodation within the framework of a basic set of inflexible units. Such simplification has concealed a fundamental mathematical reality that emerges with the algebra of magnitudes: the law of dyadic variation, presented in Section 38, which expresses the two components of change of any dimensional quantity: the isometric variation of the current paradigm and the *dysmetric* variation associated with the dynamic flexibility of magnitudes, ignored by *arithmetization*. This mathematical innovation makes it possible to represent physical domains in which magnitudes may exhibit unrestricted variation. The International System thus appears as the particular case in which such *dysmetric* variation is null, a reduction that prevents the observation of dynamic properties of magnitudes and restricts the scope of Physics to strictly isometric phenomena.

The bibliography included at the end of this synthesis, apart from the reference works used to develop this synthesis, consists of titles that are merely recommendations to readers, intended to indicate where they may find well-presented treatments of abstract algebra concerning common mathematical structures, so as to facilitate a better understanding of the origins of the specializations here formulated for physical magnitudes.

1. Postulate of Affinity. In general, every physical reality or extramental object manifests to the observer's mind multiple characteristics implicit in it. The brain gives mental form to these characteristics and, if thought defines a concept, that mental image of the real characteristic becomes associated with a symbol representing both the real and the mental aspects. Algebra is concerned only with those characteristics that can be represented by mathematical objects comparable among themselves and to which an algebraic structure can be assigned. A characteristic will be said to be algebraic if it satisfies this condition.

Furthermore, if all manifestations of an algebraic characteristic can be grouped into a set whose elements are symbolized by certain mathematical entities, which we shall here call dyads, and if it is thereby possible to define a one-to-one correspondence with the mathematical elements of another algebraic structure such that the operations on the elements of the first set can be defined in terms of the operations of the second, we shall say that it is an affine algebraic characteristic.

This algebra of magnitudes studies the affine algebraic characteristics within the domain of Physics. To this end, it takes as reference the structure of geometric segments, areas, and volumes—in as many dimensions as necessary—together with the additive and multiplicative operations of geometry, in combination with set theory. Accordingly, all operations conceived for the dyadic entities, which will be defined later, are grounded in this affine structure of geometry and mathematical sets, by virtue of what we shall call the postulate of affinity.

2. Definition of magnitude. Any affine algebraic characteristic is said to constitute a magnitude, composed of its various manifestations.

3. Definition of the quantity of a magnitude. Any manifestation of an affine algebraic characteristic is understood to constitute a quantity of a magnitude. Consequently, the various quantities of a given magnitude are represented by mathematical objects that are comparable with one another and correspond to elements of geometric structures under the determinations of abstract algebra.

4. Dimensional axiom. We acknowledge the evidence that quantities of magnitudes cannot be reduced to a single abstract number. On the contrary, they require the use of specific symbols to represent the various quantities in order to compare and operate with them, without being able to specify them explicitly. This makes it possible to form sets of these symbolic elements, whose whole or fractional groupings are understood to implicitly contain the various quantities of magnitude associated with those sets, as will be clearly illustrated in this synthesis.

5. Definitions of dyad, homogeneity, uniformity, and dyadic equality. A dyad is defined as any binary set (q, U) consisting of a pair of elements: a mathematical object q and a symbol U representing an non-numerical quantity of a magnitude. In general, U could symbolize anything, but for the purposes of this algebra we will only study the case where U represents any manifestation of an affine algebraic feature. The mathematical element q is called the primary of the dyad, and its non-numerical part U is called the secondary. The mathematical element q serves to define sets of whole or fractional elements of U. Two dyads expressing quantities of the same magnitude are said to be homogeneous. If, in addition, they share the same secondary, they are said to be uniform. Two homogeneous dyads (q_1, U_1) and (q_2, U_2) are considered equal, and we write $(q_1, U_1) = (q_2, U_2)$, if both implicitly contain the same quantity of magnitude.

6. Definition of a whole dyad. If q is an element of the set of integers Z, the dyad (q, U) symbolizes the set of $\{q$ elements $U\}$ and this identity is indicated with an equal sign: $(q, U) = \{q$ elements $U\}$.

7. Definition of a rational dyad. If q is a rational number a/b, with a and b integers in Z and $b \neq 0$, we define the rational dyad $(a/b, U)$ as the set $\{a$ elements $U_b\}$, which implicitly contains the same quantity of magnitude as the dyad (a, U_b), where U_b is such that $U = \{b$ elements $U_b\} = (b, U_b)$. The axiom of affinity justifies considering U and U_b analogous to segments and allows for the application of the geometric operations of whole division and multiplication defined for lengths, admitting that U_b is the quantity of magnitude resulting from dividing the quantity U by the number b, as if they were affine geometric segments. Moreover, we understand U as a shorthand notation for the dyad $(1, U)$. Obviously, if $b = 1$, every rational dyad is a whole dyad.

8. Definition of a vectorial dyad. We define a vectorial dyad as any dyad in which q is an element of a vector space. The quantity of magnitude implicitly contained is a vectorial dyad is determined by the magnitude of the vector paired with the unit

U; therefore, for the purpose of the implicit quantity, there is no distinction between scalar and vectorial dyads. However, to account for vectorial characteristics, operations on these dyads are performed according to the rules of the corresponding vector space structure.

9. Definition of whole dyadic addition. Given two uniform whole dyads (q_1, U) and (q_2, U), we define addition, denoted by $\oplus$, as the quantity of magnitude implicitly contained in the set formed by the union of the sets represented by the two dyads. This is expressed algebraically as follows:

$$(q_1, U) \oplus (q_2, U) = \{q_1 \text{ elements } U\} \cup \{q_2 \text{ elements } U\}$$

The number of elements U in the union set is, obviously, the addition of the integers $q_1 + q_2$, so that the right-hand side becomes the set $\{q_1 + q_2 \text{ elements } U\}$, and the previous definition is then expressed by the following equation, which already has operational properties:

$$(q_1, U) \oplus (q_2, U) = (q_1 + q_2, U)$$

The interpretation of this expression should be as follows: the quantity of magnitude implicitly contained in the union of the sets $\{q_1 \text{ elements } U\}$ and $\{q_2 \text{ elements } U\}$ is the addition of the dyads that independently represent them, a sum that is denoted by the dyad whose primary is the sum of the primaries and whose secondary is the same as that of the summands. The non-numerical reference quantity is implicitly contained in the symbol U.

10. Commutative property of whole dyadic addition. By the definition of addition, we have $(q_2, U) \oplus (q_1, U) = (q_2 + q_1, U)$. The commutative property of addition in the group of integers states that $q_2 + q_1 = q_1 + q_2$; therefore, $(q_2 + q_1, U) = (q_1 + q_2, U)$, which establishes the commutativity: $(q_2, U) \oplus (q_1, U) = (q_1, U) \oplus (q_2, U)$.

11. Associative property of whole dyadic addition. Consider the addition of three dyads $[(q_1, U) \oplus (q_2, U)] \oplus (q_3, U)$. By the definition of addition, we have $[(q_1, U) \oplus (q_2, U)] = (q_1 + q_2, U)$. Adding the third summand gives $(q_1+q_2,U)\oplus(q_3,U)=[(q_1+q_2)+q_3,U]$. By the associative property in the additive group of integers, $(q_1+q_2)+q_3=q_1+(q_2+q_3)$. Therefore, $[(q_1+q_2)+q_3,U]=[q_1+(q_2+q_3),U]$. The definition of whole dyadic addition ensures that $[q_1+(q_2+q_3),U]=(q_1,U)\oplus[(q_2+q_3),U)]$. In conclusion, we have $[(q_1,U)\oplus(q_2,U)]\oplus(q_3,U)=(q_1,U)\oplus[(q_2+q_3),U)]$, which expresses the associative property of whole dyadic addition.

12. Existence of the whole dyadic identity element. Let us consider the additive identity of the integers, denoted by 0. Consider the dyad $(0,U)$. For any whole dyad (q,U) we have $(q,U)\oplus(0,U)=(q+0,U)$. Since 0 is the identity element in the integers, $q+0=q$. Therefore, $(q,U)\oplus(0,U)=(q,U)$, showing that the dyad $(0,U)$ is a right identity for any dyad (q,U). Similarly, it can be shown that $(0,U)$ is a left identity, and thus $(0,U)$ is the identity element for all dyads (q,U), proving the existence of an identity element for whole dyadic addition.

13. Existence of the whole dyadic symmetric or opposite element. Consider the dyads (q,U) and $(-q,U)$, where $-q$ is the additive opposite of the integer q. Their addition is $(q,U)\oplus(-q,U)=(q+(-q),U)$. Since the integers $q+(-q)=0$, we have $(q+(-q),U)=(0,U)$, and thus the whole dyad $(-q,U)$ is the right opposite of (q,U). Similarly, it is also the left oppsite. Therefore, the inverse element of the whole dyad (q,U) exists and is $(-q,U)$.

14. Whole dyadic set. Given any magnitude and one of its quantities U, we form the set of all whole dyads (q,U), where $q\in Z$. The set thus formed, denoted $\{Z,U\}$ is called the whole dyadic set of the magnitude represented by the quantity U. This set represents all whole quantities of that magnitude.

15. Additive whole dyadic structure. The whole dyadic set $\{Z,U\}$, endowed with the internal composition law defined by the Cartesian product $\{Z,U\}\times\{Z,U\}$ in $\{Z,U\}$, given by whole dyadic addition $(q_1,U)\oplus(q_2,U)=(q_1+q_2,U)$, where all dyads (q_1,U), (q_2,U) and (q_1+q_2,U) elements of $\{Z,U\}$, forms an abelian additive group. We denote this structure by $\{Z,U,\oplus\}$.

16. Definition of multiplication of an integer by a whole dyad.
Given an integer a and a whole dyad (q, U), the multiplication,
denoted by $\circ$, is defined as the set formed by the union of a sets of
$\{q$ *elements* U$\}$. It can be written as $a \circ (q, U)$ on the left or $(q, U) \circ a$
on the right. It is clear that the number of elements U in this set
is the product of the integers $a \times q$. Thus, among other equalities,
we have:

$$a \circ (q, U) = (q, U) \circ a = (a \times q, U) = (q \times a, U) = q \circ (a, U) =$$
$$= q \circ (a \circ U) = (q, a \circ U) = q \circ (U \circ a) = (q, U \circ a)$$

Therefore, if the primary of a dyad (q, U) is multiplied by an
integer a, the resulting dyad is equal to the one obtained by
multiplying it secondary by the same number; that is,
$(a \times q, U) = (q, a \circ U)$. Likewise, if only the primary or the secondary
is multiplied by a, the dyad is multiplied by the same number.

It should be noted that the multiplication of a whole dyad by
an integer is commutative by definition, since $a \circ (q, U)$ and $(q, U) \circ a$
represent the same set $\{a \times q$ elements $U\}$.

17. Relations between the components of a rational dyad. Given
a rational dyad, we have $(a/b, U) = \{a$ elements $U_b\} = (a, U_b)$, where
U_b is such that $U = \{b$ elements $U_b\} = (b, U_b)$. The equality $U = (b, U_b)$
allows us to write $U = (b, U_b) = b \circ (1, U_b) = b \circ U_b$. The product $U = b \circ U_b$
can also be written using divisive notation as $U_b = U /\!/ b$. We use a
double slash to indicate that this division does not act on
numbers, but rather divides a quantity of magnitude U by an
integer b. Thus, we have $(a/b, U) = (a, U_b) = (a, U /\!/ b)$. Furthermore,
$b \circ (a/b, U) = b \circ (a, U_b) = (b \times a, U_b) = (b \times a/b, U) = (a, U)$, which in
divisive notation is $(a, U) /\!/ b = (a/b, U)$. In conclusion, the following
equalities hold: $(a/b, U) = (a, U /\!/ b) = (a, U) /\!/ b$. Therefore, this
expresion defines the division of a whole dyad (a, U) by an integer
b as the division of either its primary a in Q or its secondary U by
the number b, such that $b \circ U /\!/ b = U$.

18. Rational dyadic set. Let there be any magnitude and one of
its quantities denoted by U. The rational dyadic set $\{Q, U\}$ is
defined as the set of all rational dyads (q, U), such that $q \in Q$. This

set includes all rational dyads and represents all quantities of the magnitude indicated by U over Q.

19. Multiplication of a rational number by a rational dyad. Every rational dyad is reduced to a while dyad by its definition, since $(a/b, U) = (a, U_b)$, simply by substituting U by U_b. Therefore, when developing the product of the rational dyad $(a/b, U)$ by the rational number c/d, we can follow the reasoning:

$$c/d \circ (a/b, U) = (c/d, (a/b, U)) = (c/d, (a, U_b)) = (c, (a, U_b)_d) =$$
$$= (c, (a, U_b)/\!/d) = (c, (a, U_b/\!/d)) = (c, (a, U_{bd})) = (c \times a, U_{bd})$$
$$U = \{b \text{ elements } U_b\} = \{b \text{ elements } \{d \text{ elements } U_{bd}\}\}$$
$$U = \{b \times d \text{ elements } U_{bd}\}$$
$$U = b \times d \circ U_{bd} \text{ y } U_{bd} = U /\!/ (b \times d)$$
$$c/d \circ (a/b, U) = (c \times a, U_{bd}) = (c \times a, U /\!/ (b \times d)) = (c \times a/(b \times d), U)$$

In other words, the multiplication of a rational dyad by a rational number yields a dyad with the same secondary and primary equal to the product in Q of the dyad′s primary and the multiplier. This corresponds to an application of the Cartesian product set $Q \times \{Q, U\}$ in $\{Q, U\}$, which defines an external composition law with Q as the domain of operators.

20. Definition of addition of rational dyads. Let $(a/b, U)$ and $(c/d, U)$ be two rational dyads. We can write $(a/b, U) = (a \times d/b \times d, U)$ y es $(c/d, U) = (c \times b/d \times b, U)$, so that the primaries have the same denominator $b \times d$. Then we have:

$$(a/b, U) = \{a \times d \text{ elements } U_{bd}\}$$
$$(c/d, U) = \{c \times b \text{ elements } U_{bd}\}$$
$$(a/b, U) \oplus (c/d, U) = \{a \times d + c \times b \text{ elements } U_{bd}\} =$$
$$= \{a \times d + c \times b \text{ elements } U /\!/ (b \times d)\} = (a \times d + c \times b, U /\!/ (b \times d))$$

According to what was shown in Section 17, the dyad $(a \times d + c \times b, U /\!/ (b \times d))$ represents the same quantity of magnitude as $((a \times d + c \times b)/(b \times d), U)$. Thus, the definition of addition of rational dyads defines an internal law from $\{Q, U\} \times \{Q, U\}$ in $\{Q, U\}$:

$$(a/b, U) \oplus (c/d, U) = ((a \times d + c \times b)/(b \times d), U)$$

It is observed that in the sum dyad of the second member the primary is the rational addition of the primaries, while the secondary remains the same quantity of U.

21. Properties of addition of rational dyads. Since the addition of rational dyads is based on the addition of rational numbers, it is not necessary to repeat the proofs given for whole dyads. It is clear that this definition constitutes an internal composition law and satisfies the commutative and associative properties, as well as the existence of an identity element and of a symmetric or opposite element.

22. Additive rational dyadic structure. The rational dyadic set $\{Q,U\}$, endowed with the internal composition law defined by the Cartesian product $\{Q,U\}\times\{Q,U\}$ in $\{Q,U\}$ and specified by rational dyadic addition through the expression $(q_1,U)\oplus(q_2,U)=(q_1+q_2,U)$, where all rational dyads (q_1,U), (q_2,U) and (q_1+q_2,U) elements of $\{Q,U\}$, forms an abelian additive group. We denote this structure by $\{Q,U,\oplus\}$.

23. Generalization to other groups. Having verified that for whole and rational dyads the definition of dyadic adition given by the expression $(q_1,U)\oplus(q_2,U)=(q_1+q_2,U)$ endows their respective dyadic sets with the structure of an abelian additive group, based on the axioms of a commutative group, we are in a position to extend the definition of dyadic addition to the dyadic sets $\{R,U\}$ of the abelian group of real numbers R, and, in general, to any other commutative group G with its corresponding dyadic sets $\{G,U\}$.

Let us briefly recall some concepts from mathematical analysis, which are not developed here in detail. It is known that a real number can be conceived as the equivalence class of Cauchy sequences of rational numbers. In simpler terms, this means that every real number, whether rational or irrational, can be represented as the limit of a sequence of rational numbers. Moreover, it is established that these limits of rational sequences form a field. Thus, once the additive group structure of rational dyadic sets has been verified, it is justified to extend all their properties to the real dyadic sets $\{R, U, \oplus\}$ and, in abstract, to any other group $\{G, U, \oplus\}$. In practice, every irrational number in R can always be approximated by a rational number to any desired degree of precision. Therefore, rational numbers provide the ultimate verification needed to justify dyadic operations and their properties, since what holds for rational numbers is valid for any real number.

24. Definition of addition and substraction of non-uniform real dyads. Axiom of continuity. Up to this point, we have considered dyads whose secondary U is the same. These have been called uniform dyads. Now, working with real numbers, we can analyze the additive composition of homogeneous non-uniform dyads, that is, dyads of the same magnitude but with different units U_1 and U_2. With real numbers, we can establish the axiom of continuity, which states that there exists a real number k such that $U_2 = k \circ U_1$.

This multiplication corresponds to the generalization of the multiplication of a rational number by a rational dyad, as considered in Section 19. Here, the operation corresponds to an application of $R \times \{R, U\}$ in $\{R, U\}$, whose analytical definition is identical and is given by the expression $k \circ (q, U) = (q, U) \circ k = (k \times q, U)$, where k and q are elements of R. We can easily deduce:

$$k \circ (q, U) = (q, U) \circ k = (k \times q, U) = (q \times k, U) = q \circ (k, U) =$$
$$= q \circ (k \circ U) = (q, (k \circ U)) = q \circ (U \circ k) = (q, U \circ k)$$

These equalities follow directly from the definitions and can also be verified by observing that the symbols in all terms indicate sets with the same number of whole or fractional elements of U. Now, if we take two homogeneous non-uniform real dyads (q_1, U_1) and (q_2, U_2), the axiom of continuity allows us to relate their secondaries with $U_2 = k \circ U_1$. Y así $(q_2, U_2) = (q_2, k \circ U_1) = (q_2 \times k, U_1)$.

In this way, we have reduced the dyad (q_2, U_2) to a uniform expression with U_1, and it is now possible to calculate the addition

$(q_1, U_1) \oplus (q_2, U_2)$, because the sets represented by both dyads are referred to the same secondary U_1. It only remains to add the primaries as real numbers to obtain the sum of homogeneous non-uniform dyads:

$$(q_1, U_1) \oplus (q_2, U_2) = (q_1, U_1) \oplus (q_2 \times k, U_1) = (q_1 + q_2 \times k, U_1)$$

Once the general real homogeneous addition is defined, we can define the subtraction of homogeneous non-uniform dyads (q_1, U_1) and (q_2, U_2), denoted $(q_1, U_1) \ominus (q_2, U_2)$. The classical definition established for any group also applies to any dyadic set over R. Thus, we have:

$$(q_1, U_1) \ominus (q_2, U_2) = (q_1, U_1) \oplus [-(q_2, U_2)]$$

The opposite dyad of $(q_2, U_2) = (q_2 \times k, U_1)$ is, as in any group, denoted $-(q_2, U_2) = (-q_2, U_2) = (-q_2 \times k, U_1)$, because $(q_2, U_2) \oplus (-q_2, U_2) = (q_2 - q_2, U_2) = (0, U_2)$, which is the identity element of dyadic addition. Since in R we have $q_1 + (-q_2 \times k) = q_1 - q_2 \times k$, the final definition of substraction of non-uniform dyads is:

$$(q_1, U_1) \ominus (q_2, U_2) = (q_1 - q_2 \times k, U_1)$$

25. Properties of multiplication of a real number by a real dyad. Once this operation has been established for rational numbers and dyads, it is legitimate to extend it to real numbers, as we did in the previous section on subtraction, where we also summarized the essential associative properties of this product, which need not be repeated here. Let us examine the behavior of the multiplication of a real dyad (q, U) by two real numbers k and p. Consider $(k \times p) \circ (q, U)$. Using the definition of dyadic multiplication by a number and the associative property of multiplication in R, we obtain the following associative expressions:

$$(k \times p) \circ (q, U) = ((k \times p) \times q, U) = (k \times (p \times q), U) = k \circ (p \times q, U)$$

Next, let us examine the distributive properties, which follow from those of the real number field. We first analyze uniform dyads, since extending the results to homogeneous dyads is achieved by reducing them to the same secondary, as done previously. Consider two uniform real dyads (q_1, U) and (q_2, U), and let k be any real number. Form the product $k \circ [(q_1, U) \oplus (q_2, U)]$. We have:

$$k \circ [(q_1, U) \oplus (q_2, U)] = k \circ (q_1 + q_2, U) = (k \times (q_1 + q_2), U) =$$
$$= (k \times q_1 + k \times q_2, U) = (k \times q_1, U) \oplus (k \times q_2, U) = k \circ (q_1, U) \oplus k \circ (q_2, U)$$

The first and last terms illustrate the distributive property of the multiplication of a real number over the addition of two real dyads. Next, consider distributivity with respect to the addition of real numbers. For this, form the product $(k+p) \circ (q, U)$, where k and p are real numbers. Then we obtain:

$$(k+p) \circ (q, U) = ((k+p) \times q, U) = (k \times q + p \times q, U) =$$
$$= (k \times q, U) \oplus (p \times q, U) = k \circ (q, U) \oplus p \circ (q, U)$$

The first and last terms confirm the distributive property in this case. Any other associative or distributive relation can be verified in the same manner. Finally, consider the multiplication of the number 0 by any dyad, $0 \circ (q, U) = (0 \times q, U)$. Since $0 \times q = 0$ in R for finite q, we have $0 \circ (q, U) = (0, U)$, which is the identity element of the dyadic set $\{R, U\}$. Therefor, the zero element of R, multiplied by any dyad, produces the zero dyad.

26. Divisions derived from the multiplication of a real number by a real dyad. First, let us consider two homogeneous, non-zero quantities U_1 and U_2. The axiom of continuity ensures that there exists a real number k such that $U_2 = k \circ U_1$. Observing this multiplication, we can consider U_2 as the dividend, U_1 as the divisor, and k as the quotient. In this way, the multiplicative expression is equivalent to the divisive expression $U_2 /\!/ U_1 = k$. We indicate this division with a double slash to emphasize that it does not divide numbers, but magnitudes, in this case homogeneous quantities. The division $U_2 /\!/ U_1 = k$ tells us that the quotient of two quantities of the same magnitude is not itself a quantity of magnitude, but a real number, and therefore dimensionless.

Next, let us consider two non-zero homogeneous dyads (q_1, U_1) and (q_2, U_2). We seek their quotient $(q_1, U_1) /\!/ (q_2, U_2)$. By the axiom of continuity, there exists a real number k such that $U_2 = k \circ U_1$. It is immediate that $(q_2, U_2) = (q_2, k \circ U_1)$. Since q_1 and q_2 are elements of R, if $q_1 \neq 0$, there exists a real number p such that $q_2 = p \times q_1$. Thus:

$$(q_2, U_2) = (q_2, k \circ U_1) = (p \times q_1, k \circ U_1) = (p \times k \times q_1, U_1) = (p \times k) \circ (q_1, U_1)$$

Taking (q_2, U_2) as the dividend and (q_1, U_1) as the divisor, the quotient is $p \times k$. That is $(q_2, U_2) /\!/ (q_1, U_1) = p \times k$. Hence, the quotient resulting from dividing any two homogeneous dyads is the real number $p \times k$, where p and k are such that $q_2 = p \times q_1$ y $U_2 = k \circ U_1$.

This property is very useful, as we will see at the end of this synthesis of the First Algebra of Magnitudes, to determine the *dysmetric density* of any magnitude under given physical conditions.

The division of homogeneous dyads is described by a mapping from the Cartesian product set $\{R, U_2\} \times \{R, U_1\}$ in R defined by the opperation $(q_2, U_2) /\!/ (q_1, U_1) = p \times k$ and in therefore an external composition law over R.

Furtheremore, given two homogeneous real dyads (q_1, U_1) and (q_2, U_2), we know that $(q_2, U_2) = (p \times k) \circ (q_1, U_1)$. This means that (q_2, U_2) can be considered the dividend, (q_1, U_1) the divisor, and $p \times k$ the quotient. In orther words, dividing the dyad (q_2, U_2) by the real number $h = p \times k$, yields the homogeneous dyad (q_1, U_1), that is $(q_2, U_2) /\!/ h = (q_1, U_1)$ and we can state that the quotient of a real (q_2, U_2) by a non-zero number is the homogeneous real dyad (q_1, U_1) such that $(q_2, U_2) = h \circ (q_1, U_1)$. This operation is algebraically established as a mapping from the Cartesian product set $\{R, U_2\} \times R$ in $\{R, U_1\}$ via the relation $(q_2, U_2) /\!/ h = (q_1, U_1)$ which is an external law with R as the operator. This type of division can also be represented in classical abstract algebra using the inverse of $h \neq 0$, denoted h^{-1}. Then, if $(q_2, U_2) = h \circ (q_1, U_1)$, multiplying both sides by h^{-1}, yields $h^{-1} \circ (q_2, U_2) = (h^{-1} \times h) \circ (q_1, U_1)$. Since $h^{-1} \times h = 1$, we have $h^{-1} \circ (q_2, U_2) = 1 \circ (q_1, U_1) = (1 \times q_1, U_1)$. Since $1 \times q_1 = q_1$, yields $h^{-1} \circ (q_2, U_2) = (q_1, U_1)$. Using classical division notation, we have $(q_2, U_2) /\!/ h = (q_1, U_1)$, and it follows that $(q_2, U_2) /\!/ h = h^{-1} \circ (q_2, U_2)$.

27. Definitions and properties for vector dyads. All the previous definitions and properties extend to vector dyads over the field of real numbers or over any other structure isomorphic to it. Therefore, it is not necessary to include here an exhaustive analysis of the enormous range of possible cases, since the study of each particular instance is entirely analogous to what has already been developed above. In the case of vectors, it suffices to distinguish the operations involving vectors and numbers according to the algebraic structures involved. For example, the addition of uniform vector dyads $(\mathbf{q_1}, U)$ and $(\mathbf{q_2}, U)$, where $\mathbf{q_1}$ and $\mathbf{q_2}$ are vectors in R^n, is defined by $(\mathbf{q_1}, U) \oplus (\mathbf{q_2}, U) = (\mathbf{q_1} + \mathbf{q_2}, U)$, where the addition of the first member is the addition of vector dyads (not real dyads), and the addition on the second member is the vector sum in R^n, not addition in R. This operation corresponds to a mapping from the Cartesian product set $\{R^n, U\} \times \{R^n, U\}$ in $\{R^n, U\}$. The structure of the vector space R^n over R guarantees that vector dyadic addition satisfies the commutative and associative properties, as well as the existence of a neutral element and additive opposites.

Another example is the multiplication of a vector dyad $(\mathbf{q}, U)$ where $\mathbf{q}$ is an element of the vector space R^3 over R, by a rel number k. This is defined by $k \circ (\mathbf{q}, U) = (k \bullet \mathbf{q}, U)$, where $k \bullet \mathbf{q}$ is the multiplication of a real number k by the vector $\mathbf{q}$, in accordance with the laws of vector space structure. Apart from this distinction, the analysis of this vector operation is completely analogous to that of a real dyad. Thus, the operation is defined by $k \circ (\mathbf{q}, U) = (\mathbf{q}, U) \circ k = (k \bullet \mathbf{q}, U)$ and in this case constitutes a dyadic mapping $R \times \{R^n, U\}$ or $\{R^n, U\} \times R$ in $\{R^n, U\}$. Here again, the vector space structure of R^n over R ensures that the corresponding commutative, associative, and distributive properties hold between R and $\{R^n, U\}$.

28. Definition of multiplication of whole dyads. Generative external composition law. Up to this point we have established additive operations with quantities of homogeneous magnitudes. Included among these is the product of a number by a dyad, since that multiplication ultimately reduces, in a certain sense, to repeated addition. What follows describes multiplicative operations, which serve to multiply and divide quantities of magnitudes among themselves, whether homogeneous or not.

Let us take two generic whole dyads (q_1, U) and (q_2, V), where U and V are arbitrary quantities of the same or of different magnitudes. Whole dyads have been defined as sets such as $(q_1, U) = \{q_1 \text{ elements } U\}$ and $(q_2, V) = \{q_2 \text{ elements } V\}$, where q_1 and q_2 are integers.

The Cartesian product of these two sets is said to determine the multiplication of the indicated dyads, and we symbolize this product by $(q_1, U) * (q_2, V)$. The asterisk will denote this operation, alongside the usual cross sign used for Cartesian products of sets. Thus, the multiplication of whole dyads is defined by associating the dyadic symbol $(q_1, U) * (q_2, V)$ with the Cartesian product of the sets corresponding to the factors:

$$(q_1, U) * (q_2, V) = \{q_1 \text{ elements } U\} \times \{q_2 \text{ elements } V\}$$

This definition implies that the dyadic product $(q_1, U) * (q_2, V)$ corresponds to the quantity of magnitude implicit in a set consisting of a certain number of identical elements, each equal to the pair (U, V). To determine this quantity, we must find how many elements (U, V) are contained in the set $\{q_1 \text{ elements } U\} \times \{q_2 \text{ elements } V\}$. Clearly, this product can be arranged as a matrix with q_1 rows and q_2 columns of elements, all entries being the pair (U, V). Hence, the Cartesian product is a set which contains $q_1 \times q_2$ elements (U, V). Analytically:

$$(q_1, U) * (q_2, V) = \{q_1 \times q_2 \text{ elements } (U, V)\}$$

The pair (U, V) represents a quantity of the magnitude resulting from the multiplicative composition of U and V. To give meaning to these pairs, we invoke the postulate of affinity. Imagine U and V as ideal segments. Their geometric multiplication would consist on forming a rectangle whose sides are determined by segments U and V. The area of this ideal rectangle would be the quantity of magnitude represented by the pair (U, V). To indicate that the magnitude implicit in (U, V)

arises from multiplying the magnitudes represented by U and V, we denote it multiplicatively as $U*V$.

Therefore, we specify that the product just defined is characterized by generating a magnitude different from those of the quantities being multiplied. In the case of affine segments, multiplying lengths generates a new magnitude: area. Thus, by the postulate of affinity, the product of quantities (q_1, U) and (q_2, V) generates the quantity implicit in $q_1 \times q_2$ elements (U, V) or $U*V$ belonging to the magnitude produced by multiplication. Te symbol representing this quantity is the dyad $(q_1 \times q_2, U*V)$. This result is expressed analytically as:

$$(q_1, U)*(q_2, V) = (q_1 \times q_2, U*V)$$

This is the final form of the definition of multiplication of whole dyads. It corresponds to a mapping of the Cartesian product of the set defined by $\{Z, U\} \times \{Z, V\}$ over $\{Z, U*V\}$. The most notable features of this multiplicative operation are its external and generative character. It is external because it composes different dyadic sets. It is generative because it produces a new magnitude from those on which it operates. Such generative laws are characteristic of physical magnitudes and constitute a significant novelty with respect to abstract algebra. We have considered two factors. With three factors, affinity indicates that multiplication of affine segments would yield another magnitude: volume. Analytically, the form remains identical. With n factors, the resulting magnitude would be an n-dimensional hypervolume. The general analytic form is that given in Section 30.

29. Definition of multiplication of rational dyads. Let two dyads, homogenous or not, $(a/b, U)$ and $(c/d, V)$, where a, b, c and d are integers in Z, b and d non-zero:

$$(a/b, U) = \{a \text{ elements } U_b\} \text{ y } U = \{b \text{ elements } U_b\} = (b, U_b)$$
$$(c/d, V) = \{c \text{ elements } V_d\} \text{ y } V = \{d \text{ elements } V_d\} = (d, V_d)$$

As with whole dyads, we define the multiplication of rational dyads $(a/b, U)*(c/d, V)$ as the quantity of magnitude implicit in the Cartesian product of the sets defined by those dyads, $\{a$ elementos $U_b\} \times \{c$ elementos $V_d\}$. These sets behave as whole dyads; therefore, the product $\{a$ elements $U_b\} \times \{c$ elements $V_d\}$ is the set $\{a \times c$ elements $(U_b, V_d)\}$. Application of the postulate of affinity leads us, as in the previous case, to denote each pair of elements by $(U_b, V_d) = U_b * V_d$. That is, we consider $U_b * V_d$ to be the surface of an ideal rectangle whose sides are U_b and V_d. Now, $U = (b, U_b)$ and $V = (d, V_d)$, or equivalently, by multiplication of whole dyads by an integer, $U = b \circ U_b$ y $V = d \circ V_d$. Recall that $U_b * V_d$ represents the ideal surface of an affine rectangle with sides U_b and V_d. From geometry, we know that the area of the ideal rectangle $U * V$ is $b \times d$ times the surface of $U_b * V_d$; therefore, given the postulated affinity, we can assert that $U * V = (b \times d) \circ (U_b * V_d)$. The divisive form of this expression is $U_b * V_d = (U * V) /\!/ (b \times d)$. Thus we obtain:

$$(a/b, U) * (c/d, V) = \{a \times c \text{ elements } (U_b, V_d)\} = (a \times c, (U_b, V_d)) =$$
$$= (a \times c, U_b * V_d) = (a \times c, (U * V) /\!/ (b \times d)) = (a \times c / (b \times d), U * V)$$

The last equality $(a \times c, (U * V) /\!/ (b \times d)) = (a \times c / (b \times d), U * V)$ is a consequence of the corresponding property described in Section 17, whereby the same quantity of magnitude results from dividing only the primary or only the secondary of a dyad by the same number. Under these conditions, we can formulate the final definition of the multiplication of rational dyads:

$$(a/b, U) * (c/d, V) = (a \times c / (b \times d), U * V)$$

Since $a \times c / (b \times d)$ is the product in Q of the rationals a/b and c/d, it is legitimate to state the abstract definition of multiplication of rational dyads as follows: given two dyads (q_1, U) and (q_2, V), with q_1 and q_2 elements of Q, the dyadic product is defined by:

$$(q_1, U) * (q_2, V) = (q_1 \times q_2, U * V)$$

This abstract analytic form coincides with that obtained for the multiplication of whole dyads, which authorizes generalizing the same definition to real numbers.

30. Generic definition of dyadic multiplication. Generative external composition law. We recalled in Section 23 that irrational numbers are, in essence, limits of sequences of rational numbers, and that these limits possess field structure. Moreover, in practice, every irrational number in R can always be approximated by a rational number to any desired degree of precision. Therefore, here as well, rational numbers constitute the foundation that justifies multiplicative dyadic operations and their properties, since what is established for rational numbers is valid for any real number. Under these conditions, the general definition of dyadic multiplication for R is immediate:

Given n arbitrary dyads (q_1, U_1), (q_2, U_2), ... , (q_n, U_n), where q_1, q_2, ..., q_n are elements of R or, in abstract terms, of any other field K, and there are n arbitrary quantities U_1, U_2, ... ,U_n of any magnitudes, dyadic multiplication is defined by the expression:

$$(q_1, U_1) * (q_2, U_2) * \ ... \ * (q_n, U_n) = (q_1 \times q_2 \times \ ... \ \times q_n, U_1 * U_2 * \ ... \ * U_n)$$

We have seen that these composition laws are external and generative in character, which is evident from the nature of the mapping they define, from the Cartesian product set $\{K, U_1\} \times \{K, U_2\} \times \ ... \ \times \{K, U_n\}$ in $\{K, U_1 * U_2 * \ ... \ * U_n\}$. It is clear that the dyadic set $\{K, U_1 * U_2 * \ ... \ * U_n\}$ differs from all members of the Cartesian product, even when the sets $\{K, U_i\}$ are the same. Therefore, this composition law is always external and generative.

31. Properties of dyadic multiplication. Let two arbitrary dyads (q_1, U_1) and (q_2, U_2) be given, where q_1 and q_2 are scalars or vectors. The postulate of affinity indicates that the surfaces of the ideal rectangles $U_1 * U_2$ and $U_2 * U_1$ are equal or geometrically equivalent; therefore, one may admit that $U_1 * U_2 = U_2 * U_1$. Consequently, starting from the generic definition of dyadic multiplication, if the primaries q_1 and q_2 belong to an algebraic structure with commutative multiplication, we have:

$$(q_1, U_1) * (q_2, U_2) = (q_1 \times q_2, U_1 * U_2) = (q_2 \times q_1, U_2 * U_1) = (q_2, U_2) * (q_1, U_1);$$
$$(q_1, U_1) * (q_2, U_2) = (q_2, U_2) * (q_1, U_1)$$

424

It is thus established that, if the primaries q_1 and q_2 belong to an algebraic structure with commutative multiplication, dyadic multiplication is also commutative. The same occurs with the associative property. Given three arbitrary dyads (q_1, U_1), (q_2, U_2) and (q_3, U_3), consider the product $[(q_1, U_1) * (q_2, U_2)] * (q_3, U_3)$. As usual, the brackets indicate the preferred order of multiplication. We may imagine an affine geometric volume $U_1 * U_2 * U_3$, with dimensions U_1, U_2 and U_3. Geometry guarantees that the resulting volume is the same regardless of how the dimensions are associated. This means that $(U_1 * U_2) * U_3 = U_1 * (U_2 * U_3)$, given the postulate of affinity. If the primaries q_1, q_2 and q_3 belong to an algebraic structure with associative multiplication, we obtain:

$$[(q_1, U_1) * (q_2, U_2)] * (q_3, U_3) = (q_1 \times q_2, U_1 * U_2) * (q_3, U_3) =$$
$$= ((q_1 \times q_2) \times q_3, (U_1 * U_2) * U_3) = (q_1 \times (q_2 \times q_3), U_1 * (U_2 * U_3)) =$$
$$= (q_1, U_1) * [(q_2, U_2) * (q_3, U_3)];$$
$$[(q_1, U_1) * (q_2, U_2)] * (q_3, U_3) = (q_1, U_1) * [(q_2, U_2) * (q_3, U_3)]$$

In conclusion, if the primaries q_1, q_2 and q_3 belong to an algebraic structure with associative multiplication, dyadic multiplication is also associative.

Let us now examine the distributive property of multiplication over addition in dyadic algebra. Take three arbitrary dyads (q_1, U_1), (q_2, U_2) and (q_3, U_2). Consider the expression $(q_1, U_1) * [(q_2, U_2) \oplus (q_3, U_2)]$. By the definition of addition $(q_2, U_2) \oplus (q_3, U_2) = (q_2 + q_3, U_2)$ assuming the addends have previously been reduced to uniform dyads. Substituting and applying the definition of dyadic multiplication:

$$(q_1, U_1) * [(q_2, U_2) \oplus (q_3, U_2)] = (q_1, U_1) * (q_2 + q_3, U_2);$$
$$y \ (q_1, U_1) * (q_2 + q_3, U_2) = (q_1 \times (q_2 + q_3), U_1 * U_2)$$

If q_1, q_2 and q_3 belong to an algebraic structure satisfying the distributive property of multiplication over addition, then $q_1 \times (q_2 + q_3) = q_1 \times q_2 + q_1 \times q_3$. Substituting and operating:

$$(q_1 \times (q_2 + q_3), U_1 * U_2) = (q_1 \times q_2 + q_1 \times q_3, U_1 * U_2);$$
$$(q_1 \times q_2 + q_1 \times q_3, U_1 * U_2) = (q_1 \times q_2, U_1 * U_2) \oplus (q_1 \times q_3, U_1 * U_2);$$
$$(q_1 \times q_2, U_1 * U_2) \oplus (q_1 \times q_3, U_1 * U_2) =$$

$$=[(q_1, U_1) * (q_2, U_2)] \oplus [(q_1, U_1) * (q_3, U_2)]$$

The first and last members of these equalities express the distributive property of dyadic multiplication over the addition of uniform dyads:

$$(q_1, U_1) * [(q_2, U_2) \oplus (q_3, U_2)] = [(q_1, U_1) * (q_2, U_2)] \oplus [(q_1, U_1) * (q_3, U_2)]$$

32. Nonexistence of a unit element for dyadic multiplication. In abstract algebra, in a set where an internal multiplicative composition law is defined, the left unit element u is defined as the neutral element such that for every q belonging to the set, $u \times q = q$. If the operation is commutative, it also holds that u is a right unit element, so that $u \times q = q \times u = q$. Observe that the unit element is defined for internal composition laws. However, we have seen in Section 27 that dyadic multiplication is an external, generative composition law. Therefore, if we take any dyadic set $\{K, U\}$ over a field K and consider the quantity of magnitude given by U, if we suppose that there exists a quantity $U_u \in \{K, U\}$, non-uniform with U, such that $U * U_u = U$, it would follow that $U * U_u$ does not refer to the same magnitude as U, by definition of dyadic multiplication, and therefore $U * U_u$ would not belong to the set $\{K, U\}$. Thus, on the one hand, U would be an element of $\{K, U\}$, while on the other hand $U = U * U_u$ would indicate a magnitude different from U, which is absurd. Hence, the hypothesis of the existence of such a $U_u \in \{K, U\}$ is impossible. No such unit element can exist.

This reasoning is sufficient to prove the nonexistence of a multiplicative unit element in $\{K, U\}$. However, given the importance of this analysis, let us consider a generic dyad (q, U) of the dyadic set $\{K, U\}$ over a field K. Suppose there exists a quantity represented by the dyad $(q_u, U_u) \in \{K, U\}$ such that $(q, U) * (q_u, U_u) = (q, U)$. By the definition of multiplication, we would have $(q, U) * (q_u, U_u) = (q \times q_u, U * U_u)$. As before, on the one hand, (q, U) would be an element of $\{K, U\}$ and on the other hand $(q, U) = (q, U) * (q_u, U_u)$ would indicate a magnitude different from (q, U), which is absurd. Hence, the hypothesis of the existence of $(q_u, U_u) \in \{K, U\}$ is impossible. No such unit element can exist.

33. Nonexistence of inverse elements for dyadic multiplication.
Given any dyadic set $\{K, U\}$, where K is an algebraic structure
endowed with a multiplicative operation, we have just established
that $\{K, U\}$ contains no unit element. Therefore, it cannot contain
inverse elements for any dyad (q, U), since no dyad in $\{K, U\}$,
when multiplied by (q, U), can yield a unit element belonging to
$\{K, U\}$, because no such element exists, as proved in Section 32.

34. Division between non-homogeneous dyadic entities. Since
multiplicative inverse elements do not exist in closed dyadic sets,
it would not be appropriate to use the classical notation in the
form of multiplication by the inverse of the divisor to define
division, because this could convey the false idea that such
inverses exist. Consequently, we begin by defining division in
fractional form. Let two arbitrary dyads (q_1, U_1) and (q_2, U_2) be
given. Consider their product $(q_1, U_1) * (q_2, U_2) = (q_1 \times q_2, U_1 * U_2)$. To
transform this multiplicative expression into a divisive form, it
suffices, for example, to regard $(q_1 \times q_2, U_1 * U_2)$ as the dividend,
(q_2, U_2) as the divisor, and (q_1, U_1) as the quotient. Obviously, one
could also take (q_1, U_1) as divisor and (q_2, U_2) as quotient.

In the first case, we may write the divisive expression in the
form $(q_1 \times q_2, U_1 * U_2) /\!/ (q_2, U_2) = (q_1, U_1)$. Let us analyze the meanings
implicit in the dyad $(q_1 \times q_2, U_1 * U_2)$ of this equality. The product
$q_1 \times q_2$ of the primary may be viewed as a dividend, and each
factor q_1 and q_2 as divisors or quotients. Thus we obtain the two
divisive forms $q_1 \times q_2 / q_1 = q_2$ y $q_1 \times q_2 / q_2 = q_1$. In turn, the dyadic
product $U_1 * U_2$ of the primary may be viewed as a dividend, and
each factor U_1 y U_2 as divisors or quotients. In the simple manner
we obtain the two divisions $U_1 * U_2 /\!/ U_1 = U_2$ y $U_1 * U_2 /\!/ U_2 = U_1$. It
is unnecessary to repeat constantly that divisors or denominators
cannot be zero when dealing with divisive operations.

It is important to observe that these four divisions legitimize
operations that simplify repeated elements in the numerators and
denominators of the fractions thus formed. This rule is applied
mechanically from elementary schooling and is thereby justified
both for numbers and for quantities of magnitude. In conclusion,

we may establish that the division of two non-homogeneous dyads is another dyad whose primary is the quotient of the primaries and whose secondary is the quotient of the secondaries of dividend and divisor.

This prepares us to define dyadic division in a generic way. Let two dyads (q_1, U_1) and (q_2, U_2). Their dyadic quotient is defined as:

$$(q_1, U_1) /\!\!/ (q_2, U_2) = (q_1 / q_2, U_1 /\!\!/ U_2)$$

In Section 26, the division of homogeneous dyads was described in the context of multiplying a dyad by a number. We verify that the result obtained there also satisfies the definition above, conceived for non-homogeneous dyads. Therefore, although arising from different algebraic motivations, this latter definition is also formally valid for homogeneous dyads and thus becomes generic.

Obviously, if the elements of the primary belong to a field K, their quotient may be expressed as the product of the dividend by the inverse of the divisor, that is, $q_1 / q_2 = q_1 \times q_2^{-1}$. However, in the case of the secondary this cannot be strictly so, because we proved in Section 33 that inverse elements of dyads do not exist. In any case, if one wishes to maintain formal structural notation, one may symbolize $U_1 /\!\!/ U_2 = U_1 * U_2^{-1}$, replacing divisive symbolism with multiplicative symbolism. Rigorously, however, it must be borne in mind that this dyadic product does not exist algebraically; it is only symbolic, so that $U_1 * U_2^{-1}$ must be assigned the meaning of the dyadic division $U_1 /\!\!/ U_2$. No isolated inverse U_2^{-1} can be conceived, as in internal composition laws.

35. Powers and roots. Negative exponents. In this subject, special care must be taken to respect the external and generating condition of dyadic multiplication. Thus, as noted in the previous point, isolated negative exponents are excluded from exponentiation and root extraction, because no negative exponent has meaning on its own within the context of dyadic multiplication. Furthermore, there is no problem in relating exponentiation and root extraction using classical notation, so that, given any dyad (q, U) and a rational number a/b, with a and b non-zero positive integers in Z. Let U_b be such that $U = U_b^b = U_b * U_b * \ldots * U_b$, with b factors. If U_b represents an affine length, U will represent a b-dimensional affine volume. We can coherently formulate the following definition:

$$(q, U)^{\frac{a}{b}} = \left(\sqrt[b]{q^a}, \left(\sqrt[b]{U} \right)^a \right) = \left(\sqrt[b]{q^a}, U_b^{\ a} \right) \; ; \; U_b = \sqrt[b]{U}$$

This notation fits the definition of dyadic multiplication in section 30 and thus $(q, U)^{a/b}$ and $U^{a/b}$ are associated with affine volumes of $a \times b$ dimensions.

As for negative exponents in general, they only make sense when associated with another quantity; therefore, they are meaningless when considered in isolation. Thus, the form U^{-x}, with $x > 0$, cannot be written correctly, but $U^{-x} * V$, where V is any other quantity, can, with the meaning given by the quotient $U^{-x} * V = V /\!/ U^x$. In contrast, the expression $U^{-x} /\!/ V$ would not make sense because there cannot exist a W such that $W * V = U^{-x}$. However, the quotient $V /\!/ U^{-x}$ would make sense because a product $W * U^{-x} = W /\!/ U^x = V$ could be formed, with $V * U^x = W$.

36. Dyadic algebraic structure. Collecting the operations defined in this synthesis, we observe first that any dyadic set $\{K, U, \oplus\}$ constructed over a field K for the magnitude corresponding to U, endowed with dyadic addition, exhibits the structure of an abelian additive group. By adding to this group the external multiplication of any dyad by any element of K, this group $\{K, U, \oplus, \circ\}$ acquires the structure of a vector space over K. Let us analyze this structure from the point of view of affinity with the set of geometric segments. This set is nothing but a concrete and fundamental case of a dyadic set in which the constituent magnitude is length. It could be represented as $\{R, m\}$, where m is the standard meter. However, here we shall distinguish it as the set of all geometric segments, to emphasize length as the fundamental magnitude, the basis of affinity with all others and its geometric natures.

429

Consider the set of all geometric segments $\{S\}$. Segment addition, denoted $\oplus_S$, is defined in geometry by the operation called juxtaposition of summands. This means that the geometric addition of two given segments is the segment obtained by placing them on a line one after the other if both are positive, in opposite directions if one is positive and the other negative, or both in the opposite direction if both summands are negative. With juxtaposition defined in this way, it is elementary to verify that geometric segment addition endows the set $\{S, \oplus_S\}$ with the structure of a commutative group. This addition is commutative, associative, has a zero element S_0, the segment formed by a single point, and every segment S has its opposite $-S$, which is the segment added in the direction opposite to that defined for addition, so that $S \oplus_S (-S) = S_0$.

It is appropiate here to recall the definition of isomosphism, which anstract algebra defines as a bijective mapping ϕ between a set $\{E, \top\}$ and another $\{F, \perp\}$, where $\top$ and $\perp$ are internal composition laws in E and F, so that ϕ is such that every pair of elements a and b in $\{E, \top\}$ it is verified that $\phi(a \top b) = \phi(a) \perp \phi(b)$, which simply means that ϕ preserves th operational structure. Isomorphism is denoted $\{E, \top\} \simeq \{F, \perp\}$. It is immediate to verify that if $\{E, \top\}$ and $\{F, \perp\}$ are groups, the isomosphism preserves not only the operation but also the identity and symmetric elements. That is, if e is the null element of E, $\phi(e)$ is the neutral element of F; and, if a' is the symmetric element of a, then, $\phi(a') = [\phi(a)]'$, the image of the symmetric element is the symmetric element of the image.

For what follows, we restrict the field K to the real numbers R, since ordinary geometric operations fit only this structure, although in abstract terms one could generalize to any field K. Define the correspondence ϕ predicted by the postulate of affinity between $\{R, U, \oplus\}$ and $\{S, \oplus_S\}$. Let us see whether it satisfies the isomorphism condition. Take any dyad (q, U) and let $\phi(U) = S_U$, an element of $\{S, \oplus_S\}$ distinct from the null segment S_0. Define $\phi(q, U) = q \circ_S S_U$, where $\circ_S$ denotes the geometric operation of multiplying a segment by a real number. This is based on the

430

operations of dividing a segment into equal parts, an elementary problem, and on the addition of segments by juxtaposition, also fundamental, so that a segment can be multiplied by any rational number and, by extension, by any real number, as we have done in the preceding section with quantities of magnitudes in general, which is also valid for segments, since a segment is nothing more than a quantity of length, as already indicated. In any case, we must verify that the mapping ϕ is bijective. Let us first examine injectivity, or one-to-one mapping and check whether every element of $\{S,\oplus_S\}$ is the image of at most one element of $\{R,U,\oplus\}$. To this end, take two dyads (q_1,U) and (q_2,U). Their images are $\phi(q_1,U)=q_1\circ_S S_U$ and $\phi(q_2,U)=q_2\circ_S S_U$. If we postulate the equality $\phi(q_1,U)=\phi(q_2,U)$, then $q_1\circ_S S_U=q_2\circ_S S_U$. Adding to both members the opposite segment $-q_2\circ_S S_U$ we easily obtain $(q_1-q_2)\circ_S S_U=S_0$. The segment on the first member can be equal to the null segment only if $q_1-q_2=0$, because S_U is different of S_0 by hypothesis. Therefore $q_1=q_2$ and thus $(q_1,U)=(q_2,U)$. Hence, the mapping ϕ defined by $\phi(q,U)=q\circ_S S_U$ is injective, or *one-to-one*. Let us now see whether it is surjective or onto.

Let us check whether every element of $\{S,\oplus_S\}$ is the image of *at least* one element of $\{R,U,\oplus\}$. Let S be any element of $\{S,\oplus_S\}$. Geometry guarantees that there exists a real number q such that $q\circ_S S_U=S$. By definition of ϕ we have $\phi(q,U)=q\circ_S S_U$. Hence it follows that $\phi(q,U)=S$. Therefore, for every segment S in $\{S,\oplus_S\}$ there exists an element (q,U) in $\{R,U,\oplus\}$ such that $q\circ_S S_U=S$, which is the image of (q,U). Thus, the mapping ϕ is surjective or *onto*. Being ϕ *one-to-one* and *onto*, it is a bijection, as we intended to verify.

Now let us find the image given by $\phi[(q_1,U)\oplus(q_2,U)]$. Let (q_1,U) and (q_2,U) be any two uniform dyads of $\{R,U,\oplus\}$. If they were not uniform, it would suffice to apply the axiom of continuity to reduce them to uniformity. We have:

$$\phi[(q_1,U)\oplus(q_2,U)]=\phi[(q_1+q_2,U)]=(q_1+q_2)\circ_S S_U$$

From geometry we know that multiplication of a segment by real numbers is distributive.

Therefore, $(q_1+q_2)\circ_sS_U=q_1\circ_sS_U\oplus_Sq_2\circ_sS_U$. Obviously, by the definition of ϕ, is $q_1\circ_sS_U=\phi(q_1,U)$ and $q_2\circ_sS_U=\phi(q_2,U)$. Hence, $\phi[(q_1,U)\oplus(q_2,U)]=\phi(q_1,U)\oplus_S\phi(q_2,U)$, which means that ϕ is an idomorphism between the additive groups $\{R,U,\oplus\}$ and $\{S,\oplus_S\}$.

In the same way, it can be verified that ϕ is an isomorphism between the vector spaces $\{R,U,\oplus,\circ\}$ and $\{S,\oplus_S,\circ_S\}$ over R as the field of scalars. To see this, take any dyad (q,U) belonging to $\{R,U,\oplus,\circ\}$ and a real number λ of R. Let us find the image $\phi[\lambda\circ(q,U)]$. It is immediate to observe that $\lambda\circ(q,U)=(\lambda\times q,U)$. By definition of ϕ, we then have:

$$\phi[\lambda\circ(q,U)]=\phi(\lambda\times q,U)=(\lambda\times q)\circ_sS_U$$

The properties of the vector space $\{S,\oplus_S,\circ_S\}$ allow us to write $(\lambda\times q)\circ_sS_U=\lambda\circ_S(q\circ_sS_U)$ and therefore $\lambda\circ_S(q\circ_sS_U)=\lambda\circ_S\phi(q,U)$ and $\phi[\lambda\circ(q,U)]=\lambda\circ_S\phi(q,U)$. Hence, ϕ preserves scalar multiplication in both structures, and since we have already seen that it also preserves addition, we conclude that the structures $\{R,U,\oplus,\circ\}$ y $\{S,\oplus_S,\circ_S\}$ are isomorphic.

Let us now analyze the dyadic sets endowed with a multiplicative structure $\{R,U,V,*\}$, where U and V indicate quantities of any magnitudes, which do not need to be homogeneous. The notation $\{R,U,V,*\}$ represents the dyadic set $\{R,U\}\cup\{R,V\}\cup\{R,U*V\}$, with the dyadic multiplication $*$ already defined, which relates the elements of this set according to the definition $(q_1,U)*(q_2,V)=(q_1\times q_2,U*V)$. Let us examine its relation with the set of geometric segments with its own multiplication $\{S,*_S\}$, which represents the set $\{S\}\cup\{S*_SS\}$, that I, the union of all segments and all areas. Analogously, with three segments one would obtain volumes, and with more multiplicativ simensions, hypervolumes. By affinity, we have inspired dyadic multiplication on geometric multiplication, resulting in the fact that, in both cases, there are no unit elements or inverses within the sets $\{R,U\}$, $\{R,V\}$ or $\{S\}$, as justified in sections 32 and 33.

Let us see whether the same function ϕ defined earlier can serve to establish a valid correspondence between the sets

432

$\{R, U, V, *\}$ and $\{S, *_S\}$. Let us find the image under ϕ of the dyad represented by $(q_1, U)*(q_2, V)$; that is, let us analyze the elemnt $\phi[(q_1, U)*(q_2, V)]$, which is nothing more than the image under ϕ of a dyad symbolized by $(q_1, U)*(q_2, V)$, a multiplicative expression of two others. By definition of dyadic multiplication, we immediately have $\phi[(q_1, U)*(q_2, V)] = \phi(q_{1\times}q_2, U*V)$. If we consider $(q_{1\times}q_2, U*V)$ as a generic dyad formed by an element of R and a quantity of magnitudes given by $U*V$, we can conclude that $\phi(q_{1\times}q_2, U*V) = (q_{1\times}q_2)\circ_S(S_U *_S S_V)$, where $S_U = \phi(U)$ y $S_V = \phi(V)$. From geometry, we know that the area of a rectangle with sides $q_1\circ_S S_U$ and $q_2\circ_S S_V$ is $q_{1\times}q_2$ times the area of a rectangle with dimensions S_U and S_V. Analytically, this can be expressed by the equation $(q_{1\times}q_2)\circ_S(S_U *_S S_V) = (q_1\circ_S S_U)*_S(q_2\circ_S S_V)$. By definition of ϕ, we have $(q_1\circ_S S_U) = \phi(q_1, U)$ y $(q_2\circ_S S_V) = \phi(q_2, V)$. In conclusion, $\phi[(q_1, U)*(q_2, V)] = \phi(q_1, U)*_S\phi(q_2, V)$.

This result reveals that the correspondence between the sets $\{R, U, V, *\}$ and $\{S, *_S\}$, as defined, maintains the multiplicative operation, even though it is external and generating. However, the supposed isomorphism between the set $\{R, U, V, *\}$ and $\{R, \times\}$ cannot be properly established because these multiplications are heterogeneous: the multiplication of $\{R, \times\}$ is internal, and the multiplication of $\{R, U, V, *\}$ is external and generating. Nevertheless, for theoretical purposes, we can investigate the function ϕ_R between $\{R, U, V, *\}$ and $\{R, \times\}$ such that $\phi_R(q, U) = q$. This function clearly loses all the information about the dimensional part of the dyad (q, U) because it completely disregards U, but this is what the presumed isomorphism between magnitudes and real numbers implies. Under these conditions, let's look for the image given by:

$$\phi_R[(q_1, U)*(q_2, V)] = \phi_R(q_1\times q_2, U*V) = q_1\times q_2 = \phi_R(q_1, U)\times(q_2, V)$$

In conclusion, although ϕ_R satisfies the condition of isomorphism because it preserves the multiplicative operation, it is not capable of reflecting the essence of the multiplication of magnitudes, which is its external nature and its power to generate new magnitudes. Therefore, since the structures $\{R, U, V, *\}$ y

$\{R,\times\}$ are not homogeneous, a strict isomorphism cannot be properly defined between them. Consequently, the function ϕ_R cannot be considered valid for establishing an algebra of magnitudes. The true isomorphism is that established by ϕ, as defined above, between the algebraic structures $\{R,U,V,*\}$ y $\{S,*_S\}$.

Thus, the algebraic meaning of the affinity postulate becomes evident, because we observe that the defined composition laws endow generic dyadic sets with their own structure: isomorphic to groups for the additive operations, the internal $\oplus$ and the external $\circ$, and with the isomorphic specialty of the generating external multiplicatives $*$, in relation to the structure of geometric segments. These generic structures, which we denote in summary as $\{S,\oplus_S,\circ_S,*_S\}$ and $\{K,U_1,U_2, \ldots ,U_n,\oplus,\circ,*\}$, we will refer to collectively as the dyadic algebra of magnitudes.

37. Equivalence classes and order relations in sets of quantities of magnitudes. Let us form the set $M=\{m\}$, which represents the complete repertoire of all possible quantities m of the magnitude under consideration. With this notation, any dyad (q,U) is identified with some element m.

Take two homogeneous dyads (q_1,U_1) and (q_2,U_2). We have defined dyadic equality by the condition that both implicitly contain the same quantity of magnitude. If the above dyads are equal, we write $(q_1,U_1)=(q_2,U_2)$. We have established that $(q_1,U_1)=q_1\circ U_1$ and $(q_2,U_2)=q_2\circ U_2$. Hence, $q_1\circ U_1=q_2\circ U_2$, so q_1 is the quotient of $q_2\circ U_2$ and U_1, which we have denoted divisively as $q_1=q_2\circ U_2/\!/U_1$. Multiplying both members by q_2^{-1}, we obtain $q_1\times q_2^{-1}=U_2/\!/U_1$. Or in divisibe form, $q_1/q_2=U_2/\!/U_1$. That is, given two homogeneous dyads, the ratio of the primaries is the inverse of the ratio of the secondaries.

Let us take the set $D=\{(q,U)\}$, tha is, the set of all dyads (q,U) of a given magnitude, with $q\in R$ and $U\in M$. The set D can also be conceived as the Cartesian products $R\times\{U\}$ or $R\times M$. An analogous construction applies to vector dyads. In any case,

434

within D we ideally find all the possible dyads that can be formed to represent quantities of the given magnitude.

We shall say that two dyads (q_1, U_1) and (q_2, U_2) are equivalent of their components satisfy the relation $q_1/q_2 = U_2/\!/U_1$ and we shall write $(q_1, U_1) \sim (q_2, U_2)$ or, if preferred, $(q_1, U_1) = (q_2, U_2)$. Note that this equality sign does not denote numerical equality, as in ordinary algebraic expressions, but equality of quantities of magnitudes, which are non-numerical in nature. It is straightforward to verify that the relation thus defined satisfies the reflexive, symmetric, and transitive properties; therefore, it is an equivalence relation. It is a subset of the Cartesian product D×D and establishes in the set D a classification of the dyads $\{(q, U)\}$ in D into all its classes and the corresponding partition of this set. As with any equivalence relation, the partition of D into classes means that each quantity of magnitude m in M can be represented by any element of its corresponding class, which we may denote by:

$$m = [(q, U)] = \{\text{all dyads } (x, X) \text{ such that } q/x = X/\!/U\}$$

By forming the set whose elements are all the equivalence classes $\{[(q, U)]\}$ in D under $\sim$, we arrive at the definition fo M as the set of classes of equivalent dyads: $M = \{[(q, U)]\}$. Thus, M is the partition corresponding to the quotient set of D with respect to the equivalence relation $\sim$, which in algebra is written $D/\!\sim = M$.

In summary, a quantity of magnitude m is defined by a specific equivalence class $[(q, U)]$ of the set D of all dyads. In turn, the set of all classes $\{[(q, U)]\}$, which constitutes a partition of D, established by the equivalence relation $\sim$, is by definition the set $M = \{m\} = \{[(q, U)]\}$ of all quantities m and dyadic equivalence classes of the given magnitude. We must note that the definition of dyadic equality is synonymous with the equivalence relation $\sim$ defined on D. Equality of dyads does not require their primary and secondary elements to coincide; rather, it means that they belong to the same equivalence class. Obviously, as a particular case, the reflexive property guarantees that two dyads with identical primary and secondary elements are equal, since they belong to the same class.

Let us see how we can characterize analytically in R the equivalence classes of the set D. To this end, note that the axiom of continuity guarantees that any unit U' can be expressed in terms of another given unit U through the product $y \circ U = U'$, where y is a real number. Therefore, any dyad (x, U') can be written in the form $(x, y \circ U)$ and the set D of all dyads can be formed using a single unit U by means of numerical pairs (x, y) with $D = \{(x, y \circ U)\}$. It is easy to verify that the dyads $(x, y \circ U)$ such that $x \times y = h$, where h is any real number, are equivalent. Indeed, let us take the dyads $(x_1, y_1 \circ U)$ and $(x_2, y_2 \circ U)$ and follow this line of reasoning:

$$\frac{x_1}{x_2} = \frac{\dfrac{h}{y_1}}{\dfrac{h}{y_2}} = \frac{y_2}{y_1} = \frac{y_2 \circ U}{y_1 \circ U}$$

Therefore, the ratio of the primaries is the inverse of the ratio of the secondaries, and thus the dyads $(x_1, y_1 \circ U)$ and $(x_2, y_2 \circ U)$ satisfy the equivalence condition and belong to the same class. Since, by the initial hypothesis, $x \times y = h$, it follows that x and y are related by the function $y = h/x$. If we represent this function in a Cartesian coordinate system, then for each value $h \in R$, the function $y = h/x$ is graphed as a hyperbola, that is, a curve of inverse proportionality. Hence, the graphical form of the partition into equivalence classes established in $D = \{x, U'\} = \{(x, y \circ U)\}$ is a family of infinitely many hyperbolas, each associated with its corresponding value $h \in R$. Each hyperbola represents a set of dyads that are mutually equivalent, constituting the partition of D whose classes are the elements of the quotient set defined above.

Let us now define the order relation *less than or equal to* denoted $\leq$, for the elements of M. The criterion this relation must satisfy for any two dyads (q_1, U_1) and (q_2, U_2) is, by definition, $(q_1, U_1) \leq (q_2, U_2)$ if and only if $q_1/q_2 \leq U_2 /\!/ U_1$. Whenever this condition holds for the dyads (q_1, U_1) and (q_2, U_2), it follows immediately that any other dyads belonging to their same equivalence classes must also satisfy it. Analytically, this can be written as: $[(q, U)] \leq [(q, U)]$. Thus, the quantities of magnitudes given by $m_1 = \mathcal{C}(q_1, U_1)$ and $m_2 = \mathcal{C}(q_2, U_2)$ satisfy $m_1 \leq m_2$ if $q_1/q_2 \leq U_2 /\!/ U_1$. Since both q_1/q_2 y $U_2 /\!/ U_1$ are real numbers, the set M of quantities of any magnitude ir ordered in the same way as. Therefore, the relation $\leq$ for quantities of magnitudes is a total order relation. Indeed, the reflexive and transitive properties in M are immediate. Antisymmetry is easily verified: given two quantities m_1 and m_2 in M, if $m_1 \leq m_2$ and $m_2 \leq m_1$, then, $m_1 = m_2$. These three properties characterize *less than or equal to* as an order relation. Moreover, for any two quantities m_1 and m_2 in M, either $m_1 \leq m_2$ or $m_2 \leq m_1$, and not both unless they are equal. Hence, all elements of M are mutually comparable, and the relation $\leq$, is a total order. Therefore, for any magnitude, the set M of all possible quantities m is totally ordered by the relation $\leq$ defined by the condition $m_1 \leq m_2$ if and only if $q_1/q_2 \leq U_2 /\!/ U_1$.

The analysis of the *less than* relation, denoted $<$, is completely analogous to the $\leq$ or *less than ot equal to* relation. Since these relations reduce to those of R, the *less than* relation in M also behaves as a strict order on R. This means it is irreflexive, antisymmetric and transitive; in other words, it is no reflexive, not symmetric, a property called asymmetry, and is transitive. Specifically, for any quantities m, m_1, m_2 and m_3 in M, $m < m$ is never true, irreflexivity; if $m_1 < m_2$, then $m_1 \neq m_2$, which is equivalent to stating that $m_2 < m_1$ cannot hold, assymetrywhich; if $m_1 < m_2$ and $m_2 < m_3$, then $m_1 < m_3$, transitivity. Finally, for any $m_1 \neq m_2$, either $m_1 < m_2$ or $m_2 < m_1$, but not both. Hence, all elements of M are comparable, and the *less than* relation is a total order on M.

In conclusion, for any given magnitude, the set $M = \{m\}$ of all possible quantities m is, like R, totally ordered by the relation $<$, defined by the condition $m_1 < m_2$ if and only if $q_1/q_2 < U_2 /\!/ U_1$. Thus, the ordering of quantities of a magnitude is reduced to the usual ordering of real numbers through the ratio of their primaries and the inverse ratio of their associated units.

Moreover, in the dyadic notation $(x_1, y_1 \circ U)$ and $(x_2, y_2 \circ U)$ the equivalence condition is $x \times y = h$. That is, all dyads satisfying $x_1 \times y_1 = h_1$ belong to the same equivalence class and represent the same quantity of magnitude m_1. Analogously, $x_2 \times y_2 = h_2$ corresponds to the quantity m_2. Hence, $(x_1, y_1 \circ U) \leq (x_2, y_2 \circ U)$ implies $x_1 / x_2 \leq y_2 / y_1$. If $h > 0$, then $x_1 \times y_1 \leq x_2 \times y_2$ that is $h_1 \leq h_2$ and therefore $m_1 \leq m_2$. If $h < 0$, then $x_1 \times y_1 \geq x_2 \times y_2$ that is $h_1 \geq h_2$ and thus $m_1 \geq m_2$. Since $x \times y = h$ is the measure of m in the unit U, the order of h determines the order of m. The same conclusion holds for the strict relation *less than*. Therefore, the order in $M = \{m\}$ is determined by the order of h and by the associated hyperbolas that define the equivalence classes of m for each value of h. In this way, the intrinsically non-numerical quantity of magnitude m can be quantified by its associated real number $h \in R$. Interpreting the hyperbolas $x \times y = h$ as level curves in the (x, y) plane for a given level h, one obtains a hyperbolic surface representing all possible quantities of magnitude relative to any given magnitude. Each level curve $x \times y = h$ thus denotes the equivalence class of the quantity $(x, y \circ U)$, where U is an arbitrary unit of the magnitude considered.

Thus, we can finally conclude that this last equivalence relation —defined so that, given pairs of real numbers (x_1, y_1) and (x_2, y_2) in $R \times R$, they are related of and only if $x_1 \times y_1 = x_2 \times y_2$— establishes, on the Cartesian product $R \times R$, a partition into equivalence classes (a quotient set). This quotient set can be regarded as equivalent to the set M of all possible quantities of magnitude, and it can be identified with the dyadic set $\{x, y \circ U\}$ relative to any arbitrary unit U, where x and y are any real numbers. Accordingly, one may formally define $R \times R / \sim \equiv M = \{x, y \circ U\}$.

It follows from all the above that the set M of all quantities of a magnitude has been defined in three different —yet equivalent— ways: first, as the quotient set of all dyads D, so that $D/\sim \,=M$; second, by means of the dyadic set $\{x,y\circ U\}$, which employs an arbitrary reference unit U together with all ordered pairs (x,y) of real numbers; third, using only real numbers, as just established above, through the quotient set of the Cartesian product $R\times R$.

38. Law of dyadic variation. In this section, a brief account is given of how one arrives at the original prediction of the new *dysmetric dimension* of magnitudes.

To this end, we begin by deducing a property of addition that is required to set up the argument. Take two homogeneous dyads with the same primary, (q,U_1) and (q,U_2). This quantities are equal to $q\circ(1,U_1)$ and $q\circ(1,U_2)$. Adding them and applying the distributive property within the framework of dyadic algebra, we obtain:

$$q\circ(1,U_1)\oplus q\circ(1,U_2)=q\circ[(1,U_1)\oplus(1,U_2)]$$
$$(q,U_1)\oplus(q,U_2)=(q,U_1\oplus U_2)$$

Considering the previous property of dyadic addition, we now study the generic differential variation of a dyad, taking $U_1=U$ and $U_2=dU$. The differencial variation $d(q,U)$ must obviously have the form $d(q,U)=(q+dq,U\oplus dU)\ominus(q,U)$, and thus:

$$d(q,U)=(q+dq,U\oplus dU)\ominus(q,U)=$$
$$=(q+dq,U)\oplus(q+dq,dU)\ominus(q,U)=$$
$$=(q,U)\oplus(dq,U)\oplus(q,dU)\oplus(dq,dU)\ominus(q,U)=$$
$$=(dq,U)\oplus(q,dU)\oplus(dq,dU)$$

The term (dq,dU) is a second-order infinitesimal, and therefore it can be neglected with respect to the other two terms, which are first order. Hence, we obtain:

$$d(q,U)=(dq,U)\oplus(q,dU)$$

This result is called *the law of dyadic variation*. The term (dq,U) represents the modification of the dyad (q,U) as a consequence of the change in the primary component, and it is called metric variation. It describes the conventional framework that has always been used to analyze variations of quantities of

439

magnitudes. In turn, the innovative term (q,dU) is called *dysmetric variation*, and it determines the component attributable to this new dimension. It refers to the change experienced by the magnitude implicitly contained in every quantity U, whatever the cause may be.

39. Definition of dysmetric density. In general, the dysmetry of a magnitude consists in the prediction that a given dyad (q,U) may represent different quantities of magnitude depending on its position and the material environment. This is mathematized by relating the quantity at any point P to that at a fixed point O through the ratios $(q,U)_P /\!/(q,U)_O$, that is, the quantity of magnitudes at P divided by the quantity at O for the same dyad (q,U). In Section 25 it was concluded that the ratio of two real homogeneous dyads is a real number. These quotients are called the *dysmetric density* of the magnitude considered at point P with respect to O. That is to say, the real number

$$\delta(P) = \frac{(q,U)_P}{(q,U)_O} \in \mathrm{R}$$

indicates the *dysmetric density* of the magnitude associated with the dyad (q,U) at point P with respect to the reference point O.

40. Definition of dysmetric space. We have extracted this concept, briefly referring to the simplified case of bilinear mappings in Cartesian coordinates. Its full explanation and the general case in curvilinear coordinates are found in the reference texts.

The *dysmetric space* arises from applying the *dysmetric prediction* of length to the empty geometric space. It is mathematically configured as a set of four elements denoted $\{\mathscr{M}, \mathscr{F}, \mathcal{D}, \triangleq\}$: a mathematical space $\mathscr{M}$, a physical space $\mathscr{F}$, both with the structure of an affine point space of the same dimension n, a linear transformation between them $\mathcal{D}$, which transforms $\mathscr{M}$ into $\mathscr{F}(\mathcal{D}$:

$\mathcal{M} \to \mathcal{F}$) and a bilinear transformation $\triangle$ of the tensor product $\mathcal{M} \otimes \mathcal{F}$ into R ($\triangle: \mathcal{M} \otimes \mathcal{F} \to$ R).

Mathematical space is defined as the space in which measurements are taken; it can also be considered apparent or perceived and visible space. Physical space, on the other hand, is where phenomena occur; it represents real and invisible space. The mapping $\mathcal{D}$, or *spatial deformation tensor*, describes the difference and relationship between mathematical and physical space. The mapping $\triangle$, or *dysmetric density tensor*, reflects the *dysmetry of space*, associating each pair of homologous points $\mathcal{M}$ and $\mathcal{F}$ with a real number that represents this *dymetric density*, defined as the ratio between the quantity of a magnitude implicit in a dyad for a given point and the corresponding quantity for the same dyad at another fixed reference point.

We take any basis $\{e_i\}$ of the considered affine point space and a vector **u** from the mathematical space $\mathcal{M}$. We denote v as the image vector in the physical or deformed space $\mathcal{F}$ of the linear transformation $\mathcal{D}$, so that $\mathbf{v} = \mathcal{D}(\mathbf{u})$ and $v^j = u^i d_i^j$. The terms v^j and u^i are the contravariant coordinates of **v** and u in the basis $\{e_i\}$. For their part, the elements dij are the contravariant coordinates of each vector ei of the basis, transformed by $\mathcal{D}$. The matrix relation between all these coordinates can be formulated in matrix form with the notation $\mathbf{v} = \mathbf{u}[\mathcal{D}]$, where v and u are row matrices and $[\mathcal{D}]$ denotes a square matrix of order $n \times n$, which represents the deformation tensor $\mathcal{D}$. The elements d_i^j of this tensor can be conceived in general as functions of the coordinates ui and of time.

The bilinear mapping $\triangle$ onto the basis vectors will be given by $\triangle(e_i \otimes e_j) = \Delta^{ij}$, where Δ^{ij} denotes the corresponding real numbers. And thus it is possible to determine the action of $\triangle$ on any tensor product $\mathbf{u} \otimes \mathbf{v}$, according to the following law with summation indices and superscripts:

$$\triangle(\mathbf{u} \otimes \mathbf{v}) = u^i v^j \Delta^{ij} = u^i u^k d_k^j \Delta^{ij} = \delta(P) \in R$$

The matrix expression of the above law can be written in matrix notation as follows::

$$\triangle(\mathbf{u} \otimes \mathbf{v}) = \mathbf{u}\,[\mathcal{D}][\triangle]^{T}\mathbf{u}^{T} = \delta(P) \in \mathbf{R}$$

Where $\delta(P)$ represents the *dysmetric density* at each point P. Any set of $n \times n$ values ordered in the matrix $[\triangle]$ is called a *dysmetric density tensor* and is such that at every affine point P of the vector u it determines the *dysmetric density* of the space at that point, given by $\mathbf{u}[\mathcal{D}][\triangle]^{T}\mathbf{u}^{T} = \delta(P) \in \mathbf{R}$. The $n \times n$ elements Δ^{ij} of this tensor can generally be indicated as functions of the coordinates u^{i} and the time magnitude.

In physical applications, it is practical to identify $\mathcal{M}$ and $\mathcal{F}$ with the ordinary three-dimensional space $\mathbf{R}^{3}$ or, in general, $\mathbf{R}^{n}$.

41. Immediate consequences of *dysmetry*. In very brief terms, without aiming to be exhaustive, it is important to highlight here that in a *dysmetric space*, the mathematical constant indicated by the number pi is not constant. Nor is the speed of light in a vacuum constant, whose propagation would be curved without any gravitational or other perturbation. In both cases, these constants turn out to be the limits of their respective *dysmetric* values when the *dysmetric density of length* tends to one.

In kinematic laws, a *dysmetric* component appears, which tends to zero when the *dysmetric density* vanishes.

The *dysmetric* nature of other magnitudes besides length completes the classical formulation of all physical laws. For example, *dysmetric* effects are observed in Newton's laws or in relativistic and quantum formulations. In short, it can be easily verified that *dysmetric* is a fundamental mathematical reality for representing infinitely variable physical universes in all their dimensions. However, all *dysmetric laws* reduce to their current equivalents when the *dysmetric* components cancel out and only the isometric components remain. This occurs if the magnitudes are considered rigid, that is, when the *dysmetric densities* of all magnitudes are equal to unity in the set of real numbers in every case.

In general, any physical formulation contains a *dysmetric component*, which disappears when this effect is eliminated.

Therefore, all current laws correspond to their *general dysmetric law* when all magnitudes are considered *isometrics at* all points in space and at all times.

THEMATIC PAPER II

DYSMETRIC SPACES
Mathematical properties of empty space

Abstract

In the publication entitled Dyadic Algebra of Magnitudes and Dysmetry, we formulated an abstract synthesis of the dyadic algebra developed by J. M. Arnaiz in his text The New Physics of Dysmetric Spaces. We continue with this synthesis, this time to analyze the theory of *dysmetric spaces* conceived by the same author. Here, we examine the mathematics that arises from considering exclusively the *dysmetry of length*, without any other *dysmetric effects*, to discover how *dysmetric empty space* would behave and its effects on physical laws. Concepts are developed to describe the *dysmetry of length*, such as *dysmetric tensors*, and the metric of a generic *dysmetric space* is formulated, achieving the goal of determining the *dysmetric density* at every point and time. This lays the foundation for analyzing the *dysmetric component* of any physical law, a component that will, in any case, vanish when the dysmetry disappears completely.

I. Introduction

To develop the subject matter of this study, it is necessary to understand the laws of *dyadic algebra of magnitudes*, which we presented in the previously cited publication. We pick up where we left off to continue developing the concept of *dysmetry* in its various facets. In the following sections, we will focus on the *dysmetry* of the most fundamental magnitude of all: length. This implies that, just as *isometry* consists of admitting that every dyad (q, U) represents the same quantity of magnitude under all circumstances—that is, at any point in space, at any time, and under any material condition—*dysmetry*, on the other hand, is the most generic prediction imaginable; that is, every dyad can

445

represent different quantities of the corresponding magnitude in different positions, times, or physical circumstances.

Applying the above to length, the *isometry* of this magnitude consists of assuming that, given q as any measurement expressed with some mathematical element, and m as the meter or standard unit of length, the dyad (q,m) will implicitly include the same amount of length under any circumstances. *Dysmetry*, on the other hand, anticipates that the amount of length implicit in the dyad (q,m) can vary from point to point, from time to time, or depending on the physical environment under study.

For the purposes of this work, we will not consider any *dysmetry* other than that of length. In this way, we intend to focus on the *dysmetric properties of empty space*. Hence, our concept of *dysmetric space* is oriented towards the development of mathematical tools that allow us to represent a space of this nature in the absence of any other physical reality besides space itself.

II. Preliminary assumptions and definitions

1. Definition of the dysmetric density of a magnitude. It is worth recalling here the formulation of this concept already presented in our publication dedicated to dyadic algebra. Since the *dysmetry* of any magnitude consists of the prediction that any dyad (q,U) can represent different quantities of the magnitude depending on its position and the material environment, nothing prevents us from relating the quantity at any point P with that of another fixed point O, by means of the ratios $(q,U)_P/\!/(q,U)_O$, that is, the quantity of the magnitude at P divided by the quantity of the magnitude at O of the same dyad (q,U). *Dyadic algebra of magnitudes* teaches us that the ratio of two homogeneous real dyads—of the same magnitude—is a real number. We have called a quotient of these the *dysmetric density* of the magnitude considered at point P with respect to O. That is, the real number that represents the *dysmetric density* of the magnitude associated with the dyad (q,U) at point P with respect to the reference point O is given by the following dyadic ratio:

$$\delta(P) = \frac{(q,U)_P}{(q,U)_0} \in \mathrm{R}$$

The effect of the *dysmetry* of any magnitude is eliminated when isometric conditions are restored; that is, when every dyad (q,U) implicitly has the same quantity of the magnitude in all cases. Isometric conditions thus assume that the *dysmetric density* of the magnitude in question at every point P is equal to one, $\delta(P)=1$, under any circumstance. If this holds true for all magnitudes, the *dysmetric formulations* reduce to isometric ones.

2. The dysmetric density of length. Since we are limiting the analysis to length, the *dysmetric density of length* at P with respect to O will be given by the expression:

$$\delta(P) = \frac{(q,m)_P}{(q,m)_0} = \frac{m_P}{m_O} \in \mathrm{R}$$

The second ratio stems from the fact that the abstract mathematical element q does not change from O to P, so the dismetric density reduces to the dyadic ratio between the quantity of length m_P implicit in the standard meter at P and the corresponding quantity of the same standard meter m_0 at O. From now on, we will assume that the reference point O is the terrestrial environment, and thus we can omit the subscript O, leaving $m_0 = m$, which symbolizes the quantity of length of our standard meter. In this way, we will have both quantities of length related in the form $m_P = \delta(P) \circ m$. Recall that the operation $\circ$ is the dyadic multiplication of a number $\delta(P)$ by a quantity of magnitude m.

3. Definition of mathematical space. Starting from the isometric hypothesis of current metrology, the length standard does not change with position or time. Therefore, in R^n, every point will be defined by its n coordinates, measured by geometric congruence with the chosen unit, according to the measurement procedure we all know from elementary school. Thus, each coordinate will be represented by a dyad (u^i, m), where u^i is a real number with i being a superscript that varies from one to n. This isometric space in which measurements are carried out using the rules of geometry and a constant standard m is what we will call *mathematical space*. The position of every point P will indicate how the point is observed in space when we measure its position in this classical way. Taking into account the affine space condition of R^n, the coordinates ui are also the contravariant coordinates of the position vector u of point P.

4. Definition of physical space. If we take into account the effect of *length dysmetry*, we transform the space R^n, by definition, into a *dysmetric space*, which we call *physical space*. The same point P in this space will have associated position vector **v** with contravariant coordinates v^i in the same basis of R^n as in the previous case, because we establish this by definition. We find that every differential element of length will be given by the dyad $(dv^i, m(v^i))$, where dv^i is a differential element of the coordinate v^i at any of its points and $m(v^i)$ is the implicit quantity of length in the standard meter at that same point of the coordinate v^i. Since we are considering that *dysmetry* is active, the quantities of length implicit in m and $m(v^i)$ are generally different and are related by the *dysmetric density* $\delta(v^i)$ that corresponds to each point of the coordinate v^i. Thus, we will have $m(v^i) = \delta(v^i) \circ m$, which, as we recall, represents the dyadic multiplication of a quantity of magnitude by a real number. Under these conditions, operating on the dyad $(dv^i, m(v^i))$, we will have with the laws of *dyadic algebra*:

$$(dv^i, m(v^i)) = (dv^i, \delta(v^i) \circ m) = (\delta(v^i) \times dv^i, m)$$

We have thus reduced the differential dyad $(dv^i, m(v^i))$ to another, $(\delta(v^i) \times dv^i, m)$, where $\times$ is the multiplication of R. This dyad represents the same quantity of length expressed with the standard meter of mathematical space. Integrating all these dyads along the coordinate, we will arrive at a specific value of length, which we will symbolize by (v^i, m), thus determining the

448

corresponding coordinate of point P. In short, we can observe the effect of the *length dysmetry* as a transformation of mathematical space into physical space, such that every point P will be indicated by a set of coordinates $\{u^i\}$ in mathematical space and their corresponding $\{v^i\}$ in physical space. And, if we use the affine space R^n, the length dissymmetry is represented by the transformation of every vector $\mathbf{u}$ of mathematical space into another $\mathbf{v}$ of physical space. We thus establish that $\mathbf{u}$ and $\mathbf{v}$ associate homologous points of mathematical and physical space with respect to a point P, characterized simultaneously by $\mathbf{u}$ and $\mathbf{v}$. Under these conditions, with all measurements reduced to the standard meter m, the *spatial dysmetry* is analyzed and represented by the relationships between the primaries of the coordinates of any point in both spaces. In other words, it suffices to analyze the relationship between homologous points associated with the vectors $\mathbf{u}$ and $\mathbf{v}$ to describe how the *dysmetry* alters the distribution of lengths across space and time.

5. Correspondence between mathematical space and physical space. We say that the *length dysmetry* establishes a relationship between mathematical space and physical space that we can observe as a spatial deformation represented by a function $\mathcal{D}: R^n \to R^n$, also indicated by the classical form $\mathbf{v} = \mathcal{D}(\mathbf{u})$. When the transformation $\mathcal{D}$ is the identity function, we will have $\mathbf{v} = \mathbf{u}$ for any $\mathbf{u}$, there will be no deformation whatsoever, and the mathematical and physical spaces will coincide, canceling out and completely eliminating the *dysmetric effect*.

III. Dysmetric tensors in Cartesian coordinates

6. Dysmetric space deformation tensor. Let $\{O, \mathbf{e}_i\}$ be a basis of the affine point space R^n, where O is the origin of the system and i takes positive integer values from one to n. The vectors $\mathbf{u}$ and $\mathbf{v}$ can be expressed in terms of this basis in the forms $\mathbf{u} = u^i \mathbf{e}_i$ y $\mathbf{v} = v^j \mathbf{e}_j$. We use summation notation with superscripts and subscripts indicated by the same letter. The relationship between the vector $\mathbf{u}$ and the vector $\mathbf{v}$ is determined by the *dysmetric operator*, which we have denoted $\mathbf{v} = \mathcal{D}(\mathbf{u})$, or in terms of the basis vectors

$v^j \mathbf{e}_j = \mathcal{D}(u^i \mathbf{e}_i)$. If the function $\mathcal{D}$ is a linear transformation, $v^j \mathbf{e}_j = u^i \mathcal{D}(\mathbf{e}_i)$. The elements $\mathcal{D}(\mathbf{e}_i)$ represent the transformed versions of the basis vectors, whose contravariant coordinates we will denote d_i^j, and which satisfy $\mathcal{D}(\mathbf{e}_i) = d_i^j \mathbf{e}_j$. Thus, $v^j \mathbf{e}_j = u^i d_i^j \mathbf{e}_j$ and it follows that $v^j = u^i d_i^j$. Thus we have expressed the relationship between the coordinate sets $\{u^i\}$ and $\{v^j\}$ of point P in mathematical and physical spaces. The set of $n \times n$ real numbers d_i^j characterizes the transformation $\mathcal{D}$, which we will call the *dysmetric space deformation tensor*.

7. Spatial dissymmetric density tensor. We aim to establish a mapping that reflects *dysmetric density* in R^n. To do this, we use the tensor product $R^n \otimes R^n$, where the first factor is associated with the mathematical space and the second with the physical space. We know that in tensor algebra, the set of all linear mappings from $R^n \otimes R^n$ to R^n is called the dual space of $R^n \otimes R^n$, which is usually represented as $(R^n \otimes R^n)^*$. Every mapping $\triangle$ from $(R^n \otimes R^n)^*$ will operate on the space $(R^n \otimes R^n)^*$ by associating each pair of vectors $\mathbf{u}$ and $\mathbf{v}$ from the mathematical and physical spaces with the *dysmetric density* distribution $\{\delta(P) \in R\}$. Any mapping $\triangle$, with the underlined delta letter, will be called a *dysmetric density function*.

Let $\{\mathbf{E}_{ij}\}$ be a basis of $R^n \otimes R^n$ with $\mathbf{E}_{ij} = \mathbf{e}_i \otimes \mathbf{e}_j$ and let $\triangle$ be any function from $(R^n \otimes R^n)^*$. The action of $\triangle$ on $\mathbf{E}_{ij}$ can be written $\triangle(\mathbf{E}_{ij}) = \Delta^{ij} \in R$. Given two vectors $\mathbf{u}$ and $\mathbf{v}$ in R^n, their tensor product is $\mathbf{u} \otimes \mathbf{v} = u^i v^j \mathbf{E}_{ij}$ and thus we have the following reasoning with summation equalities:

$$\triangle(\mathbf{u} \otimes \mathbf{v}) = \triangle(u^i v^j \mathbf{E}_{ij}) = u^i v^j \triangle(\mathbf{E}_{ij}) = u^i v^j \Delta^{ij} = \delta(P) \in R$$

We know that the relationship between u^i and v^j is given by the *spatial deformation tensor* $\mathcal{D}$, which has components d_k^j, through the relations $v^j = u^k d_k^j$. Substituting, we have the following logical chain:

$$\triangle(\mathbf{u} \otimes \mathbf{v}) = u^i v^j \Delta^{ij} = u^i u^k d_k^j \Delta^{ij} = \delta(P) \in R$$

In conclusion, the *density function* $\triangle$ of images Δ^{ij} or *dysmetric densities* associated with the basis $\{\mathbf{E}_{ij}\}$, determines the generic relationship between the strain *tensor and the dissymmetric density* at any affine point simultaneously associated with the vectors **u** and **v** by the law:

$$\triangle(\mathbf{u} \otimes \mathbf{v}) = u^i u^k d_k^j \, \Delta^{ij} = \delta(P) \in \mathrm{R}$$

In the case of *n* dimensions, as corresponds to R^n, which represents both mathematical and physical space, by definition, $\mathrm{R}^n \otimes \mathrm{R}^n$ is applied to R by $\triangle$, so the matrix expression of the previous law would be developed in this way:

$$\triangle(\mathbf{u} \otimes \mathbf{v}) = \begin{bmatrix} u^1 & u^2 & \dots & u^n \end{bmatrix} \begin{bmatrix} d_1^1 & d_1^2 & \dots & d_1^n \\ d_2^1 & d_2^2 & \dots & d_1^n \\ \dots & \dots & \dots & \dots \\ d_n^1 & d_n^2 & \dots & d_n^n \end{bmatrix} \begin{bmatrix} \Delta^{11} & \Delta^{21} & \dots & \Delta^{n1} \\ \Delta^{12} & \Delta^{22} & \dots & \Delta^{n2} \\ \dots & \dots & \dots & \dots \\ \Delta^{1n} & \Delta^{2n} & \dots & \Delta^{nn} \end{bmatrix} \begin{bmatrix} u^1 \\ u^2 \\ \dots \\ u^n \end{bmatrix} = \delta(P)$$

Note that we have introduced two *dismetric tensors*, $\mathcal{D}$ and $\triangle$. The former reflects the link between mathematical and physical spaces, while the latter mathematizes the distribution of *dysmetric densities* across space.

We have called the tensor $\mathcal{D}$ the *space deformation tensor*. We can call the tensor $\triangle$ the *dysmetric density tensor*. Both tensors are related by the above law, which in concentrated matrix notation can be expressed as follows:

$$\triangle(\mathbf{u} \otimes \mathbf{v}) = \mathbf{u} \, [d][\triangle]^T \mathbf{u}^T = \delta(P) \in \mathrm{R}$$

8. Dysmetric tensor of space. The preceding mathematical law describes the *dismetric properties* of ordinary space, so we can call it the *dysmetric law of empty space*, which strictly speaking should be expressed in the plural, laws, because it determines the variable behavior of each spatial domain, as a function of the *deformation and density tensors* corresponding to each differentiated environment subjected to observation and analysis. Finally, it should be noted that the *dysmetric function* $\triangle$ can be represented by a single transformation $\mathcal{T}$ associated with the product matrix $[d][\triangle]^T$, whose elements t_{ij} constitute what we could call the *dismetric tensor of space*, a relationship that can be expressed with summation and matrix notations as follows:

$$[T]=[t_{ij}] \quad ; \quad t_{ij}=d_i^k \Delta^{jk} \quad ; \quad [T]=[d][\Delta]^T$$

IV. Dysmetric tensors in curvilinear coordinates

9. General case. In the n-dimensional affine real point space $\mathbf{R}^n$, curvilinear coordinates are defined as any system of n real variables that can be put into one-to-one correspondence with the points in space and that, therefore, can be used to represent those points. Thus, given a Cartesian reference system with origin O and basis $\{\mathbf{e}_i\}$, every point P in space can be put into one-to-one correspondence with the contravariant coordinates of the vector $\mathbf{r}(P)=\mathbf{OP}$. Let $\{r^i\}$ be the set of such n coordinates, with i varying from one to n. Any system of n variables $\{u^i\}$ that can be put into one-to-one correspondence with the Cartesian system $\{r^i\}$ is a curvilinear coordinate system, since it can be put into one-to-one correspondence with the points of the affine space. Therefore, in general, the transformations of the points of a *dysmetric space* do not necessarily have to be indicated by linear equations, so the relations between the components $\{u^i\}$ and the $\{v^j\}$ will be any functions that can be indicated in the abstract with the following broadly general forms:

$$v^j = f_j(u^1, u^2, \dots, u^n) \text{ with } j=1, 2, 3, \dots, n$$

$$u^i = f_i^{-1}(v^1, v^2, \dots, v^n) \text{ with } i=1, 2, 3, \dots, n$$

Any set of coordinates $\{u^i\}$ will be associated with another set of coordinates $\{v^j\}$ through the systems of equations described above. The theory of generic tensors is developed based on these relationships between curvilinear coordinates. If we consider the equations as a change of reference frame, then the sets $\{u^i\}$ and $\{v^j\}$ will be the coordinates of the same point in those two systems; whereas, if the sets $\{u^i\}$ and $\{v^j\}$ represent the coordinates of two homologous points —a fiction we employ to make visible the hidden *dysmetry* through the difference between the mathematical or apparent position and the physical or real position relative to the same point P affine to both—the

452

equations that link them will determine the relationship between the mathematical space and the deformed physical space, or equivalently, between the mathematical distance given by the measurement and the physical distance that affects the development of phenomena and is implicit in the measurement.

Put another way, in what precedes we have assumed that the vectors of every basis $\{\mathbf{e}_i\}$ of space are constant, and thus the transformation equations turn out to be linear. The general case will then be that in which the bases vary at each point in space, and this stems from the curvilinear coordinates with which the *dysmetric tensor* can be conceived and represented extensively. Thus we have the so-called natural vectors $\boldsymbol{\alpha}_i$ associated with the curvilinear coordinates $\{u^i\}$ at the point P, defined as follows:

$$\alpha_i = \frac{\partial \mathbf{r}(P)}{\partial u^i}$$

These n vectors $\boldsymbol{\alpha}_i$ form a distinct free system at each point P and, therefore, constitute a basis of the affine point space. In turn, if $\{u^i\}$ and $\{v^j\}$ are two curvilinear systems, their respective natural vectors $\boldsymbol{\alpha}_j$ and $\boldsymbol{\beta}_i$ at every point P are related as follows:

$$\beta_i = \frac{\partial \mathbf{r}(P)}{\partial v^i} = \frac{\partial \mathbf{r}(P)}{\partial u^j} \frac{\partial u^j}{\partial v^i} = \frac{\partial u^j}{\partial v^i} \alpha_j$$

If the functions relating the coordinates are invertible, then the direct and inverse relationships described below are obtained:

$$\beta_i = \frac{\partial u^j}{\partial v^i} \alpha_j \quad ; \quad \alpha_i = \frac{\partial v^j}{\partial u^i} \beta_j$$

For the functions $v^j = f_j(u^1, u^2, \ldots, u^n)$ to be invertible to the forms $u^i = f_i^{-1}(v^1, v^2, \ldots, v^n)$ they must be differentiable and their

Jacobian must be non-zero at each point P. Recall that the Jacobian is the determinant of the Jacobian matrix, whose elements are:

$$J_{ij} = \frac{\partial f_i}{\partial u^j}$$

Therefore, the aforementioned inversion condition is that $|J_{ij}| \neq 0$. Under these conditions, differentiating the functions $v^j = f_j(u^1, u^2, \ldots, u^n)$, we immediately arrive at the following summation expressions:

$$dv^j = \frac{\partial f_j}{\partial u^i} du^i$$

The above expressions represent the curvilinear form of the relationship between the mathematical space and the deformed physical space. Thus, the elements d_i^j of the deformation tensor $\mathcal{D}$, which in this case relates the differential coordinates du^j and dv^i of both spaces at each point P affine to them, are the transposed terms of the Jacobian matrix:

$$d_i^j = \frac{\partial f_j}{\partial u^i} \quad ; \quad dv^j = d_i^j du^i$$

The *dysmetric tensor of the space* $\mathcal{T}$ is expressed at each point P in curvilinear coordinates in a manner similar to Cartesian coordinates, simply by replacing the vectors **u** and **v** with their differential vectors in the mathematical and physical spaces, **du** and **dv**. Under these conditions, observing that the *dysmetric density* establishes a scalar field that assigns a real number to each point P of the space, in a neighborhood of the point we can determine the differential of the field, which we will denote with the notation $d\delta(P)$. Thus, considering that in the differential

454

neighborhood of the point P the natural bases function linearly like any Cartesian system, we finally have the reasoning given by the following chain of equalities:

$$\underline{\Delta}\left(d\mathbf{u}\otimes d\mathbf{v}\right)=du^{i}dv^{j}\Delta^{ij}=du^{i}du^{k}\frac{\partial f_{j}}{\partial u^{k}}\Delta^{ij}=d\delta\left(P\right)\in\mathrm{R}$$

Therefore, the *dysmetric tensor of the space* in generic curvilinear coordinates, taking into account the indifference of the letters used in the subscripts, provided that the summation links are not changed, is expressed at each point P by the law:

$$t_{ij}=\frac{\partial f_{k}}{\partial u^{i}}\Delta^{jk}$$

Just as in Cartesian coordinates the *dysmetric law of space* is described by $u^{i}u^{j}t_{ij}=\delta(P)\in\mathrm{R}$, in curvilinear coordinates its form is $du^{i}du^{j}t_{ij}=d\delta(P)\in\mathrm{R}$. From mathematical field theory we know that the differential of a scalar field such as that generated by the *dysmetric density* $\delta(P)$ is given by the expression:

$$d\delta\left(P\right)=\frac{\partial\delta\left(P\right)}{\partial u^{k}}\,du^{k}$$

Therefore, the *dysmetric law of stationary space* in curvilinear coordinates finally takes the following form:

$$du^{i}du^{j}t_{ij}=\frac{\partial\delta\left(P\right)}{\partial u^{k}}\,du^{k}$$

10. Non-stationary dysmetric density. Up to this point, we have tacitly assumed that the *dysmetric density* is permanent, independent of time. But the result must be generalized to the case where this is not so and we have a field that varies with time t. The *dysmetric density function* must therefore be configured with the time component, which can be symbolized by the form $\delta(P,t)$ to simply indicate that its value at each point P in space is not stationary, but depends on the time variable, so that t will function as another coordinate for the purposes of the *dysmetric field* in the second member of the previous formulas.

In Cartesian coordinates, the images of the basic vectors $\mathcal{D}(\mathbf{e}_i)=d_i^j\mathbf{e}_j$ and $\triangle\mathbf{e}_i,\mathbf{e}_j$ will generally vary with time, so the terms dij, ?ij, and tij will not be real numbers but real functions of the real variable time t, which can be indicated by d_i^j, Δ^{ij} and t_{ij}. Thus, the time-varying *dysmetric law* in Cartesian coordinates is obtained:

$$u^i u^j t_{ij}(t) = \delta(P,t) \in \mathbf{R}$$

It should be noted that we conceive of the *dysmetric effect* of time in such a way that it does not influence mathematical space, which remains stationary, but only modifies physical space. Thus, the previous law means that every point P with permanent coordinates ui in mathematical space will have associated coordinates $v^i(t)$ in physical space that vary with time t, and that it will have an associated *dysmetric density* also dependent on time t.

Regarding the expression of this law in curvilinear systems, we will have the functions that relate the mathematical and physical coordinates given by $v^j=f_j(u^1, u^2, \dots, u^n, t)$, where time t is an additional coordinate. Thus, each specific set of coordinates ui will be transformed into a distinct v^j for each time value t. Therefore, this fact should be reflected by the term $dv^j(t)$, or equivalently, the terms of the Jacobian matrix will be functions of the time variable t, as will the natural bases in physical space. Consequently, the terms d_i^j, Δ^{ij} and t_{ij} will also be real functions of the time variable $d_i^j(t)$, $\Delta^{ij}(t)$ and $t_{ij}(t)$. In turn, the *dysmetric density field*, also being a function of time t, will have a differential variation with the new time term. All of this leads us to formulate the following general law for curvilinear systems:

$$du^i du^j t_{ij}(t) = \frac{\partial \delta(P,t)}{\partial u^k} du^k + \frac{\partial \delta(P,t)}{\partial t} dt$$

In summary, we have formulated the *variable dysmetric law of space* in Cartesian and curvilinear coordinates. This law appears surprisingly simple and suggestive for a phenomenon of extreme complexity, and in turn reveals a physical-mathematical conclusion that arises independently from what has been presented in this section: ordinary and dismetric empty space is not well described by the three classical dimensions, but requires at least the nine coordinates of the dismetric tensor to reflect its true properties, apart from the time dimension. This is without prejudice to the possibility that there may be physical environments that require *dysmetric tensors* with an even greater number of coordinates, which will have the same formal expression as the previous law, simply by changing the extension of the subscripts.

V. Metric in a dysmetric space

11. Metric in Cartesian coordinates. Up to this point, we have not used the inner connection or scalar product of the affine point space structure of R^n. However, this operation is necessary to introduce a metric, so we recall its concept. Given a basis $\{e_i\}$ of R^n, every vector **u** will be characterized by its components or contravariant coordinates u^i in that basis, such that $\mathbf{u}=u^i e_i$. The definition of covariant coordinates uses the *scalar product* of every vector with the basis vectors, that is, $u_i=\mathbf{u}\cdot e_i$. In particular, by multiplying all the basis vectors together, we obtain $e_i\cdot e_j=g_{ij}$, where g_{ij} are scalars, in this case of R, which can be arranged in matrix form, thus resulting in the so-called Gram matrix.

The norm of any vector **u** is defined as the dot product of the vector with itself, $\mathbf{u}\cdot\mathbf{u}$. Here we will use the notation $\|\mathbf{u}\|$ to refer to the norm of **u**, and if this norm is positive, as is axiomatized in Euclidean spaces, it represents the square of the magnitude of the vector, which is the distance between the origin of coordinates O

and the point P affine to $\mathbf{u}$. It is common to find conceptions that identify the norm with the magnitude or that represent the norm with other notations such as $\mathbf{Nu} = N(\mathbf{u}) = |\mathbf{u}|^2 = (\mathbf{u})^2$, among other symbols. Here we will use the definition of the norm $\|\mathbf{u}\|$ of a vector $\mathbf{u}$ as equivalent to the dot product $\mathbf{u} \cdot \mathbf{u}$, indicative of the square of a distance given by the magnitude $|\mathbf{u}|^2$.

Establishing a metric involves defining a specific way to measure distances. Thus, given the vector $\mathbf{u}$ and its transformed vector $\mathbf{v}$, we can determine the relationship between their norms, or equivalently, between the distances from P to the origin O in mathematical and physical space. To do this, we must express the norm of the transformed vector $\|\mathbf{v}\|$ and proceed as follows:

$$\|\mathbf{v}\| = \mathbf{v} \cdot \mathbf{v} = g_{ij} v^i v^j = g_{ij} u^k d_k^i u^l d_l^j = g_{ij} d_k^i d_l^j u^k u^l$$

Analyzing the last term of the previous expression, we observe that, fictitiously assuming that the Gram matrix in mathematical space is $\gamma_{kl} = g_{ij} d_k^i d_l^j$, the norm $\|\mathbf{v}\|$ in physical space is formulated in this way:

$$\|\mathbf{v}\| = \gamma_{kl} u^k u^l = \|\mathbf{u}\|_\gamma$$

We can read this expression to mean that $\|\mathbf{u}\|_\gamma$ is the norm of $\mathbf{u}$ calculated with $\gamma_{kl} = g_{ij} d_k^i d_l^j$, which would reveal how *dysmetry* changes the norm of every vector $\mathbf{u}$, transforming it from $\|\mathbf{u}\|$ to $\|\mathbf{u}\|_\gamma$. We can also describe the *metric in a dysmetric space* with Cartesian coordinates by the relationship between the magnitudes of the vectors affine to every point P in the mathematical and physical spaces, simply by forming the ratio between the norms of the associated vectors: that of the mathematical space $\|\mathbf{u}\| = \mathbf{u} \cdot \mathbf{u} = g_{ij} u^i u^j$ and that of the physical space $\|\mathbf{v}\| = \gamma_{kl} u^k u^l$, both expressed in terms of the invariant coordinates of the vector $\mathbf{u}$ in the mathematical space. Thus, we easily obtain the following resulting law:

$$\frac{\|\mathbf{v}\|}{\|\mathbf{u}\|} = \frac{\gamma_{kl} u^k u^l}{g_{ij} u^i u^j}$$

If the *dysmetric space* is not stationary, but rather the physical transform **v** of every vector **u** in the mathematical space depends on time t, then we will have a time-dependent physical vector $\mathbf{v}(t)$. It will follow that the elements of the *deformation tensor* d_l^j will not be real numbers, but real functions of the time variable t with $d_l^j(t)$, and consequently γ_{kl} will also depend on time, denoted $\gamma_{kl}(t)$. Thus, the variable norm of every physical vector $\mathbf{v}(t)$ will be expressed by the law:

$$\|\mathbf{v}(t)\| = \gamma_{kl}(t)\, u^k u^l = \|\mathbf{u}(t)\|_\gamma$$

Considering the ratio between the norms, the relationship between the metric in mathematical and physical spaces will be reflected by the following law:

$$\left\| \mathbf{v}(t) \right\| = \frac{\gamma_{kl}\left(t\right) u^k u^l}{g_{ij}\, u^i u^j} \left\| \mathbf{u} \right\|$$

12. Métrica en coordenadas curvilíneas. Let us now describe the *dysmetric metric in curvilinear coordinates* in a stationary regime, that is, time-invariant. We must limit ourselves to a differential neighborhood of a generic point P with the natural bases $\{\boldsymbol{\alpha}_i\}$, associated with the curvilinear coordinates of the mathematical space, and the natural bases $\{\boldsymbol{\beta}_j\}$ corresponding to the physical space. Every differential vector $d\mathbf{u}$ in the mathematical space will be transformed by the *dysmetry* into its physical transformed form $d\mathbf{v}$. The norm $\|d\mathbf{u}\|$ is the square of the modulus of $d\mathbf{u}$, which we can represent as the square ds^2 of a distance ds in the mathematical space. In turn, the norm $\|d\mathbf{v}\|$ is the square of the modulus of $d\mathbf{v}$, which we can represent as $d\sigma^2$ or the square of a distance $d\sigma$ in the physical space. Denoting $g_{\beta,ij}$ as the Gram matrix for the natural basis $\{\boldsymbol{\beta}_j\}$ with $g_{\beta,ij} = \boldsymbol{\beta}_i \cdot \boldsymbol{\beta}_j$, we have:

$$\|d\mathbf{v}\| = d\mathbf{v} \cdot d\mathbf{v} = g_{\beta,ij}\, dv^i dv^j$$

If the relationship between the curvilinear coordinates of the mathematical and physical spaces is given by the functions

459

$v^j = f_j(u^1, u^2, \dots, u^n)$, in the differential neighborhood of the point P we know that the following are verified:

$$dv^j = \frac{\partial f_j}{\partial u^i}\, du^i$$

Substituting into the expression for the differential norm $\|dv\|$, we have:

$$\|dv\| = g_{\beta,ij} \frac{\partial f_i}{\partial u^k} du^k \frac{\partial f_j}{\partial u^l} du^l$$

Operating on the previous expression and grouping the non-differential factors into $\gamma_{\beta,kl}$, we have:

$$\gamma_{\beta,kl} = g_{\beta,ij} \frac{\partial f_i}{\partial u^k} \frac{\partial f_j}{\partial u^l}$$

$$d\sigma^2 = \|dv\| = \gamma_{\beta,kl}\, du^k du^l = \|du\|_{\gamma\beta}$$

Thus we have related the norms of the differential vectors in the mathematical and deformed physical spaces by means of what we could call a fictitious Gram matrix given by $\gamma_{\beta,kl}$. In turn, in the mathematical space it is verified that $ds^2 = g_{\alpha,ij}\, du^i du^j$ and coordinating with the expression $d\sigma^2 = \gamma_{\beta,kl}\, du^k du^l$ we arrive at the ratio between both metrics:

$$\frac{d\sigma^2}{ds^2} = \frac{\gamma_{\beta,kl}\, du^k du^l}{g_{\alpha,ij}\, du^i du^j}$$

In the case where the *dysmetric space* is not stationary, but depends on time t, the physical and mathematical coordinates will be related by $v^j = f_j(u^1, u^2, \dots, u^n, t)$, where time is an additional

460

component. The differential of the physical coordinates $dv^j(t)$ at each point P will be a function of t and will be given by:

$$dv^j(t) = \frac{\partial f_j}{\partial u^i} du^i + \frac{\partial f_j}{\partial t} dt$$

With these $dv^j(t)$ at each point P and instant t, the norm $\|d\mathbf{v}(t)\| = g_{\beta,kl}(t) dv^k(t) dv^l(t)$ can be calculated by substituting $dv^k(t)$ and $dv^l(t)$, thus a function of terms du^k, du^l, dt, will appear, which we can denote $\gamma(du^k, du^l, dt)$, dependent on time t, with which the norm in physical space will also be variable $d\sigma(t)^2$ in P and the following *general metric law of the dysmetric space* will result:

$$d\sigma(t)^2 = \frac{\gamma(du^k, du^l, dt)}{g_{\alpha,ij} du^i du^j} ds^2$$

In light of this expression, it is now possible to describe the *dysmetric density* at every point and time in a general space, because the ratio between $d\sigma(t)$ and ds, which are the respective distances in physical and mathematical space at every point P and time t, is precisely the *dissymmetric density* $\delta(P,t)$ at that point and time. Therefore, this ratio links the dual metric of space with the scalar field of *dysmetric densities*, according to what we might call the *densimetric law of space*:

$$\frac{d\sigma(t)}{ds} = +\sqrt{\frac{\gamma(du^k, du^l, dt)}{g_{\alpha,ij} du^i du^j}} = \delta(P,t)$$

VI. Final considerations

13. Formal definition of a dysmetric space. In concise mathematical language, a general *dysmetric space* can be conceived as a set of four elements denoted by $\{\mathcal{M}, \mathcal{F}, \mathcal{D}, \triangle\}$: a mathematical space $\mathcal{M}$, a physical space $\mathcal{F}$, both with the structure of an affine point space of the same dimension n, a mapping between them $\mathcal{D}$, which transforms $\mathcal{M}$ into $\mathcal{F}$ ($\mathcal{D}: \mathcal{M} \rightarrow \mathcal{F}$), and a mapping $\triangle$ of the tensor product $\mathcal{M} \otimes \mathcal{F}$ onto R ($\triangle: \mathcal{M} \otimes \mathcal{F} \rightarrow$ R). The mapping $\mathcal{D}$ relates the spaces $\mathcal{M}$ y $\mathcal{F}$; we call it the *deformation tensor*. The mapping $\triangle$ assigns a real number to each pair of corresponding points in the spaces $\mathcal{M}$ y $\mathcal{F}$; we call it the *dismetric density tensor*. And this for sing at instant t in non-stationary regime. When the transformation $\mathcal{D}$ is the identity function, the mathematical and physical spaces coincide, reducing the *dysmetric space* to an *isometric space*, which is the one currently used.

14. Conclusions. The properties of *dysmetric space* are described by the rigid and stationary mathematical or apparent space, where measurements of magnitudes are performed, in conjunction with and distinguishing it from the flexible and variable physical or real space, where natural phenomena occur. We have conceptualized the material link between these two entities through a tensor duality that determines numerous highly relevant and easily observable consequences: the apparent straight lines of mathematical space are generally curves in physical space; magnitudes that appear constant when measured in mathematical space, such as velocities or accelerations, are variable in physical reality; certain physical-mathematical constants, such as pi or the speed of light, are not constants in physical reality; all physical laws contain a *dysmetric component*, which disappears when the *dysmetric effect* vanishes, and many other important generalizations.

With the *densimetric law of space*, we conclude the mathematical formulation of the principles of *dysmetric prediction*. We have aimed to show that it is possible to mathematize a real physical space that is infinitely variable in all its domains, because it is possible to calculate the *dysmetric density* $\delta(P,t)$ under broadly generic conditions while maintaining all current foundations of knowledge, not contradicting them, but rather complementing them.

THEMATIC PAPER III

DIFFERENTIAL DYSMETRY
The mathematical proof of dysmetric forecasting

Abstract

In the article entitled Dismetric Spaces, we described the properties of empty space tensorially based on the *dysmetry of length*. Now it is time to address how the variations of quantities of any other magnitude over space and time can be described. We will continue to make use of the properties of homogeneous division, developed in our synthesis of the Dyadic Algebra of Magnitudes and Dysmetry. We will see how the mathematical analysis of this phenomenon leads us directly to the mathematical prediction that the *dysmetry of magnitudes* may be natural.

I. Introduction

The development of *dysmetric mathematics* is based on the laws of *dyadic algebra of magnitudes*, which must be understood to construct the developments presented here. First, we will analyze how a unit U of a certain magnitude varies. Then, we will examine the variation of any quantity given by a generic dyad (q, U).

II. Dysmetric variation of a magnitude

1. Analysis of the variation of any unit U. Let us take a reference point O in the affine point space R^n and let U_0 be the standard unit at O of the magnitude in question. By definition, U_0 is the standard used for measurement by congruence in mathematical space and, therefore, must be considered invariant at every point and at all times. However, due to the *dysmetric effect*, this same constant standard unit will implicitly present different quantities of magnitude depending on its physical position in space and time. Let U be this variable implicit

463

quantity for the same standard U_0 in physical space. We know that the dyadic ratio between these two homogeneous quantities of magnitude $U/\!/U_0$ is a real number that we have called the *dysmetric density* δ of the magnitude measured at the considered point and instant. Therefore, there is no difficulty in relating these three entities by means of $U=\delta \circ U_0$, which means composing the real number δ with the quantity of magnitude U_0 by means of the additive dyadic multiplication of a number by a magnitude, resulting as a product the variable quantity U.

Let us take an arbitrary basis $\{O, e_i\}$ of the affine point space $\mathbb{R}^n$. Let P be a generic arbitrary point and suppose that the contravariant coordinates of the vector **OP** are $\{u^i\}$ in that basis. We will assume in what follows that the dyad (q, U) is such that q and U are functions of the coordinates u^i and of time t, expressing it as $q=q(u^1, u^2, \dots, u^n, t)$ and $U=U(u^1, u^2, \dots, u^n, t)$. If the quantity in question were stationary, we would have $q=q(u^1, u^2, \dots, u^n)$ and $U=U(u^1, u^2, \dots, u^n)$.

At a general point, let's analyze the variation of U when the measure of the coordinate u^i is increased by the incremental amount ?ui, such that the other coordinates remain unchanged. The magnitude U will increase by ΔU_i. The homogeneous dyadic ratio between ΔU_i and U_0 will be a real number $\Delta \delta_i$ di that represents the variation of the *dysmetric density* at the same point and for the same variation of the coordinate measure Δu^i. Therefore, noting that in this case the subscripts are not summation, we have:

$$\frac{\Delta U_i}{U_0} = \Delta \delta_i \implies \Delta U_i = \Delta \delta_i \circ U_0 \implies \frac{\Delta U_i}{\Delta u^i} = \frac{\Delta \delta_i}{\Delta u^i} \circ U_0$$

It must be observed in the last expression that the dyadic division of the first member is the homogeneous one between a magnitude ΔU_i and a real number Δu^i, while that of the second member is the quotient of real numbers, because $\Delta \delta_i$ and Δu^i are real scalars.

464

Following the established criterion for common partial derivatives, the limit of the first member as Δu^i approaches zero is said to be the partial derivative of U with respect to u^i at the point and time considered. And thus, it will result in:

$$\lim_{\Delta u^i \to 0} \frac{\Delta U_i}{\Delta u^i} = \lim_{\Delta u^i \to 0} \frac{\Delta \delta_i}{\Delta u^i} \circ U_0 \Rightarrow \frac{\partial U}{\partial u^i} = \frac{\partial \delta}{\partial u^i} \circ U_0$$

Therefore, since U_0 is invariant by hypothesis, we arrive at this important property: *to study at a point with respect to a coordinate the unit variation of the quantity of magnitude U implicit in the standard unit U_0, it suffices to analyze at the same point the variation of the scalar field of dysmetric densities d with respect to the same coordinate and multiply it by U_0.* Or, in other words, the measure of the unit variation with respect to a coordinate at any point of the quantity of magnitude implicit in U_0 is the unit variation with respect to the same coordinate of the *dysmetric density* of the magnitude at that point.

For any quantity, the definition of *dysmmetric density* $\delta = U /\!/ U_0$ relates, at any generic point P in space, the quantity of magnitude U to the density d relative to the standard unit U_0 at a fixed reference point O. Let's see if the same relationship holds for the variations $d\delta$ and dU. To do this, we observe that in a differential neighborhood of every point P, there will be a density $\delta + d\delta$ and its corresponding quantity of magnitude $U \oplus dU$. By the definition of *dysmetric density*, we conclude the following:

$$\delta + d\delta = \frac{U \oplus dU}{U_0} \Rightarrow \frac{U}{U_0} + d\delta = \frac{U}{U_0} + \frac{dU}{U_0}$$

This easily leads to the following mathematical result, which we could call the *law of spatial dysmetric variation of magnitudes*:

$$d\delta = \frac{dU}{U_0} \Leftrightarrow \frac{dU}{d\delta} = U_0$$

This significant mathematical truth links at every point P of space-time the reference standard unit U_0 at another fixed point O, taken as the origin, with the variation of the quantity of magnitude dU for that same standard and with the *variation of the dimensionless dysmetric density dδ*, both corresponding to any position P, so that the dyadic field U and the one corresponding to the *dysmetric density δ* of a certain magnitude are linked in such a way that at every point and instant the ratio of their differential variations is precisely equal to the standard unit U_0 at the origin O.

The differential variation dU of a unit U can be deduced from the approximation using partial derivatives, once these are defined. In the case where the unit does not depend on time and only varies with position, ΔU can be approximated with an incremental error $\varepsilon\left(\Delta\right)$ using this summation index expression:

$$\Delta U = \frac{\partial U}{\partial u^i} \circ \Delta u^i + \varepsilon\left(\Delta\right)$$

The incremental error $\varepsilon\left(\Delta\right)$ must tend to zero when all the increments of the coordinates tend to zero, and in this case these increments Δu^i become by definition differential elements du^i, so we can write the following differential equation:

$$dU = \frac{\partial U}{\partial u^i} \circ du^i = \frac{\partial \delta}{\partial u^i} du^i \circ U_0$$

Obviously, this result could also have been obtained directly from the *universal dysmetric law* expressed in the form of the dyadic product $dU = d\delta \circ U_0$, since, δ being a scalar field, $d\delta$ is given by ordinary mathematics:

$$d\delta = \frac{\partial \delta}{\partial u_i} du_i$$

In non-stationary regime, it would suffice to add the term corresponding to the time coordinate, and with this the following would result:

$$dU = \left(\frac{\partial \delta}{\partial u^i} du^i + \frac{\partial \delta}{\partial t} dt \right) \circ U_0$$

2. General mathematical law of dyadic variation. Let us briefly recall that every dyad represents a quantity of magnitude that can be symbolized in many different ways: (q, U), $(q\ U)$, $q\ U$, $q \circ U$, $q \circ (1, U)$, or similar. The mathematical part q is called the primary or measure, and the dimensional part U is called the secondary. A dyad can vary because its primary q varies or because its secondary U is modified. Classical *isometric* formulations only consider the first option. However, the generality of dysmetry also admits the second variant. The general case of infinitesimal variation of a dyad can be represented by $d(q, U)$, which must be the difference between the quantities $(q+dq, U \oplus dU)$ and (q, U). Obviously, the addition of the term $U \oplus dU$ is the dyadic addition, and the indicated difference is also a dyadic addition.

We observe in $U \oplus dU$ that it is a homogeneous sum, because the quantities of the addends refer to the same magnitude; however, the addends are not uniform, because they are of different units. The *dyadic algebra of magnitudes* provides the analytical form of these singular cases of dyadic addition, based on the postulate of affinity with the geometric algebra of segments. This is a typical exception to the axiom of uniformity when the primary quantities coincide and the secondary quantities are homogeneous but not uniform. Recall that homogeneity refers to dyads of the same magnitude and uniformity to dyads with the same unit. It follows that, although the units of the addends do not coincide, nothing prevents us from analytically formulating the addition of homogeneous but non-uniform quantities. Thus, given two dyads (q, U_1) and (q, U_2)

467

of the same magnitude, with U_1 and U_2 being homogeneous, the following additive law can be described:

$$(q, U_1) \oplus (q, U_2) = (q, U_1 \oplus U_2)$$

Considering the previous property of dyadic addition, taking $U_1 = U$ and $U_2 = dU$, and substituting in the expression $d(q, U) = (q + dq, U \oplus dU) \ominus (q, U)$, we have the following reasoning:

$$d(q, U) = (q + dq, U \oplus dU) \ominus (q, U) =$$

$$= (q + dq, U) \oplus (q + dq, dU) \ominus (q, U) =$$

$$= (q, U) \oplus (dq, U) \oplus (q, dU) \oplus (dq, dU) \ominus (q, U) =$$

$$= (dq, U) \oplus (q, dU) \oplus (dq, dU)$$

The term (dq, dU) is a second-order infinitesimal, so it can be neglected compared to the other two, which are first-order, resulting in:

$$d(q, U) = (dq, U) \oplus (q, dU)$$

This conclusion could be called the *law of dyadic variation*. The term (dq, U) represents the modification of the dyad (q, U) as a consequence of the change in the primary, which could be called *metric variation* and describes the convention used to analyze variations in quantities of magnitudes throughout history. In turn, the innovative term (q, dU) could be called *dysmetric variation* and determines the component attributable to this effect, which refers to the change experienced by the quantity of magnitude implicit in every standard unit U_0, a transcendent phenomenon ignored until now.

The differential elements dq and dU are known, since dq is provided by the ordinary differential mathematics of scalar fields and the quantity of magnitude dU we have just deduced. So we have the following two expressions:

$$dq = \frac{\partial q}{\partial u^i} du^i \; ; \; dU = \frac{\partial \delta}{\partial u^j} du^j \circ U_0$$

To perform the dyadic sum $d(q,U) = (dq,U) \oplus (q,dU)$, which must be possible because the summands are quantities of the same magnitude, it is only necessary to refer them to a common unit. For this, it suffices to consider that $U = \delta \circ U_0$ and that dU is given by the last of the two previous expressions. Thus, the dyads (dq,U) and (q,dU) will be expressed in the unit U_0, standard in O, will satisfy the axiom of uniformity, and the desired differential quantity $d(q,U)$ can be obtained, given by the equation:

$$d(q,U) = \left(\delta \frac{\partial q}{\partial u^i} du^i + q \frac{\partial \delta}{\partial u^j} du^j \right) \circ U_0$$

In the preceding analysis of the infinitesimal variation of a dyad (q,U) it has been assumed that the dyadic field examined is stationary, that is to say, constant over time, so that the differential increment $d(q,U)$ depends only on the spatial coordinates.

Conversely, if the dyadic field were not stationary but time-dependent, one would simply have to add the differential elements corresponding to the time variable t, as if it were just another coordinate, resulting in:

$$d(q,U) = \left[\delta \left(\frac{\partial q}{\partial u^i} du^i + \frac{\partial q}{\partial t} dt \right) + q \left(\frac{\partial \delta}{\partial u^j} du^j + \frac{\partial \delta}{\partial t} dt \right) \right] \circ U_0$$

It is obvious that everything stated so far for scalar magnitudes with numerical measures q is applicable to vector quantities, simply by considering the corresponding vectors $\mathbf{q}$ as measures. Therefore, to analytically describe the above laws in vector form, it suffices to substitute the scalar q with the vector $\mathbf{q}$ in the formulas.

3. Law of spatial dysmetric variation of magnitudes. In section 1, we derived this law based on the definition of *dysmetric density*. Now we will obtain it from the *dyadic variation law* of point 2, which has general validity, even in non-stationary conditions. To do this, we know that $d(q,U)=(dq,U)\oplus(q,dU)$. And, by the definition of *dysmetric density*, $U=\delta \circ U_0$. We know that $\delta \circ U_0 = (\delta, U_0)$. Applying the differential operator to the dyad (δ, U_0), we have $d(\delta, U_0)=(d\delta, U_0)\oplus(\delta, dU_0)$. It is clear that $d(\delta, U_0)=dU$ and that dU_0 is a null dyad, therefore, $dU=(d\delta, U_0)$ and thus $dU=d\delta \circ U_0$. This finally yields the well-known *spatial dysmetric variation law of the magnitudes*, which we rewrite below:

$$\frac{dU}{d\delta} = U_0$$

Given the importance of this law, it is justified that we have derived it by two different methods, and we do not consider it repetitive. It is of utmost importance to state unequivocally that, in a very simple way, the law describes how the spatial distribution field of the quantities U implicit in the same pattern U_0 and the field of *dysmetric densities* δ are related in such a way that the relative variation between them remains constant at every point, and that constant is precisely the standard unit U_0.

III. Dysmetric formulation of physical phenomena

4. Dysmetric component of kinematic laws. Let's analyze a simple kinematic example to appreciate the difference between the physical laws formulated with classical arithmetic, based on *isometry*, and their corresponding *dysmetric laws*. Let's assume a space where length is the only fundamental magnitude exhibiting *dysmetric* behavior, meaning time will not. Let's evaluate the motion of a point P. Let O be the origin of coordinates. Each position of P will be characterized by the vector $\mathbf{r}=\mathbf{r}(P)=\mathbf{OP}$. Let m_0 be the standard meter at O and m the length of the standard meter at P at a given time t. Note that, to maintain consistency in the notation in this section with U and U_0, the nomenclature used for the standards differs from the conventional one, where m represented the standard meter at the origin, which we now designate m_0, reserving here the symbol m for the variable quantity of length at each point P. With this notation, by definition, $m = \delta \circ m_0$ and $dm = d\delta \circ m_0$ will hold, where δ is the *dysmetric density of length* at P at a time t. The *dyadic variation law* for a vector quantity of length $(\mathbf{r}, m)$ can be written as follows:

$$d(\mathbf{r}, m) = (d\mathbf{r}, m) \oplus (\mathbf{r}, dm)$$

Substituting $m = \delta \circ m_0$ and $dm = d\delta \circ m_0$ into the terms of the second member, we will have, operating dyadly:

$$d(\mathbf{r}, m) = (\delta \bullet d\mathbf{r}, m_0) \oplus (\mathbf{r} \bullet d\delta, m_0) = \delta \circ (d\mathbf{r}, m_0) \oplus d\delta \circ (\mathbf{r}, m_0)$$

The operation indicated by $\bullet$ is the multiplication of a scalar by a vector, which we will omit from what follows, in accordance with conventional mathematical notation.

The term $(d\mathbf{r}, m_0)$ is the infinitesimal variation of the quantity of length of **OP** with respect to the standard meter m_0 at O, which can be defined as the *differential metric variation* of the position of P. The term $(\mathbf{r}, m_0)$ is the quantity of length of the position vector **OP** referred to the standard meter m_0 at O, which could be called the mathematical position of P. The term $d(\mathbf{r}, m)$ is the infinitesimal variation of the dyad $(\mathbf{r}, m)$, that is, it is the variation of the quantity of length of the position vector **OP** referred to the standard meter m at the same point P, which we could define as the physical variation of the position of P.

Let (t, s) be any quantity of time, where s is the standard quantity of time at P. By hypothesis, time is not an *dysmetric* magnitude, so the quantity of time s_0 of the standard at O is the same as that of s at P. Thus, $ds = (0, s)$, which means that ds is a zero quantity of time, so the dyad (t, ds) will be zero, and the differential variation of any dyad (t, s) will finally be:

$$d(t, s) = (dt, s) \oplus (t, ds) = (dt, s) = (dt, s_0)$$

Dividing both sides of $d(\mathbf{r}, m) = \delta \circ (d\mathbf{r}, m_0) \oplus d\delta \circ (\mathbf{r}, m_0)$ by the infinitesimal amount of time $d(t, s)$, which in this case is equal to (dt, s), which is the same as (dt, s_0), results in:

$$\frac{d(\mathbf{r}, m)}{d(t, s)} = \delta \circ \frac{(d\mathbf{r}, m_0)}{(dt, s_0)} \oplus d\delta \circ \frac{(\mathbf{r}, m_0)}{(dt, s_0)}$$

Operating with the laws of *dyadic algebra of magnitudes*, we immediately arrive at this other dyadic equation:

$$\frac{d(\mathbf{r}, m)}{d(t, s)} = \delta \circ \left(\frac{d\mathbf{r}}{dt}, \frac{m_0}{s_0} \right) \oplus \frac{d\delta}{dt} \circ \left(\mathbf{r}, \frac{m_0}{s_0} \right)$$

The first term represents the quantity of *physical or real velocity* of P at time t, which can be symbolized as $\mathbf{w}$. The first adding of the second term is the *dysmetric density* at P for time t, multiplied by the mathematical, classical, or apparent velocity of P, which can be expressed as $\mathbf{w}_0$ and which implicitly includes the velocity unit $m_0/\!/s_0$ at O. The second adding of the second term is the derivative of the *dysmetric* density with respect to time at time t and at point P, multiplied by a quantity of velocity whose measure is the mathematical position $\mathbf{r}$ of P, which can be abbreviated as $\mathbf{w}_r$ and which we will call *complementary velocity*, also referred to the velocity unit $m_0/\!/s_0$ at O. Thus, the relationship between what we could call the *mathematical or apparent* and *physical or real motions* of point P will be given by the following abbreviated expression of dyadic algebra:

$$\mathbf{w} = \delta \circ \mathbf{w}_0 \oplus \frac{d\delta}{dt} \circ \mathbf{w}_r$$

Note that the above equation uses the condensed symbology $\mathbf{w}$, $\mathbf{w}_0$ and $\mathbf{w}_r$ to indicate dyads or quantities of magnitudes that implicitly carry their respective units, according to these definitions:

$$\mathbf{w} = \frac{d(\mathbf{r}, m)}{d(t, s)} \; ; \; \mathbf{w_0} = \left(\frac{d\mathbf{r}}{dt}, \frac{m_0}{s_0} \right) \; ; \; \mathbf{w}_r = \left(\mathbf{r}, \frac{m_0}{s_0} \right)$$

We have $\mathbf{w_0}$ and $\mathbf{w}_r$ expressed in the same unit of speed in O, that is, $m_0 /\!/ s_0$, so we can add them and calculate the vector measure $\mathbf{v}$ of the physical speed $\mathbf{w}$ in this same unit, determined by the measures $\mathbf{v_0}$ of $\mathbf{w_0}$ and that of $\mathbf{w}_r$ with the following expression of vector algebra, which only uses the dyadic primaries:

$$\mathbf{v} = \delta \, \mathbf{v}_0 + \frac{d\delta}{dt} \mathbf{r}$$

This equation tells us that this measure $\mathbf{v}$ of physical speed depends on the *dysmetric density* δ, which was to be expected, but perhaps the most striking thing is that it also depends on the situation given by the position vector $\mathbf{r}$ of point P. Specifically, if the mathematical trajectory of P were a straight line, its physical trajectory would be curved, because the complementary velocity would deviate it from rectilinear motion. And this leads us to anticipate that the *dysmetric analysis* of physical phenomena is full of surprises. For example, as explained extensively in the reference texts, we discover here again that *in an dysmetric space light cannot propagate in a straight line nor is it able to maintain a constant speed*. And it is no less important to point out that every physical or real trajectory depends on the reference origin O, since the complementary velocity is given by the position vector of point P, that is, *everything observed depends on the position of the observer*, which may be surprising at first before recognizing that in a *dysmetric space* nothing is what it seems and everything that is seen also depends on the point of view of the one who looks at it.

5. Reduction of dysmetry to isometry. In the particular case where the *dysmetric density* is equal to the unit of R at every point

473

and instant, the differential quotient or derivative $d\delta/dt=0$ will be zero in the second term, so $\mathbf{w}=\mathbf{w_0}$ and $\mathbf{v}=\mathbf{v_0}$ will be verified, so we would be in the scope of the current *isometry*, and everything would manifest itself as in the classical models.

IV. The dysmetric component of the law of dyadic variation

6. Analysis of neutral dyads. The previous example, quite simple for classical kinematics, reveals the remarkable complexity and richness of *dysmetric phenomena*, which we have thus far ignored, tacitly assuming that the quantities of magnitudes implicit in the units are invariant, due to the measurement process and the lack of a rigorous algebra for magnitudes. Hence, the most general manifestation of physical phenomena, which is *dysmetric prediction*, described in this work and in the two cited in the abstract and dedicated to the same topic, has gone unnoticed. In differential terms, it has always been believed that $d(q,U)=(dq,U)$, because it is accepted without explicitly stating that (q,dU) is a zero quantity of magnitude; that is, the quantity of magnitude dU is assumed to be zero. Note that, since dU represents a quantity of magnitude, even if it is of differential quantity, it symbolizes a dyad, so if said quantity were zero, it would have to be represented with the notations $(0,U)$ or $(0,U_0)$ or equivalents and it would be incorrect to indicate it only with the zero of the real numbers 0, according to the following analysis pertaining to the case of the differential dyadic variation.

Besides the neutral quantity $(0,U)$, there is another form of the zero element of any magnitude. Recall that, in a very simplified summary, every dyad (q,U) represents a quantity of magnitude and is formed, by definition, by a numerical primary q and a dimensional secondary U, such that the quantity of magnitude to which the dyad refers is established by the additive product $q \circ U$, or the dyadic sum of U the integer or fractional times indicated by q, and analogously if the dyad is vectorial, with the primary q. This is how the problem of the impossible objective determination of the quantity of magnitude implicit in every unit is overcome. That is why we call the quantity of length of a certain segment,

474

whose actual quantity is not determinable, the standard meter, or we call a quantity of physical matter that cannot be measured without referring it to another, the standard kilogram. And thus, nothing prevents us from defining the zero of magnitudes as a null quantity of length or mass, or, ultimately, as the absence or emptiness of all magnitude. Let U_n, or any other symbol, represent the void or non-existence of any quantity of magnitude. With this unit, we can establish dyads (k, U_n), formed with any number k in R and this zero unit, which would be equivalent to adding the empty quantity of all magnitude, denoted U_n, k times (integer or fractional). This dyadic addition must yield a result that can only be no quantity of magnitude, so every dyad of the form (k, U_n) will be zero for all k in R, even if $k \neq 0$ if it is a scalar, or $\mathbf{k} \neq \mathbf{0}$ if it is a vector. Thus, we must admit the existence of infinitely many neutral dyads lacking any magnitude. Therefore, any empty dyad of magnitude (k, U_n) constitutes an exception to the axiom of uniformity, resulting in the addition of any other dyad (q, U) without the need to transform the addends to refer to the same or secondary units. The addition is $(q, U) \oplus (k, U_n) = (q, U)$, regardless of q, k, and U, making the dyad (k, U_n) neutral. Regarding the multiplicative operation, $(q, U) * (k, U_n) = U_n$, since, due to the affinity of magnitudes with length, $U * U_n = U_n$. This acquires algebraic meaning if we consider the neutral dyad as the equivalence class of (k, U_n), according to the definition formulated in our publication entitled Dyadic Algebra of Magnitudes and Dysmetry.

7. Analysis of the differential dyad (q, dU). Having established the concept of a null dyad, we can undertake the analysis of the dyad (q, dU), which would only be null for all values of q if we assume that $dU = U_n$. Only in that case would the dyad (q, dU) be neutral for addition, and we could reasonably write that $(q, dU) = U_n$ for all q. Thus, operating with the *law of dyadic variation*, we would obtain the erroneous simplification $d(q, U) = (dq, U) \oplus (q, dU) = (dq, U)$. This would explain the classical method of evaluating the variation of quantities of magnitudes without taking into account the significant *dysmetric* addend

(q,dU), although it certainly would not justify it at all, because, mathematically, the dyad (q,dU) does not have to be null.

On the contrary, the foregoing provides conclusive proof that voluntarily or inadvertently nullifying dU is a completely unfounded restriction if not previously justified, since it is obligatory to observe and comply with the *dyadic variation law* $d(q,U)=(dq,U)\oplus(q,dU)$, respecting it without truncating it, unless, for convenience or practicality in certain cases, it must be consciously assumed that $dU=U_n$ and that dU contains no quantity of magnitude. It is obvious that the hypothesis of the nullity of dU is equivalent to assuming that every unit of magnitude is constant in space and time, with $U=U_0$ at every point and instant. However, in order not to exclude any reality, there is no doubt that the correct approach is to embrace what is presented most broadly, that is, the *dysmetry*, without omitting it through negligence or closed-mindedness, because otherwise the only result would be to disregard those pieces of evidence that are *dysmetrics* and that researchers do not perceive because they do not even look at them. Obviously, there will be physical domains where it is appropriate or practical to simplify and assume that the quantity dU is zero or empty, but there may also be other domains where the *dysmetric prediction* is correct. Therefore, the primary proposition of this work is that *dysmetry* manifests itself as the reliable way to avoid ignoring anything measurable. This is fully supported by the mathematical *law of dyadic variation* $d(q,U)=(dq,U)\oplus(q,dU)$, which proves that the general prediction is *dysmetry*, because the term representing it (q,dU) exists; it is a mathematical reality that cannot be ignored. The *isometry* that we compulsively apply without thinking does not correspond to this true law that governs physical-mathematical phenomena. Therefore, we believe that the *dysmetric term* (q,dU) should not be excluded a priori, assuming in fact that $dU=U_n$ for any U, or in terms of the standard meter that $dm=U_n$ or the standard kilogram that $dkg=U_n$. In general, in order not to exclude any possible *dysmetric phenomenon*, it should be presupposed that $dU\neq U_n$ for every unit U, specifically admitting that the variations of the units of length, mass, time, and all others may not be zero, with $dm \neq U_n$ or $dkg \neq U_n$ or $ds \neq U_n$.

V. Primary formulation of the law of dyadic variation

8. Dysmetric density and dyadic variation. We still need to analyze how dyadic variation manifests itself as a function of the *dysmetric density*, which, as we recall, is a way of mathematizing *dysmetry*. To do this, we take a standard unit U_0 at a given point O, considering that at any other point P is, $U = \delta \circ U_0$ and $dU = d\delta \circ U_0$, where U is the magnitude of U_0 at P and d is the *dysmetric density* at that same point, determined by the dyadic ratio $U /\!/ U_0 = \delta$. Thus, operating dyadically, we arrive at the following conclusion:

$$d(q,U) = (dq,U) \oplus (q,dU) = (dq, \delta \circ U_0) \oplus (q,d\delta \circ U_0) =$$

$$= (\delta \, dq, U_0) \oplus (q \, d\delta, U_0) = [(\delta \, dq + q \, d\delta), U_0]$$

We observe that the differential variation of any dyad (q,U) is the infinitesimal quantity $[(\delta \, dq + q \, d\delta), U_0]$, which, according to the nomenclatures we have been using, can be written as $d(q,U) = (\delta \, dq + q \, d\delta) \circ U_0$ or also $d(q,U) = (\delta \, dq + q \, d\delta) \circ (1, U_0)$ or any other notation that may be of interest. In any case, the important thing is that $d(q,U)$ has a measurable analytical expression and that its value is given by this fundamental law. Dividing $d(q,U)$ by the standard unit U_0, the result must be a real number, as we know from *dyadic algebra*, and it turns out that this number is $\delta \, dq + q \, d\delta$, which is the measure with U_0 of the differential dyadic variation, determined by the following expression:

$$\frac{d(q,U)}{U_0} = \delta \, dq + q \, d\delta$$

Therefore, the measure or primary of the quantity of magnitude implicit in every differential dyad $d(q,U)$ is determined in relation to the standard unit U_0 and is given by an abstract dimensionless real number, which is $\delta \, dq + q \, d\delta$, an addition that

477

is the differential of the numerical product δq, which allows us to put:

$$\delta\, dq + q\, d\delta = d(\delta q) \quad ; \quad \frac{d(q, U)}{U_0} = d(\delta q)$$

The preceding differential equations formalizes the *primary expression of the dyadic variation law*. Obviously, for vector magnitudes, the isomorphic equatiosn of the preceding scalar dyadic differentials will appear, where addition refers to vector addition and multiplication to the product of a real number by a vector, that is:

$$\delta\, dq + q\, d\delta = d(\delta q) \quad ; \quad \frac{d(q, U)}{U_0} = d(\delta q)$$

9. Isometric simplification. Currently, the *dysmetric phenomenon* is ignored in practice, both for scalar and vector quantities, which tacitly implies considering that the *dysmetric density* of every quantity is constant and equal to one, thus presupposing $d\delta = 0$ and so the scalar expression $\delta dq + q\, d\delta$ is reduced to dq, and the vector $\delta dq + q\, d\delta$ is transformed into dq, which undoubtedly appears as a very impoverished particular case of the *dysmetric prediction*, a singular situation that unduly excludes all those phenomena in which $\delta \neq 1$ or $d\delta \neq 0$.

However, the *primary formulation of the law of dyadic variation* shows us again the fact that every equation of *dysmetric mathematics* reduces to the corresponding current *isometric* when the *dysmetric effect* disappears.

VI. Final considerations

10. The dysmetry expands without refuting all current formulations. The concept of *dysmetry* is neither a theory nor a hypothesis. As we have already mentioned, it is a *mathematical*

478

prediction that arises from the *dyadic algebra of magnitudes* and manifests itself as an abstract reality when working with dyads of magnitudes, and not only with purely mathematical elements. Dyads are the entities that accurately represent quantities of any magnitude, marking a very significant difference from the classical way of working with magnitudes, which reduces operations to the composition of simple measures or primaries, represented by mathematical elements, either scalar or vectorial.

Therefore, *dysmetry* does not constitute a separate theory from the others, but rather complements them and shows them all the path they should follow to transcend the narrow limits of arithmetic and give algebraic consistency and full meaning to the formulations of natural laws, overcoming the contradictions and limitations inherent in *isometric arithmetization*. This transition toward the algebraic abstraction of operations with magnitudes is extremely enriching and has the mathematical backing of the *law of dyadic variation*.

11. Conclusions. It seems beyond doubt that the *dysmetric differential* presented here is mathematical proof that the *dysmetric prediction* should not be ignored. It is up to the scientific method of experimentation to investigate its validity under the various assumptions examined. Abstract mathematics is limited to proposing useful formulations for basic science to use in its models and theories. Here, all we have done is mathematize the *dysmetric prediction*, which is an indisputable logical variant, without affirming or denying anything about the nature of being and workings. We have, however, demonstrated that *dysmetry* is the broadest concept and *isometry* the narrowest; that *dysmetry* reduces to *isometry* when its effect is nullified; and that, in logical terms, it is not justified to admit *isometry* a priori, since there is no proof whatsoever of its universal validity. On the contrary, there is already evidence that the universe may be *dysmetric*, a matter we will not address here in order to limit our analysis to the strict realm of abstract mathematics.

BIBLIOGRAPHY

JOSEPH FOURIER. *Théorie Analitique de la Chaleur*, Gauthier Villars, París, 1888.

DAVID HILBERT. *Grundlagen der Geometrie* (*Fundamentos de la geometría*), 1899.

MAX PLANCK. *Vorlesungen über die Theorie der Wärmestrahlung*, Leipzig, 1906.

R.C. TOLMAN. *Physics Review*, 1914, 1917.

GIOVANNI GIORGI. *Sistemi e unita di mesura*, Enciclopedia delle Matematiche Elementari.

P. W. BRIDGMAN. *Dimensional Analysis*, Yale, University Press (Universidad Nacional de Tucumán, República Argentina).

RICARDO SAN JUAN. *Teoría de las magnitudes físicas y sus fundamentos algebraicos*, Revista de la Real Academia de Ciencias de Madrid, 1947.

P. W. BRIDGMAN. *British Enciclopedia*, edition 1951, article *Dimensional Analysis*.

JULIO PALACIOS. *El lenguaje de la física y su peculiar filosofía*, 1953.

U. STILE. *Messen und Rechnen in der Physic*, Vieweg, Braunschweig, 1961.

JULIO PALACIOS. *Análisis dimensional*, Espasa Calpe, segunda edición, 1964.

P. PUIG ADAM. *Geometría métrica*, Biblioteca Matemática Rey Pastor-Puig Adam, 1970.

SEARS ZEMANSKY. *Física general*, Aguilar, University Physics, 1970.

Luis A. Santaló. *Vectores y tensores y sus aplicaciones*, Editorial Universitaria de Buenos Aires, 1970.

R. Kurth. *Dimensional Analysis and Group Theory in Astrophysics*, Pergamon, 1972.

F. Catalá Moreno, *Álgebra lineal y multilineal*, Academia Iribas, Madrid, 1972.

André Lichnerowicz. *Elementos de cálculo tensorial*, Aguilar Sociedad Anónima de Ediciones, 1972.

I. Cano de la Torre. *Mecánica Racional*, Academia Luz de Madrid, 1973.

Sixto Ríos. *Métodos Estadísticos*. Ediciones del Castillo S.A. Sexta edición, 1974.

International Practical Temperature Scale of 1968, Amended Edition of 1975, *Metrology*, Comité International des Poids et Mesures, 1976.

R. M. Cooke. *The Algebra of Physical Magnitudes, Foundatios of Physics*, 1980.

José Catalán Chillerón. *Teoría de las magnitudes físicas*, Instituto Geográfico Nacional, Madrid, 1983.

Isaac Newton. *Principios matemáticos de la filosofía natural*, Alianza Editorial, 2016.

J. M. Arnaiz. *Matematizar 1 (Fundamentos), Matematizar 2 (Complementos) y Matematizar 3 (Aplicaciones)*, Ediciones Go Beyond, 2016.

Bureau International des Poids et Mesures. *The International System of Units (SI)*.